third edition

Technically—Write!

Communicating in a Technological Era

RON S. BLICQ

PRENTICE-HALL, ENGLEWOOD CLIFFS, NEW JERSEY 07632

Library of Congress Cataloging-in-Publication Data

Blicq, Ron (date)
  Technically-write!

  Includes index.
  1. Technical writing. I. Title.
T11.B62   1986   808'.0666021   85-24367
ISBN 0-13-898750-5

Editorial/production supervision and
    interior design: Barbara DeVries
Cover design: Wanda Lubelska Design
Manufacturing buyer: Harry P. Baisley

*The names of people and organizations are imaginary and
no reference to real persons is implied or intended.*

Printed in the United States of America

10 9 8 7 6 5 4 3

ISBN 0-13-898750-5 01

Prentice-Hall International (UK) Limited, *London*
Prentice-Hall of Australia Pty. Limited, *Sydney*
Prentice-Hall Canada Inc., *Toronto*
Prentice-Hall Hispanoamericana, S.A., *Mexico*
Prentice-Hall of India Private Limited, *New Delhi*
Prentice-Hall of Japan, Inc., *Tokyo*
Prentice-Hall of Southeast Asia Pte. Ltd., *Singapore*
Editora Prentice-Hall do Brasil, Ltda., *Rio de Janeiro*
Whitehall Books Limited, *Wellington, New Zealand*

Contents

2

MEET "H.L. WINMAN AND ASSOCIATES"   29

3

TECHNICAL CORRESPONDENCE   41

7

OTHER TECHNICAL DOCUMENTS   221

8

ILLUSTRATING TECHNICAL DOCUMENTS   252

9

TECHNICALLY—SPEAK!   275

10

COMMUNICATING WITH PROSPECTIVE EMPLOYERS

11

THE TECHNIQUE OF TECHNICAL WRITING

12

GLOSSARY OF TECHNICAL USAGE

Preface

This book presents all those aspects of technical communication that you, as a technician, technologist, engineer, scientist, or technical manager, are most likely to encounter in industry. It introduces you to the employees of two technically oriented companies, to the type of work they do, and to typical situations that call for them to communicate with clients, suppliers, and each other.

Since the earlier editions of *Technically—Write!* there have been some changes in management of the larger company (H. L. Winman and Associates—see Chapter 2), which has become involved in computer/electronics engineering through acquiring a company called Macro Engineering Inc. Harvey Winman—the owner of the two companies—has stepped into semiretirement and new faces and departments have been introduced into H. L. Winman and Associates to better reflect the parent company's involvement in current technology. Three other changes in this edition reflect much greater emphasis on job search techniques (now a totally new chapter: Chapter 10), the introduction of the computer as a writing tool (Chapter 1), and preparing a scientific research paper (Chapter 7).

At the end of each major chapter there is a selection of detailed assignments of varying complexity, in which you assume you are employed by one of these companies and are engaged in projects that require you to write letters and reports, and sometimes to present your findings orally. A marking control chart (at the end of the book) offers a place for you to record your writing errors when assignments are returned to you, and to check whether you have corrected your faults in successive work. It is based on a Correction Key in *Writing to be Read,* a Prentice-Hall textbook by Eleanor Newman Hutchens, and is printed here, in an altered form, with her permission.

The information on technical paper presentation (Chapter 9) was presented originally as a technical paper at the IRE (now IEEE) Sixth National Symposium on Engineering Writing and Speech, Washington, D.C. It was subsequently published by *Electronic Design* under the title "It's Your Turn at the Rostrum—Can You Deliver a Technical Paper?" and it is reprinted here through their courtesy. And some of the information on job search techniques (Chapter 10) appeared first as a technical paper in the *IEEE Transactions on Professional Communication* for December, 1984, under the title "Job Hunting: Sharpening Your Competitive Edge." These extracts also are reprinted with permission.

Finally, the following Prentice-Hall reviewers provided helpful comments during the revision process: Sid Burks and Nancy Chilson-Grzesik, National Education Corp., Sal Duarte, Manpower Training Consultants, Inc., Helen M. Loeb, Northeastern University, Stephanie K. McCulley, Michigan Technological University, James L. McKenna, San Jacinto College-Central and Kristin R. Woolever, Northeastern University.

R.S.B.

TECHNICALLY—WRITE!
Communicating in a
Technological Era

People as "Communicators"

As communicators, people are lazy and inefficient. We are equipped with a highly sophisticated communication system, yet consistently fail to use it properly. (If our predecessors had been equally lazy and failed to develop their communication "senses" to the level required to ensure survival, many species would be extinct today; indeed, it's even questionable that we would exist in our present form.) This communication system comprises a transmitter and receiver combined into a single package controlled by a computer, the brain. It accepts multiple inputs and transmits in three mediums: action, speech, and writing.

We spend much of our wakeful hours communicating, half the time as a transmitter, half as a receiver. If, as a receiver, we mentally switch off or permit ourselves to change channels while someone else is transmitting, we contribute to information loss. Similarly, if as a transmitter we permit our narrative to become disorganized, unconvincing, or simply uninteresting, we encourage frequency drift. Our listeners detune their receivers and let their computers (brains) think about the lunch that's imminent or wonder if they should take in a show tonight.

As long as a person transmits clearly, efficiently, and persuasively, the persons receiving keep their receivers "locked on" to the transmitting frequency (this applies to both written and spoken transmissions). Such conditions expedite the transfer of information, or "communication."

In direct contact, in which one person is speaking directly to another, the receiver has the opportunity to ask the transmitter to clarify vaguely presented information. But in more formal speech situations, and in all forms of written communication, the receiver no longer has this advantage. He or she cannot stop a speaker who mumbles or uses unfamiliar terminology, and ask that parts of a talk be repeated or clarified; neither can the receiver ask a writer in another city to explain an incoherent passage of a business letter.

The results of failure to communicate efficiently soon become apparent. If people fail to make themselves clear in day-to-day communication, the consequences are likely to differ from those they anticipated, as Cam Collins has discovered to his chagrin.

Cam is a junior electrical engineer at Macro Engineering Inc, and his specialty is high voltage power generation. When he first read about a recent EHV DC power conference, he wanted urgently to attend. In a memorandum to Fred Stokes, the company's chief engineer, Cam described the conference in glowing terms which he hoped would convince Fred to approve his request. This is what he wrote:

Fred:

The EHV conference described in the attached brochure is just the thing we have been looking for. Only last week you and I discussed the shortage of good technical information in this area, and now here is a conference featuring papers on many of the topics we are interested in. The cost is only $175.00 for registration, which includes a visit to the Freeling Rapids generating station. Travel and accommodation will be about $450 extra. I'm informing you of this early so you can make a decision in time for me to arrange flight bookings and accommodation.

Cam

Fred Stokes was equally enthusiastic and wrote back:

Cam:

Thanks for informing me of the EHV DC conference. I certainly don't want to miss it. Please make reservations for me as suggested in your memorandum.

Fred

Cam was the victim of his own carelessness: he had failed to communicate clearly exactly what he wanted. There was still a chance for Cam to remedy the situation, but only at the expense of extra time and effort.

Lisa Drew, on the other hand, did not realize she had missed a golden opportunity to be first with an innovative computer technique until it was too late to do anything about it. Her story stems from an incident that occurred several years ago, when she was a recently graduated engineer employed by a manufacturer of agricultural machinery. Lisa's job was to design modifications to the machinery, and then prepare the change procedure documentation for the production department, service representatives, sales staff, and customers.

"For each modification I had to coordinate three different documents," Lisa explained to me over lunch. "First, there had to be a design change notice to send out to everyone concerned. And then there had to be an 'exploded' isometric drawing showing a clear view of every part, with each part cross-referenced to a parts list. And finally there had to be the parts list itself, with every item labeled fully and accurately."

Lisa found that cross-referring a drawing to its parts list was a tedious, time-consuming task. The isometric drawing of the part was developed first, by the drafting department, and the coordinates of its key points were entered into the computer for production on the graphics printer. The parts listing was also prepared on a computer, but by a separate department. Because the two computer systems were not compatible, cross-referencing had to be done manually.

"And then I hit on a technique for interfacing the two programs," she explained. "It was simple, really, and I kept wondering why no one else had thought of it!"

Without telling anyone, Lisa modified one of the computer programs and tested her idea with five different modification kits. "It worked!" she laughed. "And, best of all, I found that the cross-referencing could be done in one tenth of the time."

Lisa felt her employer should know about her idea—possibly the company could market the program or even help her copyright it. So the following day she stopped the Engineering Manager as they passed in the hallway, and blurted out her suggestion. This is the conversation that ensued:

Lisa	Mr. Haddon
Oh! Mr. Haddon! You know how long it takes to do the documentation for a new part . . . ?	
	Yes . . s . . s . . ?
The problem is in trying to interface between the graphics computer and the parts list . . .	
	(Mr. Haddon appeared to be listening politely, but internally he was growing impatient.)
. . .It has to do done by hand, you see . . .	
	Doesn't the drafting department do all that?
Oh, yes! They do. I was just trying to help them . . . to speed up their work a bit.	
	You're working for the chief draftsman now?
Oh, no! It was just an idea I had— to modify one of the computer programs . . .	
	I don't remember issuing you a work order . . .
No. You didn't. I was doing it on my own . . . *(She meant she was doing it on her own time.)*	
	You mean the computer people asked you to do it?
Well--uh--no. Not exactly . . .	

(Reluctantly) Uh-huh.

I wanted to try . . .

But you have been modifying one of the computer programs?

Without authority?

I thought I had made it quite clear to all the staff: No projects are to be undertaken without my approval! *(His tone was cold and abrupt.)* That's final! *(And he turned on his heel and continued down the hall.)*

Lisa's simple suggestion had become lost in a web of misunderstanding. By the time she was through explaining what she had been doing, she had given up trying to offer her idea to the company. And so her idea lay dormant for two years, until a major software company came out with a comparable program. Lisa knew then that perhaps there *had* been market potential in the program modification she had designed.

If Cam Collins and Lisa Drew had paused to consider the needs of the persons who were to receive their information, they would never have launched precipitously into discourses that omitted essential facts. Cam had only to start his memorandum with a request ("May I have your approval to attend an EHV DC conference next month?"), and Lisa with a statement of purpose ("I have designed a computer program that can save us hundreds of dollars annually. May I have a few moments to describe it to you?"), to command the attention of their department heads. Both Mr. Stokes and Mr. Haddon could then have much more effectively appraised the information.

Such circumstances occur daily. They are frustrating to those who fail to communicate their ideas, and costly when the consequences are carried into business and industry.

Bill Carr recently devised and installed a monitor unit for the remote control panel at the microwave relay station where he is the resident engineering technician. As his modification greatly improved operating methods, the head office asked him to submit an installation drawing and an accompanying description. Here is part of his description:

> Some difficulty was experienced in finding a suitable location for the monitor unit. Eventually it was mounted on a special bracket attached to the left-hand upright of the control panel, as shown on the attached drawing.

On the strength of Bill's explicit mounting description and detailed list of hardware, the head office converted his description into an installation instruction, purchased materials, assembled 21 modification kits, and shipped them to the 21 other relay stations in the microwave link.

Within a week the 21 resident engineering technicians reported to the head office that it was impossible to mount the monitor unit as instructed, because of an adjoining control unit. No one at the head office had remembered that Bill Carr was located at site 22, the last relay station in the microwave link, where there was no need for an additional control unit.

Bill had assumed that the head office would be aware that the equipment layout at his station was unique. As he commented afterward: "No one said why I had to describe the modification, or told me what they planned to do with it."

In business and industry we *must* communicate clearly and understand fully the implications of failing to do so. A poorly worded order that results in the wrong part being supplied to a job site, a weak report that fails to motivate the reader to take the urgent action needed to avert a costly equipment breakdown, and even an inadequate job application that fails to sell an employer on the right person for a job, all increase the cost of doing business. Such mistakes and misunderstandings are wasteful of labor and resources. Many of them can be prevented by more effective communication—communication that is receiver-oriented rather than transmitter-oriented, and that transmits messages using the most expeditious, economical, and efficient means at our command.

1

A Technical Person's Approach to Writing

Almost every book on technical writing contains a statement that says in effect: "The key to effective writing is good organization." This rule is basically true, although many technical people who have tried to follow it too conscientiously find they have difficulty writing clearly. Frequently their trouble is caused by overorganization, or by trying to organize too early. Organizing that starts too early is self-defeating because it stifles a person's natural ability to write creatively.

Look at it this way: engineering technician Dan Skinner has a report to write on an investigation he completed seven weeks ago for his employer, H. L. Winman and Associates, a consulting engineering firm you will meet in Chapter 2. He has made several halfhearted attempts to get started, but each time never seemed to be the right moment: maybe he was interrupted to resolve a circuit problem, or it was too near lunchtime, or a meeting was called, or, when nothing else interfered, he "just wasn't in the mood." And now he's up against the wire and he hasn't yet set pen to paper. (Although Dan has a computer terminal beside his desk and uses it for engineering work, he has not yet attempted harnessing its word-processing capabilities.)

Unless Dan is one of those unusual persons who cannot produce except when under pressure, he is in danger of writing an inadequate, hastily prepared report that does not represent his true abilities. He knows he should not have left his report-writing project until the last moment, but he is human like the rest of us and constantly finds himself in situations like this. He does not realize that by leaving a writing task until it is too late to do a good job, and then frantically organizing the work, he is probably inhibiting his writing capabilities even more than necessary.

If Dan were to relax a little in the initial stages, instead of spending time trying to organize both himself and his writing task, he would find the physical process of writing reports a much more pleasant experience. But he must first change his whole approach to writing.

Every technical person, from student technician through potential scientist to practicing engineer, must recognize that he or she has the ability to write clearly and logically. (You may have to prove this to many responsible people in industry who continue to believe the old adage that "technical people just can't write.") This ability must be developed by continued practice: Only by planning and writing all types of documents will you develop the confidence that is the prerequisite to good writing.

PLANNING THE WRITING TASK

The word "planning" seems to contradict everything I have just said: It implies that report writers must start by organizing their material. However, I suggest they do so "creatively," to allow their latent writing ability to develop naturally. For example, I recommend that at first Dan Skinner do nothing about making an outline or taking any action that smacks of organization. All he has to do is work through several simple planning stages that will not be the chore he probably expects.

Normally, the first stage in planning any writing project is to gather information. This means assembling all the documents, results of tests, photographs, samples, specifications, and so on, that will be needed to write the report or that will be included with it. Dan should gather more information than he will probably need, because it is better to be selective and discard information than to look for additional facts and figures just when the writing is beginning to go well.

The next stage—and probably the most important—is for Dan Skinner to define his reader. He must conjure up an image of the actual person, or type of person, for whom he is writing. He must ask himself some pertinent questions: Who is this reader? What is his reading level? How much does he know about my project? Is he a technical person? Does he need to know all the details that I know? Where do his main interests lie? What will he do with my report? How will he use it? And who else is likely to read it? The answers to these questions will help Dan not only plan a good report, but also set the right tone when he is writing it.

When he has a reader in mind Dan can start making notes, either with pen or pencil and paper, or at a computer terminal. At this third stage he must "loosen up" enough to generate ideas. He needs to let his mind freewheel, so that it throws out ideas and pieces of information quickly and easily. He must not stop to question the relevance of this information—his role is purely to collect it.

But first, if he is handwriting, he must find a quiet place where he can work undisturbed; it's no use trying to be creative in a noisy, crowded office. (I will mention more about the need for establishing a good writing environment in the next section, when I discuss the practical aspects of settling down to write.)

Normally at this stage a technical person will write on a blank sheet of paper a set of arbitrary headings, such as "Introduction," "Initial Tests," and "Material Resources," and arrange them in logical order. (These seem to be standard headings in many reports.) But I want our report writer to be different. I want Dan Skinner to free his mind of the elementary organized headings, and even to refuse to divide his subject mentally into blocks of information. Then I want him to jot down or type into the computer a series of main topics that he will discuss, writing only brief headings rather than full sentences. He must do this in random order, making no attempt to force the topics into groups (although it is quite possible that grouping will occur naturally, since many interdependent topics are likely to occur to him in logical order). The topics that he knows best will spring readily to mind, then there will be a gradual slowdown as he encounters less-familiar topics.

When he stops his initial list, he should go back and examine each topic to see if it will suggest less obvious topics. As additional topics come to mind he must jot them down or type them, still in random order, until he finds he is straining to find new ideas. This should be a signal for him to stop before he becomes too objective.

Dan must not try to decide whether each topic is relevant during this spontaneous freewheeling session. If he does, he will immediately inhibit his creativeness because he will become too logical and organized. He must list all topics, regardless of their importance and eventual position in the final report.

At the end of this session Dan's list should look like Figure 1-1. He can now take a break, knowing that the first details of his report have been committed to paper. What he may not yet realize is that he has almost painlessly produced his first outline.

The fourth stage calls for Dan to examine his list of headings with a critical eye, dividing them into headings that bear directly on the subject and those that introduce topics of only marginal interest. His knowledge of the reader—identified in stage two—will help him decide whether each topic is really necessary so he can delete irrelevant topics, as has been done in Figure 1-2.

The headings that remain should be grouped into "topic areas" that will be discussed together. This he can do simply by coding related topics with the same symbol or letter. In Figure 1-2, letter (A) identifies one group of related topics, letter (B) another group, and so on.

Now, at last, Dan can take his first major organizational step, which is to arrange the groups of information in the most suitable order and at the same time sort out the order of the headings within each group. He must consider three factors: which order of presentation will be most interesting, which will be most logical, and which will be simplest to understand. The result will become his

Building OK — needs strengthening
Elevators — too slow, too small
Talk with YoYo — elev. mfr (10% discount)
Waiting time too long — 70 seconds
Shaft too small
How enlarge shaft?
         Remove stairs?
Talk with fire inspector
Correspondence — other elev mfrs
Talk with Merrywell — Budget $500,000
Sent out questionnaire
Tenants' preferences —

    Express elev   No stop — 2nd floor
    Executive elev   Faster service
    Prestige elev   No stop — ground flr
    Freight elev

Freight elev — takes up too much space
Shaft only 35 × 8 ft
            (when modified)
Big freight elev — omit basement
Tenants "OK" small freight elev
YoYo has office in Montrose   ← YoYo "C"
                                       (8 ft)
Basement level has loading dock
Service reputation — YoYo?
                   — Others?

Figure 1-1. Initial list of topic headings, jotted down in random order.

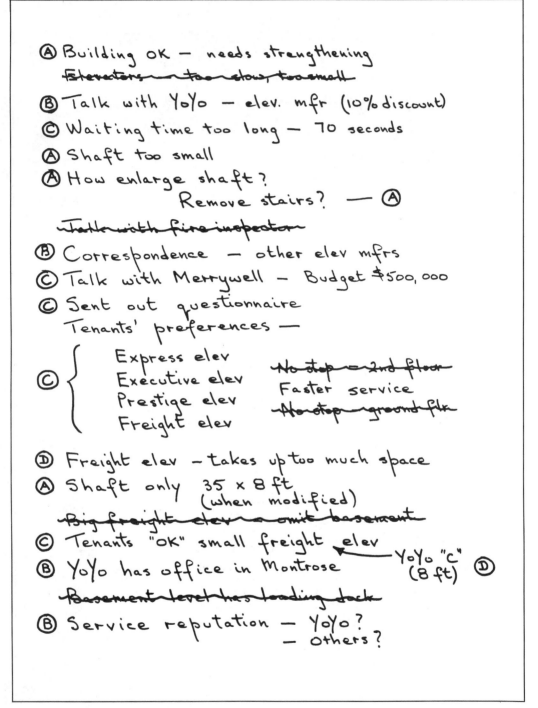

Ⓐ Building OK — needs strengthening
~~Elevators — too slow, too small~~
Ⓑ Talk with YoYo — elev. mfr (10% discount)
Ⓒ Waiting time too long — 70 seconds
Ⓐ Shaft too small
Ⓐ How enlarge shaft?
           Remove stairs? — Ⓐ
~~Talk with fire inspector~~
Ⓑ Correspondence — other elev mfrs
Ⓒ Talk with Merrywell — Budget $500,000
Ⓒ Sent out questionnaire
   Tenants' preferences —

Ⓒ { Express elev
      Executive elev
      Prestige elev
      Freight elev

~~No stop — 2nd floor~~
Faster service
~~No stop — ground flr~~

Ⓓ Freight elev — takes up too much space
Ⓐ Shaft only 35 × 8 ft
           (when modified)
~~Big freight elev — omit basement~~
Ⓒ Tenants "OK" small freight elev   ← YoYo "C" Ⓓ
Ⓑ YoYo has office in Montrose   (8 ft)
~~Basement level has loading dock~~
Ⓑ Service reputation — YoYo?
                — others?

Figure 1-2. The same list of topic headings, but with irrelevant topics deleted and remaining topics coded into subject groups (A—structural implications; B—elevator manufacturers; C—tenants' preferences; D—freight elevator).

```
    Building condition:

            OK - needs strengthening (shaft area)
            Existing elev. shaft too small
            Remove adjoining staircase
            Shaft size now 35 x 8 ft

    Tenants' needs:

            Sent out questionnaire
            Identified 5 major requests
            Requests we must meet:
                    Cut waiting time: 32 sec (max)
                    Handle freight up to 7 ft 6 in.
            Requests we should try to meet:
                    Express elev to top 4 floors
                    Deluxe models (for prestige)
                    Private elev (for executives)

    Budget: must be within $500,000

    Elevator manufacturers:

            Researched 3
            Only YoYo Co. offers discount
            Only YoYo Co. has Montrose office
```

Figure 1-3. Topic headings rearranged into a writing outline, typed at a computer terminal and printed by a dot-matrix line printer. This was part of the writing plan for the formal report on elevator selection in Chapter 6. Compare it with the final product on pages 189 and 190.

final writing plan or report outline, which may be handwritten or obtained as a computer printout as in Figure 1-3. (If Dan has typed his outline at a computer terminal, he may prefer to call the outline onto the screen at his terminal whenever he wants to refer to it, rather than work from a printed outline.)

In summary, overorganizing a report, or organizing it too early in the writing process, inhibits writing. The key to good report writing is to organize material in a spontaneous, creative manner, allowing one's mind to freewheel through the initial planning stages until the topics have been collected, scrutinized for relevance, sorted, grouped, and written into a logical outline that will appeal to the reader. This method will not necessarily suit everyone. If you already have a workable method for planning and organizing, you should continue to use it. But if you do not, or if you have difficulty starting your writing tasks, try doing it this way. The stages are simple, as illustrated in Figure 1-4, and apply to any major writing project.

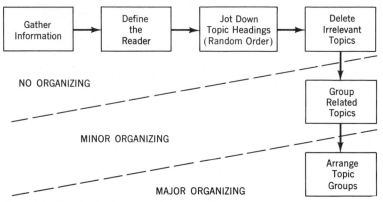

Figure 1-4. The six planning stages. In practice, these stages can overlap.

WRITING THE FIRST DRAFT

As I sit at my desk, the heading "Writing the First Draft" at the top of a clean sheet of paper, I find that I am experiencing exactly the same problem that every writer encounters from time to time: an inability to find the right words— *any* words—that can be strung together to make coherent sentences and paragraphs. The ideas are there, circling around inside my skull, and the outline is there, so I cannot excuse myself by saying I have not prepared adequately. What, then, is wrong?

The answer is simple. In the distance I can hear the percolator in the kitchen murmuring quietly. Soon its tempo will increase to a rumbling crescendo, and then it will stop. Five minutes later my wife will appear at my study door bearing two cups of coffee and a plate of English biscuits on an old-fashioned carved tray, and I will be expected to pause for a ten-minute break. I cannot concentrate on writing when I know my continuity of thought is so soon to be broken.

Continuity is the key to getting one's writing done. In my case, this means writing at fairly long sittings during which I *know* I will not be disturbed. I must be out of reach of the telephone, visiting friends, children wanting to say good night, and even my wife, so that I can write continuously. Only when I have reached a logical break in the writing, or have temporarily exhausted an easy flow of words, can I afford to stop and enjoy that cup of coffee.

It is no easier to find a quiet place to write in the business world. The average technical person who tries to write a report in a large office cannot simply ignore the surroundings. A conversation taking place a few desks away will interfere with one's creative thought processes. And even a co-worker collecting money for the pool on that night's NHL game between the New York Rangers and the L. A. Kings will interrupt writing continuity.

The problem of finding a quiet place to write can be hard to solve. In business I recommend a short walk around the premises to find a hidden corner, or a small office which is temporarily unoccupied. Then perhaps, if you are lucky, you can just "disappear" for a while. For technical students, who frequently have to work on a tiny writing area in a crowded classroom, conditions are seldom ideal. Outlining in the classroom, followed by writing in the seclusion of a library cubicle, is a possible alternative.

When Dan Skinner has found a peaceful location out of reach of the telephone, his friends, and even his boss, he can take a fresh sheet of paper and start writing. This is exactly where a second difficulty may occur. Equipped with an outline on his left, a pad of paper in front of him, and a row of sharpened pencils on his right, he may find that he does not know where to start. Or he may tackle the task enthusiastically, determined to write a really effective introduction, only to find that everything he writes sounds trite, unrealistic, or downright silly. Discouraged after many false starts, he sweeps the growing pile of discarded paper into the wastebasket or erases what he has typed from the computer's memory.

I have frequently advised technical people who have encountered this "no start" block that the best place for them to start writing is at paragraph two, or even somewhere in the middle. For example, if Dan finds that a particular part of his project interests him more than other parts, he should write about that part first. His interest and familiarity with the subject will help him to write those first few words, and keep him going once he has started. The most important thing is to *start* writing, to put any first words down, even if they are not the right words or ideas, and to let them lead naturally into the next group of ideas.

This is where continuity becomes essential. A writer who has found a quiet place to work, and has allowed sufficient time to write a sizable chunk of text, simply must not interrupt the writing process to correct a minor point of construction. If you are to maintain continuity, you must not strive to write perfect grammar, find exactly the right word, insert perfect punctuation, or construct effective sentences and paragraphs of just the right length. That can be done later, during revision. The important thing is to keep building on that rough draft (no matter how "rough" it is), so that when you do stop for a break you know you have assembled some words that can be worked up into a presentable document.

If, as he writes, Dan cannot find exactly the word he wants, he should jot down a similar word and draw a circle around it (or, at a computer terminal, type a question mark enclosed in parentheses immediately after it) as a reminder to change the word when the first draft is finished. Quite likely, when he reviews the draft, the correct word will spring to mind. Similarly, if he is not sure how to spell certain words, he should resist the temptation to turn to the dictionary, for that will disrupt the natural flow of his writing. Again, he should draw attention to the word as a reminder that he must consult his dictionary later.

I cannot stress too strongly that writers should not correct their work as they write. Writing and revising are two entirely separate functions, and they call for different approaches; they cannot be done simultaneously without the one lessening the effectiveness of the other. Writing calls for creativeness and total immersion in the subject so that the words tumble out in a constant flow. Revision calls for lucidity and logic, which will force a writer to reason and query the

Figure 1-5. Part of the author's first draft, typed at a computer terminal and then printed by a line printer. Note that the author has not stopped to hunt up minor details. Several revisions were made between this draft and the final product (see pages 189 and 190).

<u>Tenants Needs</u>

To find out what the building's tenants most needed in elevator service, we asked each company to fill out a questionaire (sp?). From their answers we were able to identify 5 factors needing consideration:

1. A major problem seems to be the length of time a person must wait for an elevator. Every tenant said we must cut out lengthy waits. A survey was carried out to find out how long people had to wait (during rush hours). This averaged out at 70 sec, more than twice the 32 sec established by Johnson (Ref?) before people get impacient (sp?). From this we calculated we would need 3 or 4 passenger elevators.

2. At first it seemed we would be forced to include a full-size freight elevator in our plan. Two companies (which?) both carry large but light displays up to their floors, but both later agreed they could hinge them, and if they did this they would need only 7 ft 6 in. width (maximum). They also said they did not need a freight elevator all the time, mainly

---

suitability of the words that have been written. The first requires exclusion of every thought but the subject; the second demands an objectivity that constantly challenges the material from the reader's point of view. Writers who try to correct their work as they write soon become frustrated, for creativity and objectivity are constantly fighting for control of their pens.

Most technical people find that, once they have written those first few difficult·sentences, their inhibitions begin to fall away and they write more naturally. Their style begins to change from the dull, stereotyped writing we expect from the average technical person, to a relaxed narrative that sounds as though the writer is telling a co-worker about the subject. As the writer becomes interested in the topic, speed builds up and he or she has difficulty writing or typing fast enough. The words may not be exactly those of the final report, but when the writer starts checking the first draft he or she will be surprised at the effectiveness of many of the sentences and paragraphs written earlier.

The length of each writing session will vary, depending on the writer's experience and the complexity of the topic. If a document is reasonably short, it should be written all at one sitting. If it is long, it should be divided into several medium-length sessions that suit the writer's particular staying power.

At the end of each session Dan should glance back over his work, note the words he has circled or questioned, and make a few necessary changes (Figure 1-5 is a page from a typical first draft). He must not yet attempt to rewrite paragraphs and sentences for better emphasis. Such major changes must be left until later, when enough time has elapsed for him to read his work objectively. Only then can he review his work as a complete document and see the relationship among its parts. Only then can he be completely critical.

TAKING A BREAK

When the final paragraph of a long report has been written, I have to resist the temptation to start revising it immediately. There are sections that I know are weak, or passages with which I am not happy, and the desire to correct them is strong. But I know it is too soon. I must take my pages, staple them together, and set them aside while I tackle a task that is completely unrelated.

Immediate reading without a suitable waiting period encourages writers to look at their work through rose-tinted glasses. Sentences which normally they would recognize as weak or too wordy appear to contain words of wisdom. Gross inaccuracies which under normal circumstances they would pounce upon go unnoticed. Paragraphs that may not be understood by a reader new to the subject, to their writers seem abundantly clear. Their familiarity with their work blinds them to its weaknesses.

The only remedy is to wait. If Dan has handwritten his report, he should arrange to have a double- or triple-spaced draft typed during this time. Not only will this make his work easier to read, but he may also be fortunate enough to find the typist has corrected some of his punctuation and spelling errors. If he has typed his report into a computer, he can run a spelling-check program and have a double-spaced copy printed out.

READING WITH A PLAN

If you have handwritten a report, always try to read from a *typed* draft so you will see the same cold, clinical type your readers will see. There is a danger in reading your own handwriting, for you may "hear" an emphasis which your readers will not notice when they read the typewritten words.

Dan Skinner's first reading should take him straight through the draft without stopping to make corrections, so that he can gain an overall impression of the report. Subsequent readings should be slower and more critical, with changes written in as he goes along. As he reads he should check for clearness, correct tone and style, and technical and grammatical accuracy.

CHECKING FOR CLEARNESS

Checking for clearness means searching for passages that are vague or ambiguous. If the following paragraph remained uncorrected, it would confuse and annoy a reader:

> *Muddled*   When the owners were contacted on April 15, the assistant manager,
> *Paragraph*   Mr. Pierson, informed the engineer that they were thinking of advertising Lot 36 for sale. He however reiterated his inability to make a definite decision by requesting this company to confirm their intentions with regard to buying the land within two months, when his boss, Mr. Davidson, general manager of the company, will have come back from a business tour in Europe. This will be June 8.

The only facts that a reader could be sure about after reading this paragraph are that the owners of the land were contacted on April 15 and that the general manager will be returning on June 8. The important information about the possible sale of Lot 36 is confusing. Probably the writer was trying to say something like this:

> *Revised*   The engineer spoke to the owners on April 15 to inquire if Lot 36 was
> *Paragraph*   for sale. He was informed by Mr. Pierson, the assistant manager, that the company was thinking of selling the lot, but that no decision would be made until after June 8, when the general manager returns from a business tour in Europe. Mr. Pierson suggested that the engineer submit a formal request to purchase the land by that date.

The more complex the topic, the more important it is to write clear paragraphs. Although the paragraph below is quite technical, it would be generally understood even by nontechnical readers:

Clear
Paragraph
A sound survey confirmed that the high noise level was caused mainly by the radar equipment blower motors, with a lesser contribution from the air-conditioning equipment. Tests showed that with the radar equipment shut down the ambient noise level at the microphone positions dropped by 10 dB, whereas with the air-conditioning equipment shut down the noise level dropped by 2.5 dB. General clatter and impact noise caused by the movement of furniture and personnel also contributed to the noisy working conditions, but could not be measured other than as sudden sporadic peaks of 2 to 5 dB.

Here paragraph unity has added much to readability. The writer has made sure that:

The topic is clearly stated in the first sentence (the topic sentence).
The topic is developed adequately by the remaining sentences.
No sentence contains information that does not substantiate the topic.

If any paragraph meets these basic requirements, its writer can feel reasonably sure that the message has been conveyed clearly.

Writers who know their subject thoroughly may find it difficult to identify paragraphs that contain ambiguities. A passage that is abundantly clear to them may be meaningless or offer alternative interpretations to a reader unfamiliar with the subject. For example:

Our examination indicates that the receiver requires both repair and recalibration, whereas the transmitter needs recalibration only, and the modulator requires the same.

This sentence plants a question in the reader's mind: Does the modulator require both repair and recalibration, or only recalibration? The technician who wrote it knows, because he has been working on the equipment, but the reader will never know unless he or she cares to write or phone and ask. The technician could have clarified the message easily, simply by rearranging the information:

Our examination indicates that the receiver requires both repair and recalibration, whereas the transmitter and modulator need recalibration only.

Sometimes ambiguities are so well buried that they are surprisingly difficult to identify, as in this excerpt from a chief draftsperson's report to a department head:

The Drafting Section will need three Model D7 drawing boards. The current price is $975 and the supplier has indicated that his quotation is "firm" for three months. We should therefore budget accordingly.

The department head took the message at face value and inserted $975 for drawing boards in the budget. Two months later he received an invoice for $2925. Unable by then to return the three boards, he had to overshoot his budget by $1950. This financial mismanagement was caused by the chief drafts-person, who omitted to state whether the price quoted applied to one drawing board or to three. If the word "each" had been inserted after $975, the message would have been clear.

Many ambiguities can be sorted out by simple deduction, although it really should not be the reader's job to interpret the author's intentions. Occasionally such ambiguities provide a humorous note, as in this extract from a field trip report:

> High grass and brush around the storage tanks impeded the technicians' progress and should be cleared before they grow too dense.

So that readers will not mistakenly think it is the technicians who are growing too dense, the two thoughts in this sentence should be separated:

> High grass and brush around the storage tanks impeded the technicians' progress. This undergrowth should be cleared before it grows too dense.

CHECKING FOR CORRECT TONE AND STYLE

How do you know when your writing has the right tone? One of the most difficult aspects of technical writing is establishing a tone that is correct for the reader, suitable for the subject, and comfortable for you, the writer. If you know your subject well and have thoroughly researched your reader you will most likely write confidently and, often, will automatically establish the correct tone. But if you try to set a tone that does not feel natural, or if you are a little uncertain about the subject and the reader, your reader will sense an unsureness in your writing. And no matter how skillfully you edit your work, the hesitancy will show up in the final sentences and paragraphs.

Finding the Best Writing Level. To check that he has set the right tone, Dan Skinner must assess whether his writing is suitable for both the subject matter and the reader. If he is writing on a specific aspect of a very technical topic, and knows that his reader is an engineer with a thorough grounding in the subject, he can use all the technical terms and abbreviations that his reader will recognize. Conversely, if he is writing on the same topic for a nontechnical reader who has little or no knowledge of the subject, Dan must alter his approach. He may have to generalize rather than state specific details, explain technical terms that he would normally expect to be understood, and generally write in a more informative manner.

Writing on a technical subject for readers who do not have the same technical knowledge as yourself is not easy. You have to be much more objective when checking your work, and try to think in the same way as the nontechnical reader. You can use only those technical words which you know will be recognized, yet you must avoid oversimplified language that may irritate the reader.

When Dan Skinner wrote this excerpt from a modification report, he knew his readers would be electronics engineers at radar-equipped airfields:

> We modified the M.T.I. by installing a K-59 double-decade circuit. This brightened moving targets by 12% and reduced ground clutter by 23%.

Although this statement would be readily understood by the readers for whom it was intended, to any reader not familiar with radar terminology it would mean very little. So when Dan reported on the same subject to the airport manager, he wrote this:

> We modified the radar set's Moving Target Indicator by installing a special circuit known as the K-59. This increased the brightness of responses from aircraft and decreased returns from fixed objects on the ground.

For this reader Dan has included more descriptive details and eliminated specific technical details that might not be meaningful to the airport manager. In their place he has made a general statement that aircraft responses were "increased" and ground returns "decreased." He also knew that the airport manager would be familiar with terms such as *Moving Target Indicator, responses,* and *returns.*

Now suppose that Dan also had to write to the local Chamber of Commerce to describe improvements in the airport's traffic control system. This time his readers would be entirely nontechnical, so he would have to avoid using *any* technical terms:

> We have modified the airfield radar system to improve its performance, which has helped us to differentiate more clearly between low-flying aircraft and high objects on the ground.

Keeping to the Subject.   Having established that he is writing at the correct level, Dan must now check that he has kept to the subject. He must take each paragraph and ask: Is this truly relevant? Is it direct? And is it to the point?

If Dan prepared his outline using the method described earlier, and followed it closely as he wrote his report, he can be reasonably sure that most of his writing is relevant. To check that his subject development follows his planned theme, he should identify the topic sentences of some paragraphs and check them against the headings in his outline. If they follow the outline, he has kept to the main theme; if they tend to diverge from the outline or if he has difficulty in identifying them, he should read the paragraphs carefully to see whether they need to be rewritten, or possibly even eliminated. (For more information about topic sentences, see Chapter 11.)

Technical writing should always be as direct and specific as possible. Technical writers should convey just enough information for their readers to understand the subject thoroughly. Technical writing, unlike literary writing, has no room for details that are not essential to the main theme. This is readily apparent in the following descriptions of the same equipment.

Literary Description   The new cabinet has a rough-textured dove gray finish that reflects the sun's rays in varying hues. Contrary to most instruments of this type, its controls are grouped artistically in one corner, where the deep black of the knobs provides an interesting contrast with the soft gray and white background. A cover plate, hardly noticeable to the layman's inexperienced eye, conceals a cluster of unsightly adjustment screws that would otherwise mar the overall appearance of the cabinet and would nullify the esthetic appeal of its surprisingly effective design.

Technical Description   The gray cabinet is extremely functional. All the controls used by the operator are grouped at the top right corner, where they can be grasped easily with one hand. Subsidiary controls and adjustment screws used by the maintenance crews are grouped at the bottom left corner, where they are hidden by a hinged cover plate.

Comparison of these examples shows how a technical description concentrates on details that are important to the reader (it tells *where* the controls are and why they have been so placed); it does not waste time being artistic. At the same time it observes the rules of good construction, using parallel structure to carry the reader easily through the description. Hence, it maintains an effective, businesslike tone.

Using Simple Words.   A writer who uses unnecessary superlatives sets an unnaturally pompous tone. The engineer who writes that a design "contains ultrasophisticated circuitry" seems to be trying to justify the importance and complexity of his or her work rather than saying that the design has a very complex circuit. The supervisor who recommends that technician Janice Smith be "given an increase in remuneration" may be understood by the company controller but will only be considered pompous by Janice. If he had simply written that Janice should be "given a raise," he would have been understood by both of them. Unnecessary use of big words, when smaller, more generally recognized, and equally effective synonyms are available, clouds technical writing and destroys the smooth flow that such writing demands.

Removing "Fat."   During the reading stage Dan should be critical of sentences and paragraphs that seem to contain too many words. He should check that he has not inserted words of low information content, that is, phrases and expressions that add little or no information. Their removal, or replacement by simpler, more descriptive words, can tighten up a sentence and add to its clarity. Low information content words and phrases are often hard to identify, because the sentences in which they appear seem to be satisfactory. Consider this sentence:

> For your information, we have tested your spectrum analyzer and are of the opinion that it needs calibration.

The words of low information content are "for your information" and "are of the opinion that." The first can be deleted, and the second replaced by "consider," so that the sentence now reads:

> We have tested your spectrum analyzer and consider it needs calibration.

Now try to identify the low information content words in this sentence:

> If you require further information, please feel free to telephone Mr. Thompson at 489-9039.

The phrase "if you require" is not entirely wrong, although it could be replaced by the single word "for," but "please feel free to" is archaic and should be eliminated. The result:

> For further information please telephone Mr. Thompson at 489-9039.

Tables 11-2 and 11-3 in Chapter 11 contain lists of low information content words and wordy expressions.

Inadvertent repetition of information can also contribute to excessive length. Although repetition can be an effective way to emphasize a point, in most cases its use is accidental. For example, Dan may explain something in one paragraph, then several paragraphs later say the same thing in different words. By welding paragraphs that repeat previously mentioned facts into a single, cohesive block of information, he will help to clarify and shorten his report. Similarly, he may repeat words in consecutive sentences:

> We tested the modem to check its compatibility with the remote terminal. After completing the modem tests we transmitted messages at low, medium, and high baud rates. The results of the transmission tests showed . . .

If he deletes the repeated words in sentence 2 ("After completing the modem tests") and sentence 3 (". . . of the transmission tests. . ."), he will achieve a much tauter paragraph:

> We tested the modem to check its compatibility with the remote terminal, and then transmitted messages at low, medium, and high baud rates. The results showed. . .

CHECKING FOR ACCURACY

Checking accuracy means examining one's work to ensure that the information is correct and that such technicalities as grammar, punctuation, and spelling have not been overlooked.

Nothing annoys readers more than to discover that they have been presented with information that is not absolutely accurate. They automatically assume that a writer knows the facts and has checked that they are correctly transcribed into the report. Errors may remain undetected for a long time, possibly until an inquiring reader starts conducting further tests and making calculations based on the author's results. Suspicion of an error in the original report can lead to tedious correspondence and wasted time until the inaccuracy is corrected. Worse, the readers' confidence in the writer, as well as the company, is downgraded.

There is no way to prevent some errors when quantities and details are copied from one document to another. These errors may be made either by the writer or by the typist. Therefore, Dan must personally check that facts, figures, equations, quantities, and extracts from other documents are all copied correctly. He must do this regardless of whether he typed the report at a computer terminal or a typist typed it from his handwritten draft. Even though he may feel the typist is responsible for proofreading the report, the ultimate responsibility for checking the accuracy of the written work is his.

This is also the time for Dan to identify poor grammar, inadequate punctuation, and incorrect spelling. He must do this with care because his knowledge of the subject may blind him to obvious errors. (How many of us have inadvertently written "their" when we intended to write "there," and "to" when we meant "too"? And these are words that a computer spelling checker will not question!) Other methods Dan can use to improve his report are suggested in Chapter 11.

REVISING ONE'S OWN WORDS

As Dan reads his work, he should make any corrections he finds necessary. Some corrections may require only minor changes to individual sentences; others may require complete revision of whole paragraphs, and even of complete sections. If Dan is working at a computer terminal he can make the changes himself and immediately see the improvements, either on the screen or on a printout. But if he is handwriting extensive revisions to a typed draft, he should have the draft retyped, and then he should carefully reread and revise it. This process should be repeated until a draft emerges which says exactly what he wants to say in as few words as possible.

As he reads, Dan must continually ask himself:

Can my readers understand me?

Will the person I am writing for be able to read my report all the way through without becoming lost?

What about other readers who might also see my report: Will they understand it?

Is the focus right?

Is my report reader-oriented?

Are the important points clearly visible?

Have I summarized the main points in an opening statement which the reader will see right away?

Is my information correct?

Is it accurate?

Is it complete?

Is all of it necessary?

Is my language good?

Is it clear, definite, and unambiguous?

Are there any grammatical, punctuation, or spelling errors?

Does every paragraph have a topic sentence (preferably at the start of the paragraph)?

Have I used any big "overblown" words where simpler words would do a better job?

Are there any low information content words and phrases?

Have I kept my report as short as possible while still meeting my readers' needs and covering the topic adequately?

By now Dan's draft should be in good shape. Any further reading and revising will be final polishing, and the amount will depend on the importance of the report. If his report is for limited or in-company distribution, a standard-quality job probably will suffice. But if it is to be distributed outside the company, or submitted to an important client, then Dan will spend as much time as necessary to ensure that it conveys a good image of both himself and H. L. Winman and Associates (his employer).

REVIEWING THE FINAL DRAFT

The final step occurs when Dan feels that his report is ready for final typing. Before charging across to the typing pool, or instructing the computer to print out a letter-quality copy, he should ask himself three more questions:

1. Would I want to receive what I have written?
2. What reaction will it incur from the intended reader?
3. Is this the reaction I want?

Now he must be extremely self-critical, ready to doubt his ability to be truly objective. If his answers are at all hesitant, he should ask an independent reviewer to read his report. Ideally, this person will be technically and mentally equivalent to the eventual reader, but not too familiar with the project. The reviewer must be able to criticize constructively, reading the report completely rather than scanning it, and making notes describing possible weaknesses and ambiguities. The reviewer should take time to discuss the report with Dan, and explain to him why certain sentences or paragraphs need to be reworked.

Dan Skinner will now be able to issue his report with confidence, knowing that he has fashioned a good product. The approach described here will not have made report writing a simple task for him, but it will have assisted him through the difficult conceptual stages, and helped him to read and revise more efficiently. When Dan has to write his next report, he will be much less likely to put it off until it is so late that he has to do a rush job.

COMPOSING AT A COMPUTER TERMINAL

Most of this chapter has been written assuming that you will be handwriting your letters and reports, which you will give to a typist. Except for short pieces, the typist will type your work twice: first as a rough draft for you to check over and make changes on, and then in the final form that you will sign and send to the ultimate reader. This scenario, however, is gradually changing.

Increasingly, engineers and engineering technicians such as Dan Skinner have a computer on or beside their desks. The computer may be a stand-alone unit (often referred to as a personal computer), or it may be part of a much larger system known as a "mainframe." In the stand-alone configuration the computer comes complete with a keyboard, video terminal, central processing unit (CPU), disk drive(s), and sometimes a line printer (a typewriter-like machine, driven by the computer). When part of a large system, only the keyboard and video terminal are at the user's position; the CPU, disk drives, and line printer are part of the mainframe—often in a remote location—and are shared by many users.

When Dan first had the computer at his desk he used it solely for solving technical problems and for mathematical calculations. More recently his computer has been equipped with a word-processing program, so now he is able to write (type) his reports directly into the computer and then call them back to his screen for editing. This saves him a lot of time, since he does not have to wait for a typed draft to come back from the typist and neither does he have to send his revised draft back for retyping. With the computer, the changes he makes between the first and second drafts are incorporated immediately, so that he has a final copy available as soon as his "editing" is complete.

WORD-PROCESSING SYSTEMS

A true word-processing system has been designed by its manufacturer specifically for writing. Known as a "dedicated" word processor, it provides powerful methods for setting up a letter or report, correcting it quickly and easily, moving paragraphs around, centering and underlining words, justifying (making straight) the right-hand margin, and so on. All the writing software is built into the system, so that additional programs normally are not needed. Its limitation, however, is that a dedicated word processor has much less computing capability, and needs additional programs to perform many computing functions.

Secretaries and stenographers normally have dedicated word processors at their desks, since their role primarily is to provide typing and secretarial services.

A true computer, whether it is a stand-alone unit or a large mainframe system, treats word processing as a secondary operation. Stand-alone units normally require a special program if they are to be used as word processors, but the programs are not as powerful as those in a dedicated system; the user has fewer commands and capabilities available. Large mainframe computers often have powerful word processing programs built into them, but even then they are not as convenient as, and lack some of the more sophisticated capabilities of, a dedicated word processor. The gap, however, is closing. As better word processing software is developed the distinction between dedicated word processors and computer word processors is gradually being eroded.

As an engineer, technologist, or engineering technician, you are far more likely to have a computer than a dedicated word processor at your desk.

PREPARING TO WRITE

The six pre-writing planning stages described earlier in this chapter (see Planning the Writing Task) apply also to writing with a word processor. But your method may differ: where previously I suggested you write random notes on a sheet of paper, now you can sit at the keyboard and *type* the random headings for stage three. And you can—with practice—just as easily insert additional headings, delete random headings, and shuffle topic headings around until you have developed a usable outline. Then from the line printer you can obtain a printout (known as hard copy) and use it as your writing guide.

But before you start writing you have to consider other aspects which were not necessarily a concern when you were writing with a pen or pencil. For example, you will have to type some specific instructions into the computer to define:

- The number of lines you want on a page, and the width of your planned typing lines (the number of characters or letters you want in a line).
- The width of margin you want on either side of the text.
- Whether you want the right margin to be justified (straight) or ragged.
- Where you want the page numbers to be positioned (top or bottom of the page, and either centered or to one side of the page); on most systems page numbers are printed automatically, but you can select where they are to appear.
- The line spacing you want (single or double), and how many blank lines you want between paragraphs (normally one or two).

- Whether the first line of each paragraph is to be indented or set "flush" with the left margin; and, if indented, how many spaces the indention is to be.
- For long words at the end of a line, whether you or the computer is to decide where the word is to be hyphenated (you can also elect no hyphenation.)

These are the basic instructions required for most word processors. Depending on the sophistication of the word processor you use, you should be able to give additional instructions for formatting your work.

WRITING (TYPING) AND EDITING

Just as with ordinary writing, I recommend that you settle down and type the first draft of your letter or report without stopping to go back and make corrections one paragraph at a time. This time, however, you cannot walk around and look for a secluded place to work, since your work position will be dictated by the location of the computer terminal (unless you are fortunate enough to have a portable unit to work with). Consequently you will need to establish an understanding between yourself and your associates that one does not interrupt a person who is composing at a computer terminal.

There are two ways you can revise and edit. You can call the whole document back to your screen, read it, and make corrections without using a paper printout. Or, you can request a printout and read and correct the hard copy, returning to the keyboard and screen to make final corrections. Both methods are valid.

Many word processing programs now have a built-in spelling checker, or your company may have purchased a spelling-check program to be used with the system. When using a spelling checker remember that it is not infallible. You may use a word—particularly a technical word—which is not in the checker's memory, or you may mistype a word but accidentally form another word the checker recognizes. For example, you may plan to type "word" but accidentally type "work," which the checker would not recognize as an error. (Remember that the checker does not read what you have written; it merely examines every word and compares each word to its master list.)

WORKING WITH A SECRETARY/TYPIST

In the more traditional setting, a letter- or report-writing team comprises two people: the person who writes the letter or report (the author) and the person who types it (the secretary). Good communication between author and secretary is extremely important, otherwise both will become frustrated and irritated; the author, because the work is not being produced in the intended form, and the secretary, because the work repeatedly needs to be retyped. Consequently much time is unnecessarily wasted.

You can help a secretary produce your letters and reports in the form you want them if you *remember to keep the secretary in the picture*. These are details secretaries need to know:

- Whether the work is to be typed as a draft or in final form, ready for your signature. (Nothing annoys a secretary more than to painstakingly prepare a business letter in its final form, and then to see you writing all over it because you *assumed* that the secretary understood you only wanted a draft.)
- Whether the original or a copy of the report or letter will be sent to the addressee. If the original will be mailed, then the secretary must make very careful corrections; but if a copy will be mailed, then the secretary can use white correction fluid fairly freely. (Of course if your secretary is working at a word processor terminal, rather than on an electric or electronic typewriter, corrections on the final copy are eliminated and so this aspect can be ignored.)
- Any "mechanical" considerations you have in mind, such as the size of the typing area (the number of lines vertically), the width of the column, the spacing between lines (single spacing, 1½ line spacing, or double spacing), and the "pitch" of the type (10 or 12 characters per inch).
- Where spaces are to be left for illustrations, with exact dimensions for each.
- The date the typing is required.

Assignments

Exercise 1.   Which do you feel is the better way for *you* to develop an outline for a report: the organized method or the "random" method? Explain why.

Exercise 2.   (a) What are the six stages advocated for planning a report? (b) Must the stages be followed exactly in the sequence listed?

Exercise 3.   If you had a long project report to write next month, describe where you would write it and why you would write it there.

Exercise 4.   Is it better to write a report without stopping to "clean up" the construction along the way, or to write a page at a time and to edit it before going on to the next page? Explain why.

Exercise 5.   What two factors will help you write more confidently, and probably help you set the right tone?

Exercise 6.   Explain why one of the following two sentences is the better conveyor of information.

(a)   The filter system was complex and difficult to understand.

(b)   The complexity of the filter system made it inherently difficult to understand.

Exercise 7.   From the list of five main questions that you, as a writer, should ask yourself during the revision stage (see the boldface questions on pages 22 and 23), which do you think is most important? Explain why.

Exercise 8.   What kind of person is best equipped to do a final review of a report you have written?

2

Meet "H. L. Winman and Associates"

Almost every project you read or write about in this book is based upon events affecting two companies: H. L. Winman and Associates, a Cleveland firm of consulting engineers, and its affiliate, Macro Engineering Inc, of Phoenix, Arizona. To help you identify yourself in these working environments, the following pages contain short histories and partial organization charts of the companies, plus brief biographical sketches of the men and women who have brought them to their current position and who most influence how their engineers, scientists, technologists, and technicians write. There is also a description of what your role might be if you were employed by either company.

CORPORATE STRUCTURE

H. L. Winman and Associates was founded by Harvey Winman some 30 years ago. Starting with a handful of engineers working in a small office in one of the older sections of Cleveland, Harvey gradually built up his company until now it has 860 employees: 380 at head office, 340 at the Southern affiliate in Phoenix, and 140 in smaller offices throughout the United States, Canada, and Mexico. Some are branches of H. L. Winman and Associates (there may be one in your area), while others are local engineering firms like Macro Engineering Inc that Harvey has purchased. Figure 2-1 illustrates the company's two main locations.

HEAD OFFICE: H. L. WINMAN AND ASSOCIATES

The head office of H. L. Winman and Associates is located in a modern building at 475 Reston Avenue, a major thoroughfare in the business district of Cleveland. A partial organization chart in Figure 2-2 shows Martin Dawes as

Photo A. Harvey L. Winman. Photo: Anthony Simmonds.

president, with nine department heads, who are responsible for the day-to-day operation of the company, reporting directly to him.

Over the years H. L. Winman and Associates has developed an excellent reputation among its clients for the management of large construction projects in remote areas of the United States and Canada. It has managed the construction of whole towns, airfields, and dams for large hydroelectric power generating stations; engineered, designed, and supervised construction of major traffic intersections, shopping complexes, bridges, and industrial parks; and designed and supervised numerous large buildings such as hotels, schools, arenas, and manufacturing plants. More recently, the company has entered the systems engineering field. Some of its current studies involve problems in air pollution, protection of the environment, and the effects of oil exploration and pipeline construction on the Alaskan arctic tundra. To handle so many diverse tasks, Harvey has built up an extremely adaptable and flexible staff with representation from many scientific and technical disciplines.

Although he is past retirement age, Harvey maintains an active interest in his company's business. Short, with a round face topped by a bald head with a fringe of hair, he is often seen strolling from department to department. He stops frequently to talk to his staff (most of whom he knows by first name) and to inquire about both the projects they are working on and their home lives. Well-liked and thoroughly respected by all, from senior management down to the newest clerk, he is a manager of the old school, a type seldom seen in today's brusque business atmosphere.

The company's chief executive is Martin Dawes, to whom Harvey, anticipating his coming retirement, has handed full operating control. Martin is a civil engineer who developed a fine reputation for designing and constructing hydro-

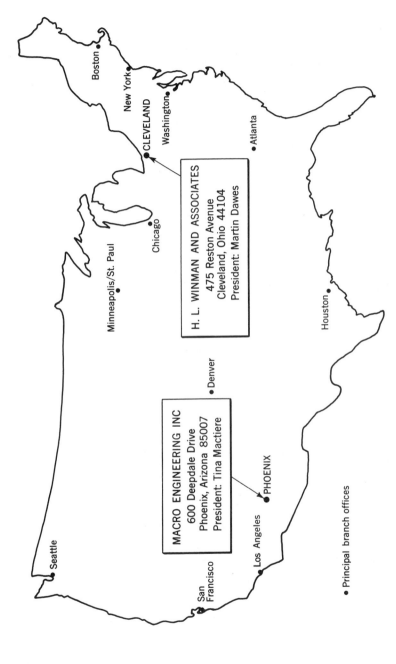

H. L. WINMAN AND ASSOCIATES
475 Reston Avenue
Cleveland, Ohio 44104
President: Martin Dawes

MACRO ENGINEERING INC
600 Deepdale Drive
Phoenix, Arizona 85007
President: Tina Mactiere

• Principal branch offices

Boston

New York

CLEVELAND

Washington

Atlanta

Chicago

Minneapolis/St. Paul

Houston

Denver

PHOENIX

Los Angeles

San Francisco

Seattle

Figure 2-1. Locations of H. L. Winman and Associates' offices.

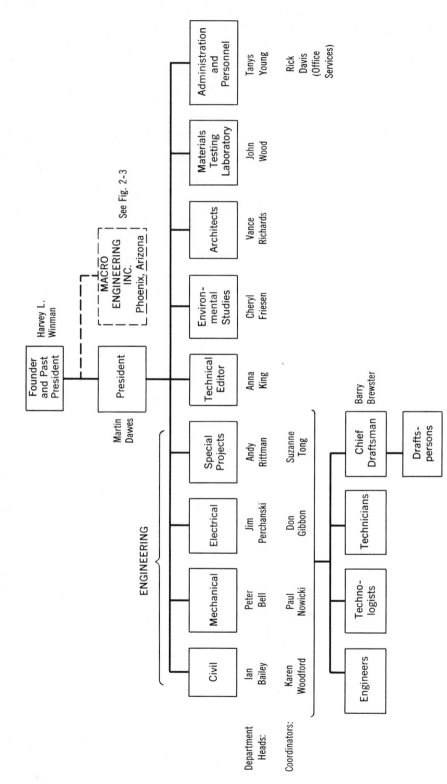

Figure 2-2. Partial organization chart—H. L. Winman and Associates.

electric dams in remote, mountainous country. An ardent fisherman, he has often combined field projects with trips into the surrounding countryside, from which he has returned laden with salmon or trout. Before becoming president he was vice-president for four years and, before that, Head of the Civil Engineering Department for 15 years.

Both Harvey Winman and Martin Dawes write well, and they expect their staff to do likewise. They recognize that the written word is the means that most often conveys an image of their company to customers. So, to ensure that all letters, reports, and company publications are of top quality, they have introduced a technical editor into their management team.

The editor is Anna King. Born in New Orleans of an English father and French mother, she grew up fully bilingual. Although her strength was languages, she wanted to be an engineer like her father and graduated with a Bachelor of Science in Electrical Engineering. She worked for the South Georgia Power and Light Company for nine years, first in the relay department and later in development engineering, where her capabilities both as engineer and report writer soon became evident.

Eventually Anna decided to move to ski country, and so accepted an offer from Harvey Winman to become H. L. Winman and Associates' technical editor. "Yours will be an exacting job," Harvey said on her first day. "You will be setting writing standards for the whole company. I don't want my engineers to think you are here simply to do their writing for them. Your role is to encourage them to become better writers."

SOUTHERN AFFILIATE: MACRO ENGINEERING INC

To increase his company's presence in the southern United States, Harvey Winman purchased two companies in the Phoenix area and amalgamated them to form Macro Engineering Inc (MEI). At the same time he achieved a second aim, which was to increase H. L. Winman and Associates' expertise in electronic and computer engineering. The first company he acquired was Mactiere Microelectronics, which specialized in designing computer interface systems. The second was Robertson Engineering Company, which had a well-established reputation for designing and custom-manufacturing sophisticated electronic, nucleonic, and electromechanical instruments. This combination, Harvey felt, would introduce some high-tech involvement into H. L. Winman and Associates' activities.

Mactiere Microelectronics' president was then Tina R. Mactiere, who had acquired full ownership of the firm two years earlier. A computer engineering graduate, she worked at first in the research and development department of Compuchip Corporation in California's Silicon Valley, and then responded to an uncle's invitation to join the firm he owned in Phoenix. She worked for him as a senior design engineer for five years, during which he discovered she was not only a highly competent designer but also a top-notch manager. And so, two years before his retirement, he appointed her as chief executive officer.

Photo B. Tina R. Mactiere, President, Macro Engineering Inc. Photo: Maria Cerullo.

Photo C. Wayne D. Robertson, General Manager, Macro Engineering Inc. Photo: David Portigal.

Wayne G. Robertson formed Robertson Engineering Company after a stint as an avionics officer with the United States Air Force. Gathering several former Air Force electronics specialists together, he set up a small equipment repair and maintenance service in Phoenix. He managed and marketed his business well, and the firm was quickly recognized for its competence and reliability. It grew steadily, with the emphasis gradually placed on research and development of complex communications systems and their installation and maintenance.

When the two companies amalgamated, Harvey Winman recommended combining the first three letters of Tina's company name and the first two letters of Wayne's, to form the new identity of Macro Engineering Inc. Tina and Wayne readily agreed to head the new company jointly, and that Tina would become President and Chief Executive Officer, and Wayne would be General Manager. Because Wayne had developed strong marketing skills, frequently travelling to South America, Australasia, and Europe to describe the technical capabilities of his company's equipment, they decided he should also retain his role as marketing manager. A partial organization chart in Figure 2-3 shows the overall structure of the company and its Engineering Department.

Macro Engineering Inc has approximately 340 employees, of whom 215 are engineers, technologists, and engineering technicians. The company occupies the first two floors of the Wilshire Building, an office and light manufacturing complex in Phoenix, owned by the Wilshire Insurance Company.

Like Harvey Winman, both Tina and Wayne recognize the importance of communication in their business, and particularly stress that the operating instructions and training manuals that accompany their instruments be abundantly clear. To this end, they employ a small technical writing group that can write manuals not only in English, but also in French, German, and Spanish, and occasionally have Anna King fly in from Cleveland when particularly important technical proposals have to be prepared.

YOUR ROLE AS A TECHNICAL EMPLOYEE

As you read the remaining chapters in this book, you will now be able to identify your "employer" and visualize yourself as an "employee." But you still need to know what your role will be as an engineer, technologist, or engineering technician working for either H. L. Winman and Associates or Macro Engineering Inc.

WORKING FOR A CONSULTING FIRM

Consultants provide specialist help to other companies or government organizations. Most clothing manufacturers, for example, are unlikely to have engineering help on staff, other than perhaps an electrical and a mechanical technician to deal with day-to-day maintenance. A clothing manufacturer who

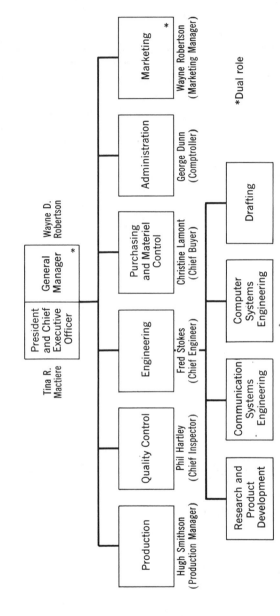

Figure 2-3. Partial organization chart showing Engineering Department of Macro Engineering Inc.

wants to build an extension to the plant, introduce robotics into the manufacturing operation, or correct an in-plant air pollution problem, does not have the expertise to do the work. Conceivably, the manufacturer could go directly to a builder and ask for a design and construction estimate, or to a manufacturer of robotics equipment and ask for a proposal, or to an air-conditioning company and request improved air circulation, but in each case the clothing manufacturer would be buying limited, rather biased, expertise. The builder, robotics manufacturer, and air-conditioning installer would each tend to recommend a design, a system, or materials that primarily employ that company's wares.

By calling in an engineering consultant the manufacturer is in effect saying, "We have a problem and we want you to use all the resources at your command to investigate it, recommend how it can be corrected, and (sometimes) manage the corrective action for us." Because consultants are not "tied in" with particular builders, robotics manufacturers, air-conditioning installers, and so on, they have an unbiased perspective from which to view their clients' needs. Some consultants specialize in a particular field, such as civil engineering or environmental control, while others offer a fairly wide range of consulting services, as do H. L. Winman and Associates. And if a general consultant does not specialize in a particular area in which expertise is needed, the consultant will collaborate with a specialist consultant to provide the necessary services for a client. Often, too, a client simply does not want to be involved in handling an investigation, issuing specifications to contractors, and coordinating and supervising contractors' work. "That can be one big headache," most clients agree. And they would rather leave all the details for a consultant to handle.

The steps in a client-engineering consultant relationship are approximately as described below and illustrated in Figure 2-4. There are two phases: the initial contact between the client and consultant, followed by a study or investigation; and coordination of a building or repair contract that may follow as a result of the investigation.

1. The client approaches the consultant, describes what is required or the problem that needs to be corrected and, if the project is likely to be extensive, asks for a preliminary proposal and a cost estimate. (If a project is small or straightforward, the client will more often dispense with the proposal and instruct the consultant to go ahead with the project.)
2. The consultant prepares a proposal outlining:
   - The requirements or the problem in detail (to demonstrate to the client that the consultant fully understands the client's needs and has interpreted them correctly).
   - How the project will be done (often presented as a series of steps or stages).
   - What human and material resources will be used.
   - The report that will be prepared at the end of the investigation.
   - Previous experience on similar projects (to demonstrate the consultant's capability).
   - The costs for: (a) carrying out the investigation and preparing the report; and (b) carrying out the steps that will be recommended in the report.

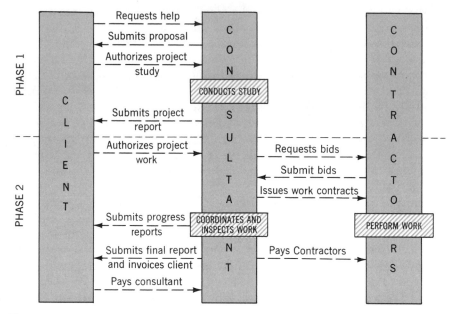

Figure 2-4. A simplified view of communication between (1) the client and consultant, and (2) the consultant and contractors.

3. If the consultant's proposal is accepted, the client writes a letter or issues a purchase order authorizing the consultant to go ahead with the project.
4. The consultant conducts the investigation and then prepares a detailed report outlining what was found out and what corrective action needs to be taken. (If the project is lengthy, periodically the consultant will send a progress report to the client.)

Many client-consultant arrangements stop here. If, however, a subsequent project calls for contractors to renovate the premises, install or modify equipment, or construct an extension to the building, the client-consultant relationship continues:

5. The client authorizes the consultant to manage the project through to completion.
6. The consultant then:
   - Prepares a work plan and job specification.
   - Calls for bids from contractors.
   - Issues contracts to the most suitable contractors.
   - Coordinates and monitors the contractors' work.
   - Deals with problems.
   - Submits regular progress reports to the client.
   - Inspects completed work, asks for correction of unsatisfactory work, and then pays the contractors.

- Prepares a job completion report and invoice for the whole project, and submits both to the client.

As an engineer, technologist, or engineering technician you may be involved in any of these stages, and you will be expected to write reports at the appropriate times both for your own management and for your company's clients. Good communication is an essential component in any client-consultant relationship, because clients who are "kept in the picture" subconsciously feel that the consultant is effectively looking after their interests. Conversely, if a consultant does not keep a client adequately informed of progress, the client may feel that the consultant is not really concerned about the client's needs, and the client may seek help elsewhere for the next consulting job.

WORKING FOR AN ENGINEERING OR TECHNICAL FIRM

An engineer or engineering technician employed by a technical firm may still provide services to clients and customers. You may be a technical service representative who installs, repairs, modifies, and maintains equipment leased or sold by your company to the client. As such you will be expected to be not only an expert on the technical aspects but also an expert in interpersonal relations. The impression you convey to a client's employees can create either a positive or a negative image of your company. Whereas a positive image may convince clients that they should do further business with your company, a negative image may make them hesitate and possibly look elsewhere when they next need to purchase new equipment. Hence, a technical service representative is both a salesperson and a technical specialist.

As a technical service representative you will write reports describing both routine maintenance and special conditions. After each service visit you will have to describe what you observed about the equipment, what repairs or adjustments you made, and what components you installed or replaced. Generally, much of this information will be written onto a preprinted form that has appropriate spaces for filling in details. But when special visits are made or problems occur, you will more likely have to write a short investigation report in which you describe a problem and how you resolved it, or an inspection report in which you describe equipment condition. (For examples of such reports, see Chapter 4.)

Large contractors keep engineers and engineering technicians on staff to direct, coordinate, and supervise the work of their installation teams, and to provide engineering assistance when problems occur. Much of the time you will work at the job site, and will keep management at head office informed of progress by submitting reports describing the teams' progress and problems encountered. (In this capacity, you may be on the payroll of one of the contractors providing services to the consultants described earlier.)

If you are employed by a design company or a manufacturer, your role will be to provide engineering or technical services. As an engineer or engineering technician you may:

- Design new products, processes, or procedures, or modifications to existing systems.
- Investigate manufacturing problems or assembly line "glitches."
- Propose improved manufacturing methods.
- Install and test prototype equipment and components.
- Instruct operators and users of new or modified equipment.
- Inspect work performed by assemblers and junior technicians.

Your reporting will have to be comprehensive, for you will be expected to describe what you have seen, to present management with an analysis or justification, and sometimes to recommend corrective action. You will have to use all the report-writing skills suggested in Chapters 4 through 7 for written reports, and Chapter 9 for spoken reports.

3

Technical Correspondence

Because the letter is our most common means of written communication, we frequently overlook its importance. We scribble down a few words and hope that someone will get the message; we bury the message under a mountain of unnecessary words; or we adopt an unnatural style that conceals our character and robs the message of humanity. Many of us are so accustomed to writing letters to friends and relatives that we find it difficult to transfer from the easygoing style of the personal letter to the more formal style of the business letter. This is where we err. Instead of letting our business letters be friendly, interesting, and persuasive, we force them to be stilted and dull, thinking that we are being businesslike.

CLARITY AND CONCISENESS

When we write personal letters to friends and relatives, we do so partly to be sociable and partly to convey a message. The length of our letters, and the amount of additional information we include, is unimportant. We can ramble along with very little organization, inserting interesting sidelights and comments as they occur to us. We assume our reader is pleased to have a letter from us, and so we use it mainly to portray what has happened since we last wrote.

But when we write a business letter we have to be much more direct. Our readers are busy people who are interested only in facts; information they do not need irks them. For these people we must keep strictly to the point. We must check that our letters are in focus, well-planned, brief, and clear, and that our facts are simply but effectively presented.

FOCUS

Letter writing is like photography: you have to center the image in the viewfinder and then focus the image sharply before you snap the picture. But first you have to decide exactly what you want to depict; otherwise viewers will wonder why they are looking at the picture.

Identify the Message.   Before you pick up a pen to write a letter, identify clearly in your mind why you are writing and what your reader most needs to hear from you. Then focus sharply on this information by placing it right up front, where it will be seen immediately. Your opening words should capture the reader's attention.

To open a letter with background information rather than the main point makes a reader wonder why you are writing. This happened to Jim Connaught as he started to read this letter from Don McKelvey:

Dear Mr. Connaught:

I refer to our purchase order No. 21438 dated April 26, 19xx, for a Vancourt microcopier model 3000, which was installed on May 14. During tests following its installation your technician discovered that some components had been damaged in transit. He ordered replacements and in a letter dated May 20 informed me that they would be shipped to us on May 27 and that he would return here to install them shortly thereafter.

It is now June 10, and I have neither received the parts nor heard from your technician. I would like to know when the replacement parts will be installed and when we can expect to use the microcopier.

Sincerely,
Don McKelvey

Jim had to read more than 70 words before he discovered what Don wanted him to do. He would have known immediately if Don had started with his main point:

Dear Mr. Connaught:

We are still unable to use the Vancourt 3000 microcopier we purchased from you on April 26. Please inform me when I can expect it to be in service.

And Don would have found his explanation was a little shorter and easier to write:

The microcopier was ordered on P.O. 21438 and installed on May 14. During tests, your technician discovered that some components had been damaged in transit. He ordered replacements and in a letter dated May 20 informed me that they would be shipped to us on May 27, and that he would return here to install them. To date, I have neither received the parts nor heard from your technician.

Sincerely,

Focus the Message. Although Don knows he should always start a letter or report with the main point, he rarely does so because it makes him feel he is being too abrupt or "pushy." If you feel the same way, you can borrow a technique used by Anna King, H. L. Winman and Associates' technical editor. Whenever Anna starts a letter, she first writes these six words:

I want to tell you that . . .

and then she finishes the sentence with the main point, like this:

I want to tell you that . . . the environmental data you submitted to us on October 8 will have to be substantiated if it is to be included in the TACMORE study.

Anna then deletes the first six words (the *I want to tell you that* phrase), to create an in-focus opening statement:

The environmental data you submitted to us on October 8 will have to be substantiated if it is to be included in the TACMORE study.

Here are three more examples of opening statements, all of which originally contained the opening words *I want to tell you that:*

Seven defective castings were received in shipment No. 308.
We will complete the XRS modification on June 14, eight days earlier than originally scheduled.
Your excellent paper "Export Engineering" arrived too late to be included in the Midwest Conference program.

If an opening statement seems too harsh or abrupt when the *I want to tell you that* phrase is removed, Anna inserts one or two additional words to soften its impact. For example, she would probably write "I regret that" in front of the third opening statement:

I regret that your excellent paper . . . (etc.)

Avoid False Starts. Anna King cautions H. L. Winman engineers against starting their letters with awkward expressions which can lead them into writing long, rambling sentences that say very little. Anna equates such expressions with "spinning one's wheels" (see Figure 3-1). The problem is that we become accustomed to seeing them at the start of other people's letters, and so we use them ourselves rather than coming to grips with the real subject. For example:

In answer to your inquiry of December 7, concerning erroneous readouts you are experiencing with your Mark 1 Analyzer, and our subsequent telephone conversation of December 18, during which we tried to pinpoint the fault . . .

WHEN YOU WRITE A LETTER . . .

Never start with a word that ends in "ing":

> *Referring . . .*
> *Replying . . .*

Never start with a phrase that ends with the preposition "to":

> *With reference to . . .*
> *In answer to . . .*
> *Pursuant to . . .*

Never start with a redundant expression:

> *I am writing . . .*
> *For your information . . .*
> *This is to inform you . . .*
> *The purpose of this letter is . . .*
> *We have received your letter . . .*
> *Enclosed please find . . .*
> *Attached herewith . . .*

IN OTHER WORDS . . .

DON'T SPIN YOUR WHEELS!

Figure 3-1. Anna King's warning to H. L. Winman engineers.

When a sentence grows as long and complicated as this, there is always a danger that its writer may stop at this point and thus write an incomplete sentence. A much better opening sentence leads straight into the problem and the probable result, and then stops:

> *I want to tell you that . .*

> I believe the problem with your Mark 1 Analyzer exists in the extrapolator circuit. Following your inquiry of December 7, in which you described the erroneous readouts you are experiencing, we examined . . .

If you can start a letter with an informative, meaningful opening statement, you are also using the pyramid technique recommended in Chapter 11.

PLAN

When you have identified and written the main message, you next have to select, sort, and arrange the remaining information you will convey to your reader. These details should amplify the bare message you have already presented, and frequently provide evidence of its validity. For example, this main message presents its reader with costly news:

> Dear Mr. Larsen:
>
> Tests of the environmental monitoring station at Wickens Peak show that 60% of the instruments need to be repaired and recalibrated at a cost of $5265.

Upon reading this opening sentence, Mr. Larsen would expect the remainder of the letter to tell him why the repairs are necessary, exactly what needs to be done and, briefly, how the total cost was derived. Simply providing him with a list of the instruments that require repairs and their individual repair costs would not be enough to satisfy his curiosity.

Develop the Message. If you are to provide your readers with the exact information they need, you first have to identify what questions will be foremost in their minds after they have read your main message. This means asking yourself what questions a reader is *likely* to ask, and then providing your own answers. Generally, these questions can be formed from six basic questions:

> *Who?*, *Where?*, *When?*, *Why?*, *What?*, and *How?*

You then apply these questions to the main message, and select which are applicable in your particular case. For example, Mr. Larsen probably would want to know:

> *Why* (are the repairs necessary)?
> *What* (repairs are needed)?
> *How* (were the costs calculated)?

He would not want to know (because the questions are irrelevant):

> *Who* (was involved)?
> *Where* (did this happen)?
> *When* (did this happen)?

The answers to the first three questions provide the details that you will use to develop the remainder of the letter to Mr. Larsen.

Arrange the Parts. The answers to the questions that you assume your reader will pose now have to be arranged in a logical, coherent sequence. A three-part arrangement is suitable for most letters.

Part 1 is a **Background** section which covers factors such as: what has happened previously; who was involved; where and when the event occurred or the facts were gathered; and, sometimes, the purpose of the letter:

> Our electronics technicians examined the Wickens Peak monitoring station on May 16 and 17, in response to your May 10 request to Mr. Patrick Friesen.

Part 2 contains **Facts** which amplify the main message. They provide the specific details the reader needs to fully understand the situation or to be convinced of the need to take further action:

> Most of the damage was caused by a tree northwest of the site that fell onto the station during a storm on April 23 and damaged parts of the roof and north and west walls. Instruments along these walls were impact-damaged and then soaked by rain. Other instruments in the station also were affected by the moisture.

> Major repairs and recalibration are required for the 16 instruments listed in attachment 1, which describes the damage and estimated repair cost for each instrument. This work will be done at our Shepperton repair depot for a total cost of $3405. Minor repairs, which can be performed on site, are necessary for the 27 instruments listed in attachment 2. These on-site repairs will cost $1860.

These two paragraphs clearly answer the *Why?*, *What?*, and *How?* questions. Note that, rather than clutter the middle of the letter with a long list, the writer has placed all the details in two attachments and summarized only the main points (the quantities and the total costs) in the body of the letter.

Part 3 contains the **Outcome,** or result. If the writer wants the reader to *do* something, then the required action should be stated here:

> If these repair costs are acceptable, please telephone me or telex your approval so that I can send in our repair crew.

> Sincerely,

But if the reader is not required to take any action, the **Outcome** simply sums up the main information contained in the letter, often by providing a result:

> I have obtained Ms. Korton's approval to perform the repairs and a crew was sent in on May 23. They should complete their work by May 31.

> Sincerely,

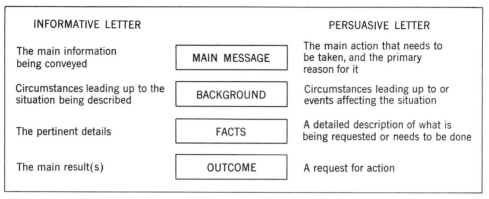

INFORMATIVE LETTER		PERSUASIVE LETTER
The main information being conveyed	**MAIN MESSAGE**	The main action that needs to be taken, and the primary reason for it
Circumstances leading up to the situation being described	**BACKGROUND**	Circumstances leading up to or events affecting the situation
The pertinent details	**FACTS**	A detailed description of what is being requested or needs to be done
The main result(s)	**OUTCOME**	A request for action

Figure 3-2. Writing plan for a business letter or memorandum.

These three parts, preceded by the main message, are illustrated in Figure 3-2. The figure also shows that all business letters and memorandums fall into one of two categories:

1. Letters that purely *inform,* for which no response or action is required from the reader. A typical informative memorandum is shown in Figure 3-3.
2. Letters that *persuade,* which require the reader to act or respond. The letter to Mr. Larsen, shown above as a summary followed by three parts, is a typical persuasive letter *because it asks for his approval.*

The four parts of a technical business letter may not necessarily represent exactly four paragraphs. The letter to Mr. Larsen has five paragraphs, while the memorandum to Ms. Mactiere in Figure 3-3 has only two paragraphs.

Close on a Strong Note. You may feel you should always end a letter with a polite closing remark, such as: *I look forward to hearing from you at your earliest convenience,* or *Thanking you in advance for your kind cooperation.* In contemporary business correspondence—and particularly in technical correspondence—such closing statements are not only outmoded but also weaken your impact on the reader. Today, you should close with a strong, definite statement.

In effect, the **Outcome** part of the letter provides a natural, positive close, as illustrated by the final sentences in the letters to Mr. Larsen and Ms. Mactiere. You should resist the temptation to add a polite but uninformative and ineffective closing remark. Simply sign off with "Regards," "Sincerely," or "Cordially."

BREVITY

Brevity means selecting and arranging information so that it fits into the smallest possible space. When packing a small suitcase for a short business trip, you have to be selective and pick out only the clothing for the environment you will encounter, and then arrange the items in the suitcase so they will arrive in

---

MACRO ENGINEERING INC

From: _____Kevin Toshak_____   Date: _____October 22, 19xx_____

To: _____Tina Mactiere_____   Subject: _____Installation of TL-680_____

_____   _____Monitor at WRC_____

I have installed a TL-680 monitor unit in room 215 at the ◄── *Main Message*
Wollaston Research Center, as instructed by your memorandum
of October 15, 19xx.   ◄── *Background*

The unit was installed without major difficulties,
although I had to modify the equipment rack to accept it ◄── *Facts*
as shown in the attached sketch. Post-installation tests
showed that the unit was accepting signals from both the ◄── *Outcome*
control center and the remote site.

K.T.

Figure 3-3. An informative letter in memorandum format.

good condition. In business writing you have to be equally selective, limiting your choice only to the essential elements and then arranging them in a coherent sequence so they will be recognized easily when they arrive on the reader's desk.

Brevity means keeping the message short: short letters, short paragraphs, short sentences, and short words.

Short Letters.   A business reader will tend to react readily to a short letter, viewing its writer as an efficient purveyor of information. In contrast, the same reader may view a long letter as "heavy going" (even before reading it) and tend to put it aside to deal with later. A short letter introduces its topic quickly, discusses it in sufficient depth, and then closes with a concluding statement. Its length should be dictated by only one factor: the number of words its writer needs to convey the message adequately and coherently.

I know of a company in which the managing director has ruled that no letters issued by the organization should exceed one page. This is an effective means for forcing the staff to be brief, but it must be unbearably severe on some writers. On occasion, letters have to extend to a second and even a third page. Limiting one's imagination and creativity at such times might reduce a convincing argument to a terse statement that fails to satisfy the reader's interest. Brevity at the expense of lucidity is false economy.

Very long letters can be shortened, however, by borrowing a technique from report writing. Instead of placing all your information in a letter, consider shaping it into a semiformal report and then summarizing the highlights (particularly the purpose and the outcome) into a short letter to be placed at the front of the report, so that the report becomes *an attachment* to the letter. Or, if a letter contains supporting information such as a series of test results, a cost analysis, an excerpt from another document, or similar data that is not immediately relevant, place this data on a separate sheet and label it as an attachment. Always refer to each attachment in the body of the letter, preferably by drawing a main conclusion from it, as has been done here:

> During our investigation we took a series of noise readings to determine sound level variations at night, during the day, and on weekends. These readings (see attachment) show that a maximum of 55 dB was recorded on weekdays, and 49 dB on weekends. In both cases these peaks were recorded between 5 and 6 p.m.

Short Paragraphs.   Novelists can afford to write long paragraphs because they assume they will have their readers' attention, and their readers have the time and patience to wend their way through leisurely description. But in business and industry readers are working against the clock: their time is limited and they want to read no more than is necessary. They need bite-size paragraphs of easy-to-digest information.

Let the first sentence of each paragraph introduce just one idea, then make sure that subsequent sentences in that paragraph develop the idea adequately and do not introduce any other ideas. In technical business writing the first sentence of each paragraph should be a "topic sentence," so that the reader will know immediately the main idea you are conveying. That is, the first sentence of each paragraph should summarize the paragraph's contents, while the remaining sentences support the first sentence by providing additional information:

> *We have tested your 15 Vancourt 801 disk drives and find that 11 require repair and recalibration.* Only minor repairs will be necessary for 6 of these drives, which will be returned to you next week. Of the 5 remaining drives, 3 require major repairs which will take approximately 20 days, and 2 are so badly damaged that repairs will cost $180 each. Since this is more than the maximum repair cost imposed by you, no work will be done on these drives.

By reading the topic sentence (italicized in this example), the reader learns immediately what he or she most wants to know: the number of disk drives that need repair. The reader can then determine the extent of repairs from the supporting sentences.

If an idea you are developing breeds an overly long paragraph, try dividing the information into a short introductory paragraph and a series of subparagraphs as has been done here:

> I have analyzed our present capabilities and estimate that we can increase our commercial business by at least $60,000 per month. To meet this objective we will have to:
>
> 1. Shift the emphasis from purely local customers to clients in major centers. To increase business from local customers alone will require an intensive sales effort for only a small increase in revenue, whereas a similar sales effort in a major center will attract a 30–40% increase in revenue.
> 2. Increase our staff and manufacturing facilities. The cost of additional personnel and new equipment will in turn have to be offset by an even larger increase in business. Properly administered, such a program should result in an ever-increasing workload.
> 3. Create a separate department for handling commercial business. If we remove the department from the existing production organization it will carry a lower overhead, which will result in products that are more competitively priced.

Subparagraphing offers a useful way to maintain continuity through a series of related points, although care must be taken to ensure that the subparagraphs are presented in parallel form. In the above example each subparagraph flows naturally from the lead-in words of the main paragraph, and the verbs not only are in the same form but also add emphasis:

> . . . we will have to:
>
> 1. *Shift* the emphasis . . .
> 2. *Increase* our staff . . .
> 3. *Create* a separate . . .

Without this *parallelism* there would be poor continuity between the subparagraphs. (For more information on parallelism, see Chapter 11.)

I am not suggesting that your letters should contain a series of small, evenly sized paragraphs, which would appear dull and stereotyped. Paragraphs should vary from quite short to medium-long to give the reader variety. How you can adjust paragraph and sentence length to suit both reader and topic, and also to place emphasis correctly, also is covered in Chapter 11.

Short Sentences.   If you write short, uncomplicated sentences, you will be helping your readers quickly grasp and readily understand each thought you present. Sometimes expert literary writers can successfully build sentences that develop more than one thought, but such sentences are confusing and out of place in the business world. Compare these two examples of the same information:

Complicated   There has been intermittent trouble with the vacuum pumps, although the flow valves and meters seem to be recording normal output, and the 5 inch pipe to the storage tanks has twice become clogged, causing backup in the system.

Clear   There has been intermittent trouble with the vacuum pumps, and twice the 5 inch pipe to the storage tank has become clogged and caused backup in the system. The flow valves and meters, however, seem to be recording normal output.

The first example is confusing because it jumps back and forth between trouble and satisfactory operation. The second example is clear because it uses two sentences to express the two different thoughts.

If you present only one thought in each sentence, then your sentences are likely to be reasonably short. I hesitate to stipulate a maximum sentence length, but 25 words might be realistic. Even this can vary, depending on the technical level of the reader and complexity of the information. As a general rule, the more complex the subject, the shorter the sentences should be.

But you should also recognize the danger in stringing together too many very short sentences. Each may be complete and clear in itself, but together they may create the effect of all starts and stops with inadequate subject development. Your objective should be to obtain rhythm by mixing short and medium-length sentences. Rhythm in sentence structure will lead the reader smoothly through each paragraph.

Short Words.   Some engineers and engineering technicians feel that the technical environment in which they work, and the complex topics they have to write about, demand that they use long, complex words in their correspondence. They write "an error of considerable magnitude was perpetrated," rather than simply "we made a large error." In so doing, they make a reader's job unnecessarily difficult.

The engineering and scientific worlds encompass many long and complex technical terms that have to be used in their original form. "Diphygenic" (to have two modes of development) is an example. If you surround these terms with simple words you will make your correspondence more readable. This does not mean using only four- and five-letter words, unless you want your work to read like a third grade spelling text. The goal is to use the right word in the right place, striking a happy balance between oversimplification and ponderousness.

CLARITY

A clear letter conveys information simply and effectively, so that the reader readily understands its message. To write clearly demands ingenuity and attention to detail. As a writer you must consider not only how you will write your letters, but also how you will present them.

Create a Good Visual Impression. Experienced writers know that a nicely laid out letter impresses a reader. Subconsciously it seems to be saying "my neat appearance demonstrates that I contain quality information that is logically and clearly organized."

The person you are writing to is unlikely to be doing nothing but waiting for your letter to come in, just so he or she can read it! Your letter will be slipped into a pile of similar-looking letters, each crying out for attention. If yours *looks* appealing, the reader will be attracted to it and may place it among those to be read first.

The appearance of a letter tells much about the writer and the company he or she represents. If a letter is sloppily arranged or contains strikeovers, visible erasures, or spelling errors, then I imagine a careless individual working in a disorganized office. But if I am presented with a neat letter, tastefully placed in the middle of the page and carefully typed with a reasonably new ribbon, then I imagine a well-organized individual working for a forward-thinking company noted for the quality of its service. I would prefer to deal with the latter company and I will read its correspondence first.

Many technical people whose correspondence is typed for them feel that once they have scribbled out or dictated a letter their responsibility for its issue is over, apart from signing it. This is not true. A typist's job is to type a neat-looking letter that presents a good image of the company; in signing the letter, the writer endorses this image. In effect, a writer has to exercise quality control and set the standards of presentation he or she considers correct for conveying the company's image. If you are presented with a letter that fails to meet these standards, you should reject it.

A good typist will recognize poor quality work and never offer a low-quality letter for signature; an inexperienced typist may need to learn what standards are acceptable. But unless you know how you want your letters to look (see page 60), and explain your requirements clearly (see Chapter 11), you may wait a long time before typists do the kind of work you want.

Develop the Subject Carefully. The key to effective subject development is to present the material logically, progressing gradually from a clear, understood point to one that is more complex. This means developing and consolidating each idea for the reader's full understanding before attempting to present the next idea. To do this you must have a thorough knowledge of your intended readers. Otherwise you may confuse them through inadequate development or annoy them by overstressing each point.

Insert Headings as Signposts. In longer letters, particularly those discussing several aspects of a situation, a writer can help the reader by inserting headings that are indicators of specific information. Such headings must be informative, summarizing clearly what is covered in the paragraphs they precede (see Chapter 11). If, for instance, I had replaced the heading to this paragraph with the single word *Headings,* I would not have summarized what this paragraph is about. The same would be true in the next paragraph if I replaced the three-word heading with the single word *Language.*

Use Good Language.   It hardly seems necessary to tell you to use good language, but in this case I mean language that you know readers will understand. Use only those technical terms and abbreviations they will recognize immediately. If you are in doubt, define the term or abbreviation, or use a simple expression in its place.

The factors I have described so far are mostly manipulative details that can be learned. Armed with this knowledge and the basic letter formats illustrated later in this chapter, the inexperienced writer has some ground rules on the practical aspects of business letter writing. Still to be acquired is the more difficult technique of letting one's character appear in letters without letting it become too obvious.

SINCERITY AND TONE

Sincerity and tone are intangibles that defy close analysis. There is no quick and easy method that will make your letters sound sincere, nor is there a checklist that will tell you when you have imparted the right tone. Both qualities are extensions of your own personality that cannot be taught. They can only be shaped and sharpened through knowledge of yourself and which of your attributes you most need to develop.

SINCERITY

At one time it was considered good manners not to permit one's personality to creep into business correspondence. Today, business letters are much less formal and, as a result, much more effective. Personal qualities that experienced letter writers develop to convey their messages efficiently are:

Enthusiasm
Humanity
Directness
Definiteness

These four qualities, together with knowledge of one's subject, form the five basic ingredients that new writers can use to inject sincerity into their correspondence.

Be Enthusiastic.   Sincerity is the gift of making your readers feel that you are personally interested in them and their problems. You convey this by the words you use and the way you use them. A reader would be unlikely to believe you if you came straight out and said, "I am genuinely interested in your project." The secret is to be so involved in the subject, so interested by it, that you automatically convey the ring of enthusiasm that would appear in your voice if you were talking about it.

Be Human. Too many letters lack humanity. They are written from one company to another, without any indication that there is one human being at the firing end and another at the receiving end. The letters might just as well be from computer to computer.

Do not be afraid to use the personal pronouns "I," "you," "he," "she," "we," and "they." Let your reader believe you are personally involved by using "I" or "we," and that you know he or she is there by using "you." Contrary to what many of us were told in school, letters may be started in the first person. If you know the reader personally, or you have corresponded with each other before, or if your topic is informal, let a personal flavor appear in your letters by using "I":

> Dear Mr. Wicks:
> I read your report with interest and agree with all but one of your conclusions. . . .

If you do not know your reader personally and are writing formally as a representative of your company, then use the first person plural:

> Dear Mr. Wicks:
> We read your report with interest and agree with all but one of your conclusions. . . .

Be Direct. Very few people have acquired the knack of being naturally direct. Anna King tells H. L. Winman engineers to state their case immediately when they start a letter, but to do so carefully so they will not sound too abrupt. The following example demonstrates her approach.

A construction company wants to know if a supply of steel it has in stock can be used to build a bridge in northern Alaska, and has submitted a small sample to H. L. Winman and Associates for analysis. When tests are complete, Mechanical Engineering Coordinator Paul Nowicki writes this letter to the client:

> Dear Mr. Sparkes:
> We conducted extensive tests on the sample of steel you sent to us for analysis. A Charpy impact test told us that the steel has a low transition temperature, while a tension test demonstrated that the steel failed in a ductile manner, with a yield point of 44,000 psi.
>
> From these tests we determined the steel to be type G40.12. Although widely used as structural steel throughout the United States and southern Canada, it is

less satisfactory for structures that are to be subjected to high loading in the far north. Its low temperature characteristics are only moderately good, which would make it unsuitable for a bridge at Peele Bay, Alaska.

Realizing that his indirect start would force Mr. Sparkes to wade through all the technical details before he finds the answer he wants, Paul heeds Anna's previous suggestions and rewrites the letter so that it opens with the main message. But this time his directness makes him seem too abrupt:

> Dear Mr. Sparkes:
>
> You will not be able to use the type of steel you sent to us for analysis to build a bridge at Peele Bay, Alaska. We found your steel to be type G40.12, which has poor low temperature characteristics. A better type would be G40.8.

So finally Paul enlists Anna's help, and between them they write this naturally direct letter:

> Dear Mr. Sparkes:
>
> The sample of steel you sent to us for analysis is G40.12, a type that would not be suitable for the short span bridge you propose to build at Peele Bay, Alaska. A more suitable type would be G40.8, which has better low temperature characteristics than type G40.12.
>
> We analyzed your sample by conducting both a Charpy impact test and a tension test, details of which appear in attachment 1. These tests demonstrate that your sample has a low transition temperature and fails in a ductile manner.

This last version tells Mr. Sparkes immediately what he most wants to know: whether his supply of steel can be used for the bridge. It also helps him read the technical details that follow more intelligently.

A writer who fails to analyze exactly what the reader wants to know can easily be led into writing an indirect letter. The writer must ask him or herself, "What basic fact is most important to my reader?" and then introduce it as early as possible. If there are conditions or interpretations that need to be described, or additional factors that have strongly influenced the results of a project, they must be brought in afterwards. To introduce them before stating the main theme weakens the effectiveness of a letter, and may even confuse or annoy its reader.

Be Definite.   Being definite means deciding exactly what you want to say, then saying it. Persons who think better with pen in hand sometimes make decisions as they write, producing indecisive letters that are irritating to read. Their writers seem to examine and discard points without really grappling with the problem. By the time they have finished a letter, they have decided what they want to say, but it has been at the readers' expense.

Decision-making does not come easily to many people. Those of us who hesitate before making a decision, who evaluate its implications from all possible angles and weigh its pros and cons, may allow our indecisiveness to creep into our writing. We hedge a little, explain too much, or try to say how or why we reached a decision before we tell our reader what the decision is. This is particularly true when we have to tell readers something unfavorable or contrary to their expectations.

Reg Wasalusky, a field employee of H. L. Winman and Associates, has written from a project office in Alaska to ask for a raise. His supervisor, Andy Rittman, writes this indefinite letter in answer to Reg's request:

An Indefinite Letter

Dear Reg:

I could not answer your letter immediately because I wanted time to consider your request for a raise very carefully. We fully appreciate down here that you are working in rough climatic conditions and that much of your work takes place out of doors. On that count we feel that your request is justified.

On the other hand, we have to equate your position with those of all the other field staff, since to increase your salary when others are not similarly considered would be unfair. We have therefore also assessed your position and seniority, since we recognize that you have been with the company for some time.

As a result of our deliberations we find that it would not be possible to increase your salary at the moment. We would like to add, however, that your request will be reviewed again in three months, when a general salary updating program is to be inaugurated.

Andy Rittman

Andy and Reg are good friends, so Andy does not like to come right out and say "No." But in trying to demonstrate that he has not treated Reg's request lightly, he has used too many wishy-washy statements that have a hollow ring:

"We fully appreciate" (para 1)
"On the other hand" (para 2)
"As a result of our deliberations" (para 3)
"We find that it would not be possible" (para 3)

The letter is also misleading because at the end of both the first and second paragraphs Reg is led to believe he is being considered, but then in each case the next paragraph contradicts this impression. This "yes-no-yes-no" tone is disconcerting and may irritate Reg more than the denial of the request. So will Andy's sudden shift from the personal "I" to the company "we," which he adopts as soon as he starts explaining why the request is going to be turned down.

If Andy Rittman wants to be definite, he must be *decisive*. He must state his decision immediately, then bring in his reasons. If supporting information must be included, it must be brief and to the point. Compare this rewritten version with the original letter:

A Decisive Letter

Dear Reg:

I'm sorry, but I cannot approve your request for a salary increase at the moment because the company will be reviewing all departmental salaries in three months. We plan then to make pay adjustments for field staff who have seniority and work in remote locations under arduous conditions.

As soon as the salary updating has been finalized, I'll write again to let you know of any changes.

Andy

Reg may not like hearing "No," but at least he will know exactly where he stands and what is being done. If he knows he has seniority and a good working record, then he can reasonably anticipate receiving a raise in three months. (Note that Andy does not say anything that will make Reg assume his request will be granted; to do so would result in a very unhappy employee if a raise failed to come through.)

Know Your Facts.   Insufficient background knowledge and inadequate preparation are the first steps to evasive, indefinite writing full of overblown words that avoid the issue. The big words act as camouflage to conceal uncertainty, but a discerning reader can easily penetrate them. Be absolutely convinced of your facts before you start to write, or you will find yourself filling in gaps with big words and many unnecessary adjectives and adverbs.

TONE

It is not easy to assess whether you are setting the right tone as you write. You will probably be doing so if you have successfully developed the personal qualities that contribute to sincerity; but you cannot assume this. To achieve the right tone, your correspondence should be simple and dignified, but friendly. Approach your readers on a person-to-person basis, following the five suggestions below.

Know Your Reader.   A writer cannot hope to set the right tone without having satisfactorily identified the reader. If the writer knows a subject well and the reader does not, the writer must take care not to be condescending by implying that the subject is so complex that the reader will be able to understand only a very simple description. Even if this is true, the careful writer will select just the right terminology to hold the reader's interest and perhaps offer a mild challenge. By letting readers feel that they are grasping some of the complexities of a subject (by using analogies within their range of knowledge), a writer can often present technical information without confusing or upsetting nontechnical readers.

Neither should inexperienced writers attempt to "write up" to the level of a technical reader whose experience and knowledge are far greater than their own. In trying to match the reader's intellect they may be tempted to insert long, important-sounding words that are incorrect, or even faintly amusing. They should write at their own level, using sensible technical terms and expressions that fit the subject.

Take It Easy with Apologies. When apologizing in writing do so quickly and cleanly, just as you would when apologizing to a person for stepping on his or her foot: "I'm sorry!" is all you need say. If you have been wrong, admit it and then consider the subject closed. To keep on apologizing, to say too much, will make your apologies seem insincere. These excerpts show the difference between an overstated apology and a brief one:

> *Too* *Wordy* We were mortified to hear of the damage caused to your carpeting by our repair technician's carelessly placed soldering iron, and are most sincerely sorry that this should have happened. The person involved has been reprimanded and we can assure you that such an occurrence will not happen again. We trust that this accident will not affect our business relationship, particularly since this has never happened before and we extremely regret that it should have happened to you. Please accept our apologies and our guarantee of top-notch service in the future. In the meantime we have arranged for a service representative to call on you shortly to discuss carpet repairs.

> *Brief* I was very sorry to hear that one of our technicians has damaged your carpet, and have asked service representative Brent Shelborne to visit you this week to arrange for repairs.

Keep Compliments Simple. Some people have a natural gift for paying compliments so that they always sound sincere. They have mastered an art that most of us find difficult: the ability to tread the narrow line between understatement and overstatement.

To pay a compliment effectively in writing is equally difficult. Most of us lean toward overstatement, forgetting that the best written compliments are stated very simply. To overstate a compliment, to gush, to overwhelm your reader with intensity, is to make the compliment seem insincere. This has happened in the following example:

> We have always received excellent service from your organization in the past, so it was only natural that we should turn to you again in the hour of our need. The assistance you provided in helping us to identify an improved transducer for phasing in our standby generator was overwhelming, and we would like to extend our heartfelt thanks to all concerned for their help. The prompt attention you gave to help resolve our problem was very deeply appreciated.

A simple "thank you" would have been much more believable:

Thank you for your prompt assistance in identifying an improved transducer for phasing in our standby generator.

Criticize Effectively.   A reader will usually accept criticism if it is presented "straight from the hip." Any attempt to soften the criticism by writing a lengthy preamble or by hiding it behind big words will reduce its effectiveness. To harp on the criticism once it has been stated, or to imply it without coming directly and clearly to the point, will lower the reader's opinion of you as a person who knows what has to be said.

When Bill Champagne complained to a manufacturer that its packing methods were inadequate, his letter said (in part):

> I regret that for the third time I must write to you to report still another of your calibrators has been damaged in shipment. It is becoming increasingly obvious that the handling methods used by the air express agents are too rough for such delicate equipment, even though the containers are clearly marked as such. I have spoken to the airline about this but their only comment is that the instrument must have been packed inadequately. To prevent damage to further shipments, I therefore suggest that you use a more resilient type of packing material. The unit in question was shipped on . . .

The trouble with this letter is that its writer did not criticize directly. Instead he shifted the blame onto the air express handlers, only *implying* that the manufacturer's packing methods should be improved. If Bill had placed the blame where it belonged, the manufacturer would have given his letter more serious attention:

> For the third time I am returning a Telemetric Calibrator that has been damaged in shipment, and am requesting that you improve your packing methods for these delicate instruments. The particular unit was shipped on . . .

Occasionally a writer has to both criticize and pay a compliment in the same letter. If this happens, always pay the compliment first, to make the criticism easier to accept. This is not the same as softening the criticism; it simply means that the writer recognizes the good work that has been done. Placing the criticism first can so reduce the effectiveness of the compliment that the reader may not even notice that it is there. Rebuked by the criticism, the reader may feel that the compliment has been added only as a sop to make him or her feel better. Thus neither criticism nor compliment does its job effectively.

Avoid Words that Antagonize.   Any statement that implies that a reader is wrong, has not tried to understand your point of view, or has failed to make himself or herself understood, will immediately place the reader on the defensive. For example, if field technician Dennis Tanski omits to send motel receipts with his expense account, Andy Rittman (his supervisor) will have to write and ask for them. If Andy writes:

> You have failed to include motel receipts with your expense account.

he will antagonize Dennis, because the words *you have failed* subtly imply that Dennis is a failure! Instead, Andy should write:

> Please send motel receipts to support your expense account.

Expressions that may annoy readers or put them on the defensive are listed below.

You have *neglected* . . .
You have *failed* . . .
You *ought to know* . . .   } *Words that make a reader feel guilty*
You seem to have *overlooked* . . .
You have *not understood* . . .

You *must* return the instrument . . .
Your *demand* for warranty service . . .   } *Words that provoke a reader*
We *insist* that you . . .
I am sure *you will agree* . . .

We *have to assume* . . .
I *must request* . . .
I *simply do not understand your* . . .   } *Words that "talk down" to a reader*
You must understand our position . . .
Undoubtedly you will . . .

When a reader has to be corrected, the words you use should clear the air rather than electrify it. Tell readers gently if they are wrong, and demonstrate why; reiterate your point of view in clear terms, to clarify any possibility of misunderstanding; or ask for further explanation of an ambiguous statement, refraining from pointing out that the writing is vague.

BUSINESS LETTER FORMAT

There are many opinions of what comprises the "correct" format for business letters. The two illustrated in Figures 3-4 and 3-5 are most frequently used by technical business organizations. The former is a conservative format known as the Modified Block, which once was the more common arrangement. The latter, a modern style known as the Full Block, is popular with organizations in which contemporary design plays an important role. Both are acceptable in today's business world. However, a business organization should be consistent and use only one format for all correspondence.

Anna King has adopted the full block format for H. L. Winman and Associates' correspondence (see the letter reports in Chapter 5 and the cover letter preceding the first report at the end of Chapter 6). Anna is also aware

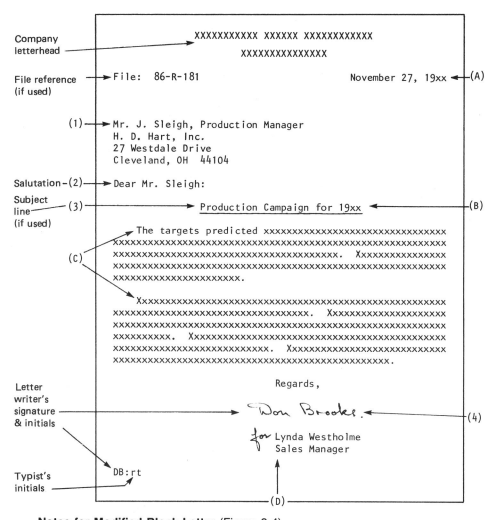

Notes for Modified Block Letter (Figure 3-4)

(A) If a file number is used, the date is offset to the right; if there is no file number, the date starts at the centerline.

(B) The subject line (if used) should be centered and underlined.

(C) The first word of each paragraph may be indented about five spaces, or started flush with the left-hand margin.

(D) The signature block should start at the centerline.

Notes for Full Block Letter (Figure 3-5)

(E) Every line starts against the left margin (simple for typist to set up). The parts otherwise are the same as for the modified block letter.

Figure 3-4. Modified block letter format.

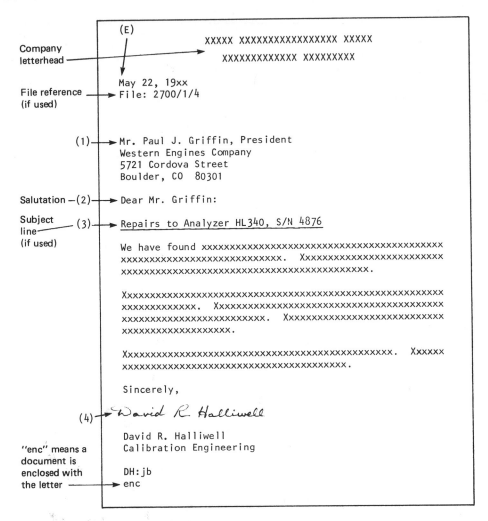

Company letterhead

File reference (if used)

(1)

Salutation —(2)

Subject line (if used) (3)

(4)

"enc" means a document is enclosed with the letter

(E)

XXXXX XXXXXXXXXXXXXXXXX XXXXX
XXXXXXXXXXXXX XXXXXXXXX

May 22, 19xx
File: 2700/1/4

Mr. Paul J. Griffin, President
Western Engines Company
5721 Cordova Street
Boulder, CO 80301

Dear Mr. Griffin:

Repairs to Analyzer HL340, S/N 4876

We have found xxxxxxxxxxxxxxxxxxxxxxxxxxxxxxxxxxxxxxxx
xxxxxxxxxxxxxxxxxxxxxxxxxxxxxx. Xxxxxxxxxxxxxxxxxxxxxxxxx
xxxxxxxxxxxxxxxxxxxxxxxxxxxxxxxxxxxxxxxxxx.

Xxxxxxxxxxxxxxxxxxxxxxxxxxxxxxxxxxxxxxxxxxxxxxxxxxxxx
xxxxxxxxxxxxx. Xxxxxxxxxxxxxxxxxxxxxxxxxxxxxxxxxxxxxxxx
xxxxxxxxxxxxxxxxxxxxxxxxxxx. Xxxxxxxxxxxxxxxxxxxxxxxxxxx
xxxxxxxxxxxxxxxxxxx.

Xxxxxxxxxxxxxxxxxxxxxxxxxxxxxxxxxxxxxxxxxxxxx. Xxxxxx
xxxxxxxxxxxxxxxxxxxxxxxxxxxxxxxxxxxxxxxx.

Sincerely,

David R. Halliwell

David R. Halliwell
Calibration Engineering

DH:jb
enc

Notes for Both Letter Styles

(1) In modern correspondence the name and title of the addressee are placed ahead of the company name. If the addressee's name is not known, a suitable title (such as Purchasing Agent) should be inserted. All but essential punctuation is eliminated from the address, salutation, and signature block.

(2) Today's trend toward informality encourages writers to use first names in the salutation: "Dear Jack" instead of "Dear Mr. Sleigh."

(3) Subject lines should be *informative* (not just "Production Plan" or " Spectrum Analyzer"); they may be preceded by *Subject:, Ref:, or Re:*.

(4) The letter writer may sign "for" the Department Head.

Figure 3-5. Full block letter format.

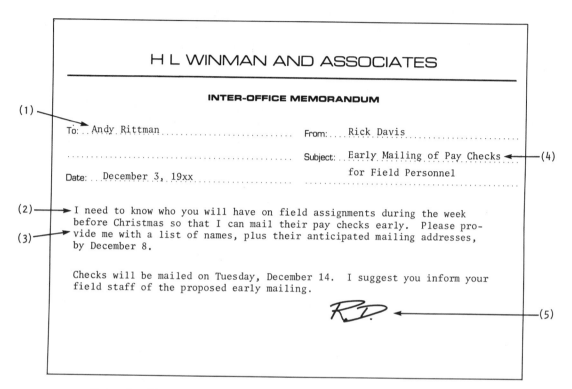

(1)

(2)

(3)

(4)

(5)

H L WINMAN AND ASSOCIATES

INTER-OFFICE MEMORANDUM

To: .. Andy Rittman ....   From: .... Rick Davis ....

....   Subject: .. Early Mailing of Pay Checks

Date: ... December 3, 19xx ....   for Field Personnel

I need to know who you will have on field assignments during the week before Christmas so that I can mail their pay checks early. Please provide me with a list of names, plus their anticipated mailing addresses, by December 8.

Checks will be mailed on Tuesday, December 14. I suggest you inform your field staff of the proposed early mailing.

Notes for Memorandum

(1) The informality of an interoffice memorandum means titles of individuals (such as Office Manager and Senior Project Engineer) may be omitted.

(2) No salutation or identification is necessary. The writer can jump straight into the subject.

(3) Paragraphs and sentences are developed properly. The informality of the memorandum is *not* an invitation to omit words so that sentences seem like extracts from telegrams.

(4) The subject line should offer the reader some information; a subject entry such as "Paychecks" would be insufficient.

(5) The writer's initials are sufficient to finish the memorandum (although some organizations repeat the name in type beneath the initials). Some people prefer to write their initials beside their name, on the "From" line, instead of signing at the foot of the memo.

Figure 3-6.   Interoffice memorandum.

that letter styles are continually changing. Some companies now omit the salutation and complimentary close (such as "Sincerely"), write dates military style (day-month-year: "27 January 1986"), may omit all punctuation from names and addresses (as in "Ms Jayne K Tooke"), and even use interoffice memorandums or quick-reply letters for informal correspondence. If these trends become more firmly established, Anna will adapt H. L. Winman Associates' letter format to reflect the changes.

The memorandum is a flexible document normally written on a prepared form similar to that shown in Figure 3-6. Formats vary according to the preference of individual companies, although the basic information at the head of the form is generally similar. Examples of memorandums appear throughout Chapters 4 and 5.

A quick-reply letter, like an interoffice memorandum, is handwritten or typed on a prepared form. The form has spaces for the originator to write a message and the recipient to write a reply (see Figure 3-7), and comprises three different-colored sheets interleaved with carbon paper. The originator detaches and retains the middle sheet. When the recipient has written a reply, he or she detaches and keeps the third sheet and returns the front sheet to the originator.

ELECTRONIC MAIL

Electronic mail has virtually replaced the telegram, which was a popular and relatively fast means of communication from the 1900s through to the 1960s, and is rapidly superseding telex and similar typewriter-based direct-communication systems. The criterion for writing telegrams, telex-type messages, and inter-computer messages, however, remains the same: keep the message brief but clear. Never allow an over-zealous desire for brevity to cloud your message: queries that an obscure communication prompts are usually more costly than writing or typing a longer but clearer message in the first place.

When Mike Toller in Topeka, Kansas opened a shipment of parts from Carlson Distributors, he found the order was incomplete and contained some items he had not ordered. He made a note of the deficiencies, sat in front of his computer terminal, and typed this message:

> TO: CARLSON DISTRIBUTORS, NEW YORK. YOUR INV 216875 OCT 19, OUR P.O. W 1634. SHORT-SHIPPED 10 TOOLSETS MKV, 4 801 SOCKETS PLUS 2 DOZ MOD 280A LATHE BITS UNORDERED. ADVISE. M. TOLLER.

In New York, Christina Baird puzzled over the message on her video screen, and then typed this reply:

☑ First Class Mail
☐ Inter-Office

TIME-SAVER LETTER

This form is designed to make it easy for you to reply. Simply note your answer in the space provided. Keep one copy for your files and return the other to us. Thank you.

MACRO ENGINEERING INC

600 Deepdale Drive
Phoenix, AZ
85007

Elmer Dubienski, Purchasing Manager
Morgan Electronics Wholesale Inc.
2160 Wavell Road
Atlanta, GA 30332

MESSAGE   DATE May 2, 19xx

Re: Your P.O. 2837

Elmer:

To reduce shipping costs, I plan to ship only the first 6 Microbar Environmental Control units by air express. This will be done on May 24.

I'll then ship the remaining 24 units by rail on June 3. Is this acceptable?

Judy P.

Judy Paulson
Production Coordinator

REPLY   DATE May 10, 19xx

Good plan, Judy!

But please increase the air express shipment to 8 units, to meet our immediate commitments.

Thanks,

Elmer.

Elmer Dubienski

©DAY-TIMERS RE-ORDER No. 23810.

SENDER - KEEP YELLOW COPY FOR YOUR FILE. MAIL WHITE AND PINK COPIES.

Figure 3-7. A quick-reply letter. (Courtesy Day-Timers Inc.) When the form is folded correctly, the recipient's name will show through the window of a standard No. 10 window envelope.

65

TO: M. TOLLER, CROWN MANUFACTURING, TOPEKA. YOUR MES-
SAGE RE OUR INV 216875 AND YOUR P.O. W1634 NOT UNDERSTOOD.
PLEASE EXPLAIN. C. BAIRD.

Mike was surprised; he thought his message was crystal clear. So he again sat at
his terminal and typed:

TO: C. BAIRD, CARLSON DISTRIBUTORS, NEW YORK. MESSAGE
QUITE CLEAR: YOU SHORT-SHIPPED 10 TOOLSETS TYPE MKV AND 4
NO. 801 SOCKETS. YOU ALSO SHIPPED 2 DOZ MODEL 280A LATHE
BITS WE DID NOT ORDER. PLEASE SHIP MISSING ITEMS AND ADVISE
HOW YOU WANT THE BITS RETURNED. M. TOLLER.

Christina replied in three words:

THANKS. WILL INVESTIGATE.

Both Mike and Christina would have been saved time had Mike been
more explicit when he typed the original message.

In electronic mail, words must be abbreviated particularly carefully.
Well-known abbreviations such as "No." and "approx" will be readily recognized. But locally coined abbreviations such as "ex" (which can mean "for example" or "coming from"), "w/o" (for "without"), and "rec" (for "received" or
"recommended") should be avoided.

Assignments

Most of the letter-writing projects below include all the details you will
need to write the assignment. You are encouraged, however, to introduce additional factors if you feel they will increase the depth or scope of your letter.

PROJECT 1: A FAULTY HOME ENTERTAINMENT CENTER

Assume that you were recently in Waverly (1100 miles from home),
where you visited friends Martin and Joan Tong. Martin gave Joan a PAM 78
"Home Entertainment Center" last Christmas, and you are impressed by its
tone, appearance, and features. Martin tells you privately that he bought it from

Craven's Discount Center at 1837 Kelly Street in Waverly, and offers to go with you if you are interested in buying one.

You are, but you are disappointed to discover that Craven's has sold all its PAM 78 entertainment centers, that no more are on order, and that no one else in Waverly carries them. However, Harry Craven, the store owner, has a suggestion: there is a demonstrator which he could sell to you at 5% off his regular discount price. You test it, and it seems okay. Martin suggests a 15% price reduction would be more normal for a demonstrator, but Mr. Craven won't budge; he adds, however, that he'll give it a good check-over if you'll leave it with him. You agree. Two days later you pick it up, pay $388.35, and receive Craven's invoice No. C7214 stamped "Paid."

But when you arrive home, you find that the tape section of the PAM 78 doesn't work. You also discover that there is no local service center for the PAM line, so you take the entertainment center to Modern TV and Radio at 280 Waltham Avenue. When you pick it up the following day, store manager Jim Williams hands you a printed circuit board (PCB) with several bent and twisted pins.

"There's your problem," Jim says. "Craven's in Waverly must have replaced this PCB—you can tell it's one of theirs because the name CRAVEN is stamped on it." He explains that whoever inserted the PCB didn't align the pins properly and bent them by forcing it into its socket.

You pay $53.00 for the repair job on Modern TV and Radio's invoice No. 2656, and take both the PAM 78 and the ruined PCB with you.

Write to Harry Craven, tell him what has happened, and ask for a refund of $ . . . (you decide how much). You may assume you attach copies of the two invoices to your letter.

Note: The PAM 78 is made by VICOM in Korea. It contains an AM/FM stereo radio receiver, a record turntable, a cassette tapedeck, and two eight-inch speakers.

PROJECT 2: TRACING A LOST SUITCASE

For the past three months you have been assessing the environmental impact of a proposed gas pipeline to be built some 450 miles (725 km) from your home office. You were based at Lake Winthrop, where you stayed at the Clock Inn Motel. Whenever you went on a field trip of more than two days' duration, you vacated your room and left unwanted baggage with the motel receptionist.

The final field trip occurred on the last three days of the project, after which you flew home. Unfortunately, you forgot to pick up one of your cases at the motel.

Part 1. Write a telegram, telex, or intercomputer message, in which you describe your case and ask for it to be shipped to you by Remick Airlines.

Address the message to Marrianne Dubray, the receptionist at the Clock Inn. Describe a case of your own.

Part 2. One week later your case arrives, but you realize it is not your case. The size and color are correct, but it has another person's initials engraved on it: KRZ. Write a second telegram, telex, or intercomputer message to the Clock Inn. Refer to your first message and explain that you have been sent the wrong case.

Part 3. Today you receive this message in reply: YOUR CASE NOT HERE. SORRY.

So you telephone the motel and ask to speak to Marrianne. No luck: she is not there. You ask when she will be in.

"She won't," says a strange voice. "She quit."

You try describing the problem, but the new receptionist cannot help. He can't even put you through to Mrs. Kravetz, the manager, because she is away for three days. You ask him to check if there are any other bags or cases behind the desk or in the office, and he does so.

"There aren't any here," he says. "Why don't you write to the manager and tell her what happened? Maybe she will remember your case if you describe it for her."

Write to the manager of the Clock Inn, Lake Winthrop. Her full name is Lara G. Kravetz.

PROJECT 3: FREE REPAIRS FOR A MICROWAVE OVEN

Assume that you are an engineering technologist employed by Delta Manufacturing Inc. at 2135 Boundary Road of your city, that a two-shift operation is in effect, and that you are a supervisor on the evening (4:30 to midnight) shift. Three years ago, at your request, the company authorized purchase of a microwave oven so that the night workers in your area could heat up snacks or meals they bring with them. Details of the purchase and the events that occurred afterward are:

1. The microwave oven is a Trident model L210 and it was purchased from Multiple Industries at 2215 Grantham Street of your city on September 30 three years ago. The sales invoice issued by Multiple Industries was No. M5628 and your company paid $695 for the oven.
2. The microwave oven has since broken down three times:
   - The first time was on December 12, fifteen months after the unit was purchased. The repair cost was $225.74 on Multiple Industries' invoice No. R2806.
   - The second time was on May 16 last year, when the repair cost was $167.30 on invoice No. R5512.

- The third time was last September 5, when the unit was repaired on invoice No. R7740 for $188.50
3. Today the microwave oven has broken down for the fourth time.

Considering the total cost of repairs against the original purchase price, you feel that the company which sold the microwave oven to Delta Manufacturing should repair the oven this time at no cost to your company. Write to the service manager at Multiple Industries. Tell him what has happened, and tell him what you want Multiple Industries to do. His name is Victor Kolinsky.

PROJECT 4: RESPONDING TO AN ANGRY CUSTOMER

You are an engineering technologist employed by a manufacturer of technical equipment at 435 Manor Road of your city. (You may name the company yourself or assume it is the local branch of H. L. Winman and Associates.) You are the supervisor of the Service and Repair Department, and three service technicians report to you. Some of the service work is done in your plant, but more is done at customers' plants since most of the products your company manufactures are heavy industrial equipment.

Today you receive an exceedingly angry letter from Mr. Frank Rudge, owner-manager of Rudge Machine Shop at 1620 Bridlington Boulevard of your city. In his letter—which is dated two days ago—Mr. Rudge claims that the Model 440 Neostat one of your technicians repaired in Mr. Rudge's plant three months ago does not work properly. He says the machine was not used for the first two months after it had been repaired, and now that his employees have started using it the Neostat is tearing the work instead of making a straight, clean cut.

You drive to Rudge Machine Shop to examine the Neostat. Immediately you notice that the cutting bars are out of alignment and the cutting edges are nicked and scarred. This seems to indicate that the machine has been used to cut hard metals rather than the plastics, fiberglass, and similar nonferrous materials for which it was designed. Worse still, you notice that the seal your technician placed over the lock of the box containing the relay mechanism has been broken, which means that one of Mr. Rudge's employees has been fiddling around inside, adjusting the relays. You talk briefly with the area supervisor—Carl Hinkstrom—but he claims no knowledge of the problem, except that ". . . the machine doesn't work properly." You ask to see the technician's repair report of three months ago, but the supervisor cannot find it.

Mr. Rudge is not in the plant during your visit, which is fortunate because you want to check the carbon copy of your technician's repair report before you reply to Mr. Rudge's complaint.

Back at the office you dig out the report (No. 0317A) and note that Wally Devries, who serviced the equipment, wrote on it:

1. Adjusted relay cut-in point, took up slack in linkage, and tested operation of cutting bars. Clearance: 0.015 in. (within specs). All satisfactory.
2. Oiled/greased all bearings and checked cutting edges: no pits or breaks.
3. Sealed relay control box (seal No. B21).

The foot of the repair report is signed "Carl Hinkstrom," who acknowledged completion of the repair job.

Normally your company's repair warranty for industrial equipment is six months. But the repair report contains several printed conditions affecting the warranty. Clause 3 says:

3. Breakage of seals over operating components automatically voids this warranty.

Now write a reply to Mr. Rudge's letter of complaint.

PROJECT 5: DAMAGE TO BORROWED EQUIPMENT

As part of an instrumentation project you are working on at Interstate Oil Consolidated (IOC), you are developing an automatic monitoring system for recording pipeline conditions at remote locations. A particular item you use to calibrate and test the remote sensing modules is a Model 212A Analyzer manufactured by Macro Engineering Inc (MEI) of Phoenix, Arizona.

Last Monday morning the 212A suddenly became unserviceable, so you telephoned MEI and described the fault to design engineer Shannon Cooke.

"Oh, that will be the comparator circuit," she replied. "Microchip NE27801."

You asked Shannon to ship you a replacement microchip air express, but she said that that particular microchip is in extremely short supply and the best delivery time she could promise you is 12 weeks. She suggested, however, that you talk to Vern Rogers, the branch manager of the H. L. Winman and Associates office in your city. "They have a 212A," she said, "and possibly they will lend it to you."

You telephoned Vern Rogers and, after checking with his staff, he said he could lend their 212A Analyzer to you for 10 days. "After that, we will need it for a study we're doing for the department of highways," he added.

You picked up the 212A on Monday afternoon and have used it full time since then, cramming all the work you can into the 10 days you will have the analyzer. But this morning at 8:10 it went dead as technician Wendell Harriock was setting up the equipment. The cause was immediately apparent: Wendell had inadvertently reversed the polarity of the connections to the 400 Hz power supply!

You again call Shannon Cooke. "It's the same circuit that is faulty on your own machine," she says. "Reversing the polarity has blown the NE27801 microchip."

Part 1. You call Vern Rogers but he is out of town for the remainder of the week. As you will be out of town next week, write to him explaining the circumstances.

Part 2. Technician Wendell Harriock comes to you with a circuit drawing. "It's a reverse-polarity protection circuit," he explains. "If you connect the terminals wrongly, a relay trips and prevents power from flowing into the equipment. Had Macro Engineering included this circuit in their analyzer, we would still be operating right now!"

Write to Shannon Cooke at Macro Engineering Inc and suggest the circuit be included in future 212A Analyzers. Tell her the cost will be about $32 per unit. Give credit to the technician who designed the circuit. Assume you enclose a copy of the circuit drawing.

PROJECT 6: REQUEST TO ATTEND A COURSE

Assume that today is Monday (the third day of the current month), and that you are employed in the Engineering Department of H. L. Winman and Associates. For the past four weeks you have been on a field assignment to Duluth, where you have been conducting an engineering study for the RamSort Corporation. You have been assisted by two competent junior technicians (Ray Hicks and Len Dominie), and you are now three days ahead of schedule. The task is to be completed by the 28th of this month.

Today you have received a folder from the University of Minnesota in Minneapolis advertising a one-week course. Details are:

Course title:	Management for Engineering and the Sciences
Course dates:	Monday the 17th to Friday the 21st inclusive (of this month)
Type of course:	Maximum immersion: 8 A.M. to 5 P.M. daily plus 7 to 10 P.M. Wednesday evening; approximately 20 hours of home assignments
Cost:	$400; includes materials, books, and lunches, with a guest speaker from industry at each lunch
Registration:	No later than noon Wednesday the 12th; telephone registrations accepted
Participants:	Limited to 16

You are impressed by the technical standard of the course described in the folder and wish to attend. (Because of previous field assignments you missed a similar Extension Department evening course offered at your local university last winter. Your company sponsored four engineers to attend that course, for which the fee was $125 each.)

Write an interoffice memorandum to your project coordinator:

1. Describe the course (convince him or her it is a good one).
2. Ask if you can attend.

3. Ask if the company will pay the tuition fee, plus travel and lodging costs.
4. Ask to be spared from the RamSort task for one week (be convincing).
5. Ask for a quick reply (because time is short).

Assume that your project coordinator can give technical approval for you to attend but must go to the department head for financial approval. Also assume that the project coordinator is a very busy person who is notorious for placing items in an *Action* tray, where they are quickly covered by other important items and often overlooked.

PROJECT 7: LETTER OF THANKS

As a member of the Environmental Studies Department of H. L. Winman and Associates, you are assisting in a three-month technical/environmental project in the Manitowac Wildlife Reservation, through which a proposed pipeline will pass. At the end of your study you are invited by Delta Marsh Project Engineer Rudi Kane to spend a long weekend with him and his family at their summer cottage at Lake Freeling, some 130 miles west of the reservation.

With Rudi, his wife Jean, and son and daughter Leon and Tracy, you swim, water-ski, golf, ride horseback along wild trails, and play tennis during the day, and then barbecue steaks beside the shore in the soft evening twilight—an idyllic existence.

On Monday evening you are to catch the 6:30 bus back to Manitowac in preparation for Tuesday's flight home. You have been waterskiing on the north side of the lake but have returned ahead of the Kane family to put your things together at the cottage (you have driven back with a neighbor).

You wait an hour and a half for their return, unaware that they have had car trouble. You want to say good-bye and thank them for having you for the weekend, but by 6:15 you can wait no longer and so trudge, with your suitcase, down to the bus depot.

Now you are back at Manitowac, ready to leave for home tomorrow morning. Write to the Kanes to thank them for their hospitality. Address your letter to: General Delivery, Lake Freeling Post Office.

PROJECT 8: ACKNOWLEDGING A COLLEGE
AWARD

Assume that you are in the second year of the course you are enrolled in, and that three weeks ago the head of the department came to you and announced that you have been selected to be this year's winner of the Vancourt Business Systems Inc. scholarship for "proficiency in computer studies." Yesterday you attended an awards luncheon attended by other scholarship winners and representatives of the firms donating the scholarships. You sat next to

Myrna Jost, Vancourt's manager of computer services, who presented the award to you.

Today, you write to Vancourt Business Systems Inc. to thank the company for the award. Use these details:

1. Address your letter to Sheenah McIvor, Vice President, Operations.
2. Vancourt's address is 1830 Mapleton Crescent of your city.
3. The award is a wall plaque inscribed with your name and a check for $300.

4

Short
Informal Reports

When you hear that someone has just finished writing a technical report, you may imagine a nicely bound formal document, tastefully typed and printed. In some cases you would be correct, but most of the time you would be wrong. Far fewer formal reports are issued than informal reports, which reach their readers as letters and memorandums. This chapter describes the short informal reports you are likely to write as a technologist, engineer, or engineering technician.

THE MEMORANDUM REPORT

The memorandum report is the least formal technical report. Normally an interoffice or interdepartmental communication, its length and tone can vary considerably. It can be very direct, it can develop its topic in great detail to present a convincing case, or it can lie anywhere in between.

Anna King is alone in the H. L. Winman office, working late one Tuesday evening, when the telephone rings. The caller is Bob Walton, a member of the electrical engineering staff who is on a field trip to Tangwell with a second staff member. He tells Anna they have had an automobile accident near Hadashville, and some of their equipment is damaged. He wants Jim Perchanski, his department head, to send out replacement items by air express.

Anna jots down notes while Bob talks. As she plans to be out of town the following day, she types out a report of the conversation and leaves it on Jim Perchanski's desk. She prepares it as a memorandum (Figure 4-1) which tells Jim Perchanski what has happened to two members of his staff, where they are now, how soon they will be able to move on, and that one of them is injured. It also tells him that equipment is damaged and replacements are needed.

Because she will not be available to answer questions the following morn-

H L WINMAN AND ASSOCIATES

INTER-OFFICE MEMORANDUM

To: Jim Perchanski

From: Anna King

Subject: Accident Report and Request for Spare Parts

Date: November 18, 19xx

(1)

Bob Walton and Pete Crandell have been involved in a highway accident which will delay their inspection of the Sledgers Control project at Tangwell. They need replacement parts shipped to them tomorrow (Wednesday, November 19).

(2)

Bob telephoned from Hadashville at 7:35 p.m. to report the accident, which occurred at 5:15 p.m. some two miles north of Hadashville. Pete has been hospitalized with a fractured left knee and a suspected concussion. Bob was unhurt. The panel van and some of their equipment were damaged.

(3)

Bob wants you to ship the following items to him by air express on the Wednesday evening flight to Montrose and to mark the shipment HOLD FOR PICKUP BY R. WALTON NOV 20:

1 Spectrum Analyzer, HK7741

1 Calibrator, Vancourt Model 23R

24 Glass phials, 30 cm long x 5 cm dia

(4)

He has arranged to rent a van and will drive to Montrose to pick up the items Thursday morning. He will then drive on to Tangwell and will arrive there about 4:00 p.m. He has informed Tangwell of the delay.

(5)

While at Hadashville Bob is staying at Hunter's Motel (tel 453 6671), where he is preparing an accident report for you.

(6)

Anna

Figure 4-1. A memorandum report.

75

ing, Anna takes care to describe the situation clearly (the paragraph numbers below are keyed to the memorandum):

(1) Anna knows that a subject line must be informative; it must tell what the memorandum is about and stress its importance to the reader. If she had simply written "Transcript of Telephone Call from R. Walton," she would not have captured Jim Perchanski's attention nearly as sharply.

(2) This brief summary gets right to the point by immediately telling Jim Perchanski in general terms what he most needs to know:

> Why the memo was written.
> What happened.
> What action has been taken.
> What action he has to take.

(3) In this paragraph Anna tells what she knows about the accident and its effects. It serves as background to the important facts that follow.

(4) Anna knows that Jim Perchanski must act quickly to ship the replacement items, so she uses a list to help him identify them. The indented list also is an attention-getter.

(5) Instructions and movement details should be explicit, otherwise the equipment and Bob Walton may not meet at Montrose. Anna has taken care throughout to identify specific days, and once even the date, to make sure that no misunderstandings occur. To state "tomorrow" or "the day after tomorrow" would be simple but might cause Jim Perchanski to assume a wrong date, since he will be reading the memorandum one day later than it was written.

(6) In this brief closing paragraph Anna indicates what further action is being taken and where Bob Walton can be contacted if more information is needed or if a message needs to be relayed to him.

When Jim Perchanski walks in on Wednesday morning, he will know immediately what has happened and what action he has to take. He does not need to ask questions because he has been placed fully in the picture.

THE LETTER REPORT

The letter report, although still basically informal, can vary in formality according to its purpose, the type of reader, and the subject being discussed. Some letter reports may be as informal as a memorandum report, particularly if they are conveying information between organizations whose members know each other well or have corresponded frequently. Others may be almost severely formal, although such reports are increasingly rare. Most often they are simple, dignified business letters that convey technical information from one company to another. As such, they serve their purpose by avoiding the rigidity of style and format demanded by the full formal report described in Chapter 6. The letter in Figure 5-2 of Chapter 5 is a typical informal letter report written to a client.

Although there are many types of informal reports, most appear in the memorandum or letter form. There are variations in format, content, and writing approach, but all follow the basic pattern shown in Figure 4-2: (1) a short

SUMMARY		A brief statement of the report's main features (often written last, but always placed *first*).
1.	**INTRODUCTION**	**Background:** information that "sets the scene."
2.	**DISCUSSION**	**Facts:** data, details; what has been and is being done.
3.	**CONCLUSIONS**	**Outcome:** results and effects; it may also suggest what needs to be done.

Figure 4-2. Basic report compartments.

introduction to the topic or problem; (2) a discussion of the data, situation, or problem, and what has been done or could be done about it; and (3) a conclusion which sums up results and possibly recommends what should be done next.

These three compartments are preceded by a summary statement, which tells readers in as few words as possible what they *most* need to know about the topic (for more about summaries and summary statements, see Chapter 11). This pattern is evident in almost every report, whether it is a simple incident report, a field trip, inspection, or progress report, an investigation report similar to that in Figure 5-2, a feasibility study, a technical proposal, or a full formal report.

WRITING STYLE

The reports described in this chapter deal with specific information. Their purpose is mainly *to inform*, so they are written in a direct, informative style that is crisp and to the point. Their writers are usually describing events that have already occurred, so they write mostly in the past tense, which helps them to be consistent. They shift gear into the present or future tense only when they have to describe something which is presently occurring, outline what will happen in the future, or suggest what needs to be done. All three tenses occur in the following report from Engineering Technician Dan Skinner to Coordinator Don Gibbon:

SUMMARY	The indoor/outdoor carpet we installed in station DMON-TV has corrected the noise problem but is "pilling" badly. I plan to examine the carpet with the manufacturer's representative to find the cause and suggest a remedy.
BACKGROUND	The carpet was installed in the satellite studio control room during the night of January 8–9.

	Sound level readings taken at 12 different locations on January 10 showed that the ambient noise level had decreased an average 3.6 dB.
Past Tense FACTS	Operating staff also noticed a marked decrease in clatter caused by impact noises. At the station manager's request, I returned to the control room today and checked the carpet's condition. After only two weeks of use it has tight little balls of carpet material adhering to its
Mainly Past Tense	surface. I called the manufacturer's representative, who said that the condition is not unusual and does not mean that the carpet is wearing quickly. He suggested that it may be caused by improper carpet-cleaning techniques and probably can be easily corrected. However, it is un-
Present Tense OUTCOME	sightly and our client is not pleased with the carpet's appearance. The manufacturer's representative and I will return to the control room between midnight and 2:00 a.m. on January 31 to study the carpet-cleaning techniques used by maintenance staff. I
Future Tense	will telephone our findings to you later in the day.

Dan has written the Background and most of the Facts paragraph mainly in the past tense because they deal with what has already been done. At the end of the Facts he has shifted into the present tense to report how the station manager feels *now*. Then for the Outcome he has jumped into the future tense to outline what he *plans* to do. This past-present-future arrangement is natural and logical; reader Don Gibbon will feel comfortable making the transitions from one tense to the next. Dan's Summary even follows the same pattern.

INCIDENT REPORT

An incident report (also known as an occurrence report) simply describes an event or incident that has occurred. Anna King's memo to Jim Perchanski in Figure 4-1 is an incident report. So is Bob Walton's report of the accident (Figure 4-3), which he is writing at Hunter's Motel in Hadashville. Although its topic is essentially the same as Anna's, its focus and emphasis are different.

Bob uses the three basic compartments in Figure 4-2 to organize his report (the letters below are keyed to the report in Figure 4-3):

(A) This is his summary: it takes a main piece of information from each compartment that follows.

(B) This is compartment 1, **Background.** By clearly describing the situation (*who? where? when?*) Bob is helping Jim more easily understand what happened. Notice how he:
   • Establishes where they were, how they happened to be there, their direction of travel, and who else was involved.

H L WINMAN AND ASSOCIATES

INTER-OFFICE MEMORANDUM

To: Jim Perchanski

From: Bob Walton

Subject: Report of Auto Accident, Hadashville, Ohio

Date: November 19, 19xx

Pete Crandell and I were involved in a multiple-vehicle accident on November 18, which resulted in injuries to Pete, damage to our panel truck and some equipment, and a two-day delay in our inspection of the Sledgers Control project. Ⓐ

The accident occurred at 5:15 p.m. on highway 44, about two miles northeast of Hadashville. We were traveling north in company panel van T2711, on our way to site RJ-17 at Tangwell. Pete was driving and we were approaching the intersection with highway 201.

Other vehicles involved in the accident were: Ⓑ

> Dodge Aries, Lic AHJ 735, driven by D. Varlick.
> Ford truck, Lic T4851, driven by F. Zabetts.
> Pontiac Catalina, Lic MD3720, driven by K. Schmitt.

Positions of these vehicles and our panel van immediately before the accident are shown on the attached sketch.

As the Dodge attempted a right turn into highway 201 it skidded into the Ford truck, which was standing at the intersection waiting to enter the highway. The impact caused the Dodge's rear end to swing into our lane, where Pete could not prevent our van from colliding with it. This in turn caused our panel van to slide broadside into the southbound lane, where the Pontiac approaching from the opposite direction collided with its left side. Ⓒ

Pete was taken to Hadashville hospital with a broken left knee and a suspected concussion; he is likely to be there for several days. The panel van was extensively damaged and was towed to Art's Autobody, 1330 Kirby St., Hadashville; I have arranged to rent a replacement van from Budget. Some of our equipment was damaged or shaken out of calibration, so I telephoned the office on Tuesday evening and requested replacements (Anna King took the message and has a list for you.) Ⓓ

I also telephoned the duty engineer at site RJ-17 and told him that my inspection of the Sledgers Control project will not start until Friday November 21, two days later than planned.

Figure 4-3. An incident report.

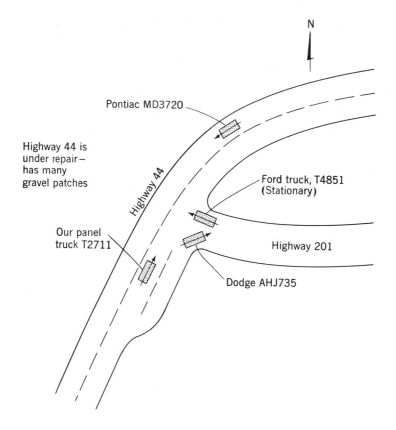

- Itemizes vehicles, license numbers, and drivers' names in an easy-to-read list.
- Mentions that he is enclosing a sketch (so Jim can look at it *before* he reads on).

(C) Because his background information is complete, Bob's compartment 2, **Facts,** can be concise. He simply provides a chronological description of what happened from the time the Dodge started to slide until all vehicles stopped moving.

(D) In compartment 3, **Outcome,** Bob describes the *results* of the accident (injuries, damage) and what he has done since (rented a van, requested replacement equipment). He closes on a strong point: What is being done about the project, which was his reason for passing through Hadashville.

Bob knows his role is to be an informative but objective (unbiased) reporter. No doubt he has an opinion of who is at fault, but to state it probably would have injected subjectivity into his report.

FIELD TRIP REPORT

After returning from a field assignment, you will be expected to write a field trip report describing what you have done. You may have been absent only a few hours, inspecting cracks in a local water reservoir; or you may have been away for two months, overhauling communications equipment at a remote defense site. Regardless of the length and complexity of the assignment, you will have to remember and transcribe many details into a logical, coherent, and factual report. To help you, carry a pocket notebook in which to jot down daily occurrences. Without such a record to rely on, you may write a disorganized report that omits many details and emphasizes the wrong parts of the project.

The simplest way to write a trip report is to answer three basic questions:

1. **Who went where?**
   Why and when did you go?
2. **What did you do?**
   What did you set out to do? Were you able to do all of it?
   What problems did you run into? What additional work did you do?
3. **What remains to be done?**
   What work could you not complete, and who should now complete it? When and how should it be done?

The answers to these questions provide the information for the three compartments in Figure 4-2:

1. INTRODUCTION: Who went where? (**Background**)
2. DISCUSSION: What did you do? (**Facts**)
3. CONCLUSIONS: What remains to be done? (**Outcome**)

Short trip reports do not need headings. A brief narrative following the three-question pattern normally can carry the story:

Compartment 1	Dave Makepiece and I visited site RJ-17 from January 15 to 17 to install a prototype automatic alarm system for the Roper Corporation.
Compartment 2	The installation was completed without difficulty and no major problems were encountered. Installation instruction W27 was followed throughout with the exception of step 33, which called for connections to be made to the remote control panel. This step was omitted since the remote panel has been permanently disconnected.
Compartment 3	The prototype alarm will be field-evaluated for one month. It will be removed in mid-March by M. Tutanne, who will be visiting the site to discuss summer survey plans.

Long trip reports require headings to help their readers identify the compartments. Typical headings might be:

1. **Assignment Details** (Background).
2. **Work Accomplished** and **Problems Encountered** (Facts; best treated as two separate headings).
3. **Suggested Follow-up** or **Follow-up Action Required** (Outcome).

The whole report should be preceded by a short summary that states very generally what was and was not achieved.

H L WINMAN AND ASSOCIATES

INTER-OFFICE MEMORANDUM

To: Systems, Project, and Field Engineers

Date: February 20, 19xx

From: Anna King

Subject: Guidelines for Writing Field Trip Reports

A standard format is to be used for all field trip reports. These notes suggest how you can organize your information under five main headings: Summary, Assignment Details, Work Accomplished, Problems, and Follow-up Action. The headings may be omitted from very short reports.

Summary

Make your summary a short opening statement that says what was and was not accomplished, and highlights any significant outcome.

Assignment Details

Under this main heading state the purpose of the trip and include any other information that the reader may want to know. If the information is too cumbersome to write as a single paragraph, use subheadings such as:

> Purpose of Trip
> Background
> Project No./Authority
> Personnel Involved
> Date(s) of Trip

Work Accomplished

In this section describe all work that was done. Normally present it in chronological order unless more than one project is involved, in which case describe each project separately. Keep it short: don't describe at great length routine work that ran smoothly. Whenever possible refer to your work instruction or specifications, and attach a copy to your report.

> The manual control was disconnected as described in steps 6 to 13 of modification instruction MI1403.

Go into more detail only if difficulty is encountered, or if work is necessary beyond that anticipated by the job specification:

> At the request of the site maintenance staff, the manual control was installed in the power house as a temporary replacement for a defective GG20 control. Parts removed from the panel, together with instructions for returning the panel to its original configuration, were left with Frank Mason, the senior power house engineer.

Figure 4-4. How to write a long trip report.

If parts of the assignment could not be completed, identify them and explain why the work was not done:

> Test No. 46 was omitted because the RamSort equipment was being overhauled.

Problems

In contrast to the conciseness recommended for the previous section, you should describe problems in detail. Knowledge of problems you encountered and how you overcame them can be invaluable to the engineering department, which may be able to prevent similar problems from occurring elsewhere.

A statement that does not tell the reader what the problem was or how it was overcome is virtually useless:

> Considerable time was spent in trying to mount the miniature control panel. Only by fabricating extra parts were we able to complete step 17.

If the information is to be used by the engineering department, it must be much more specific:

> Considerable time was spent in trying to mount the miniature control panel according to the instructions in step 17. We found that the main frame had additional equipment mounted on it, which prevented us from using most of the parts supplied. To overcome this problem we fabricated a small sheetmetal extension to the main frame and mounted it, with the miniature control panel, as shown in the attached drawing.

Follow-up Action

This section of the trip report ties up any loose ends by telling readers what still remains to be done. If any work has not been completed, draw attention to it here even though you may already have mentioned it under "Work Accomplished." Identify what needs to be done, if possible indicate how and when it should be done, and say whose responsibility it now becomes:

> The manual control mounted as a temporary replacement in the power house is to be removed when a new GG20 control is received. This will be done by F. Mason, who will use special instructions and spare parts left with him for this purpose.

In some cases you may direct follow-up action to someone else in your department:

> The manual control will be removed from the power house by R. Walton, who will be returning to the site on October 17.

If your report is very long, I suggest you insert frequent subheadings and use a paragraph numbering system to increase its readability.

Anna King

Anna King's memo at Figure 4-4 tells H. L. Winman and Associates' engineers how to organize their longer trip reports, describes the type of information that normally would follow each heading, and includes excerpts and sample paragraphs.

With the exception of the Outcome section, trip reports should be written entirely in the past tense.

OCCASIONAL PROGRESS REPORT

Progress reports keep management aware of what its project groups are doing. Even for a short-term project, management wants to hear how the project is progressing, especially if problems are affecting its schedule. Because delays can have a marked effect on costs, management needs to know about them early.

Jack Binscarth, one of Macro Engineering Inc's senior technicians, has been assigned to Cantor Petroleums at a semiremote site in Texas, where he has

Figure 4-5. An occasional progress report.

MACRO ENGINEERING INC

From: J. Binscarth (at Cantor Petroleums)   Date: October 14, 19xx

To: F. Stokes, Chief Engineer   Subject: Delay in Analysis of
Head Office   Oil Samples

My analysis of oil samples for Cantor Petroleums has been delayed by problems at the refinery. I now expect to complete the project on October 25, nine days later than planned.

The first problem occurred on September 23, when a strike of refinery personnel set the project back four working days. I had hoped to recover all of this lost time by working a partial overtime schedule, but failure of the refinery's spectrophotometer on October 13 again stopped my work. To date, I have analyzed 111 samples and have 21 more to do.

The spectrophotometer is being repaired by the manufacturer, who has promised to return it to the refinery on October 19. Today I informed the refinery manager of the delay, and he agreed to an increase in the project price to offset the additional time I will spend here. He will call you about this.

Providing there are no further delays I will analyze the remaining samples between October 20 and 24, then submit my report to the client the following morning. This means I should be back in the office on October 26.

Jack

to analyze oil samples. The job is expected to take five weeks, but problems develop that prevent Jack from completing the work on time. To let his chief know what is happening, he writes the brief progress report in Figure 4-5.

Occasional progress reports like this are similar to incident reports. They have an event or events to relate, and they do this best by adapting the standard Background-Facts-Outcome arrangement into a past-present-future pattern:

1. *Summary*   A brief description of the overall situation.

2. *Progress*   The work that has been done, the problems that have been encountered, and the effect these problems have had on progress.

3. *Situation Now*   What is being done at present.

4. *Future Plans*   What will be done to complete the project, and when it will be done.

PERIODIC PROGRESS REPORT

If a project is to continue for several months, management normally will specify that progress reports be submitted at regular intervals.

A periodic progress report may be no more than a one-paragraph statement describing progress of a simple design task, or it may be a multipage document covering many facets of a large construction project. (There are also form-type progress reports, which call for simple entries of quantities consumed, yards of concrete poured, and so on, with cryptic comments.) Regardless of its size, the report should answer four main questions that the reader is likely to ask:

Will your project be completed on schedule?
What progress have you made?
Have you had any problems?
What are your plans/expectations?

To answer these questions, a periodic progress report can readily use the standard Summary-Background-Facts-Outcome arrangement:

Summary   A brief overview of the project schedule, progress made, and plans (*answers the first question*).

Background   The situation at the start of the report period.

Facts   Progress made (*answers the second question*) and problems encountered (*answers the third question*).

Outcome   Plans/expectations for the next period (*answers the last question*).

Figure 4-6 shows how survey crew chief Pat Fraser used these four compartments to write an effective progress report (the numbers in parentheses below are keyed to parts of the report):

H L WINMAN AND ASSOCIATES

INTER-OFFICE MEMORANDUM

To: Karen Woodford, Coordinator
Civil Engineering Department

Date: May 31, 19xx

From: Pat Fraser, Survey Crew Chief

Subject: Progress Report No. 4:
Allardyce Survey Project

The Allardyce Route survey has progressed well during the May 16-31 period. The survey crew has regained two days, and now is only four days behind schedule. We expect to be back on schedule by June 30. ①

Project plan AR-51 shows we should have surveyed mileposts 30 to 34 during this period. But, as stated in my May 15 report, we were six days behind schedule at the end of the previous period, having surveyed only as far as milepost 28. Consequently, we expected to survey only to milepost 32 by May 31. ②

Dry, clear weather from May 18 to 23 enabled us to progress faster than anticipated. We reached milepost 31 on May 23, carried out a terrain analysis for the Catherine Lake diversion scheme on May 24 and 25, resumed surveying on May 26, and reached milepost 32 at 9 a.m. on May 29, two days earlier than expected. We then continued to within 300 yards of milepost 33, which we reached at 6 p.m. on May 31. Survey results are attached. ③A

Two problems affected the project during this period:

1. The electrical fault in the EDM equipment, which delayed us several times early in the project, recurred on May 23. I had the unit repaired at Fort Wilson on May 24 and 25, while we conducted the terrain analysis, and it has since worked satisfactorily. ③B

2. I have had difficulty hiring reliable people to clear brush along the route. Most remain with us for only a few days and then quit, so that I have had to waste time hiring replacements. This problem will continue until mid-June, when the college students we interviewed in March will join the crew.

We plan to advance to milepost 37 by June 15, which should place us only two days behind schedule, and hope to make up the remaining two days during the June 16-30 period. ④

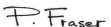

Figure 4-6. A periodic progress report.

(1) The **Summary** tells Civil Engineering Coordinator Karen Woodford how closely the survey project is adhering to schedule, and predicts future progress. This is the information she wants to read first.

(2) The **Background** section reminds Karen of the situation at the end of the previous reporting period and predicts what Pat expected to accomplish during this period. Background should always be stated briefly.

(3) The **Facts** (or **Discussion**) section is broken into two parts:
Work done during the period (3A)
Problems affecting the project (3B)
Pat Fraser opens each paragraph of this compartment with a topic sentence (a summary statement) which states the main point of the paragraph in general terms:

- *Dry, clear weather . . . enabled us to progress faster than anticipated.*
- *The electrical fault in the EDM equipment . . . recurred on May 23.*
- *I have had difficulty hiring reliable people to clear brush along the route.*

Pat then describes what happened in detail, using *facts* (exact dates and milepost numbers, for example) to support each topic sentence. To prevent the report from becoming too long, Pat attaches the survey results to it and simply refers to them in the narrative. (Because of their length, they have not been printed with Figure 4-6).

(4) In the **Outcome** paragraph Pat tells Karen what the crew expects to accomplish during the forthcoming period, and even suggests when they may eventually get back on schedule. This final statement clearly supports the opening paragraph, and so brings the report to a logical close.

Other factors you should consider when writing periodic reports are:

- If a progress report is long, use headings such as these to help readers see your organization:
Adherence to Schedule (*This is your* **Summary**).
Progress During Period (*These are your* **Facts; Background** *information should go at the front of the Progress section.*)
Problems Encountered
Projection for Next Period (*This is the* **Outcome.**)
- For lengthy progress or problems sections, start with a summarizing statement describing general progress, then write several subparagraphs each giving details of a particular aspect of the project:
 4. Interior construction work progressed rapidly but exterior work was hampered by heavy rain:
    4.1 In the east wing, all partitions were erected, 80% of the floor tiles were laid, and 20% of the light fixtures were installed.
    4.2 In the west wing, all remaining floor tiles were laid, all light fixtures were installed, and 16 of the 24 benches were bolted down; 8 of the benches were also connected to the water supply and drains.
    4.3 Landscaping started on September 16, but had to be abandoned from September 18 to 23 when heavy rains turned the soil into a quagmire. By the end of the month only the outer areas of the parking lot had been completed.
- Be as brief as possible, particularly for routine work. Use specific information (*facts* rather than generalizations) in the report narrative, and place lengthy details in an

attachment. If, for example, you are reporting an extensive analysis, in your progress section you might write:

> We analyzed 142 samples, 88 (62%) of which met specifications.
> Results of our analyses are shown in attachment 1.

Attachment 1 would contain several pages of tabular data (numbers, quantities, measurements) which, if included as part of the report narrative, would have inhibited reading continuity.

- Describe problems, difficulties, and unusual circumstances in depth. State clearly what the problem was, how it affected your project, what measures you took to overcome it, and whether the remedial measures were successful:

Summary Statement
Facts
Outcome

> Juvenile vandalism has proved to be a petty but time-consuming problem. On September 3 (Labor Day) several youths scaled the fence around the materials compound and made off with about $170 worth of building supplies. On September 16 they started a front end loader, drove it into the excavation, then got it stuck in the mud and burned out the clutch while attempting to move it. From September 18 the night watch has been doubled and the site has been policed by a patrol dog. There have been no further attempts at vandalism.

Remember that management wants to hear about problems and how they were overcome. Such knowledge can be used to avert difficulties on future projects, or can indicate project trends. A series of problems encountered during a design project may indicate that insufficient time has been allowed for it, that more engineering skills are needed, or that funds budgeted for the project are inadequate. While the Progress section shows how much work has been done, the Problems section often acts as a warning that progress may slow down unless immediate steps are taken.

- Forewarn management of any situation which, although it may not yet affect your project, may become a future problem. With such knowledge, management may be able to help you prevent a costly work stoppage or equipment breakdown. Here's a typical situation:

> 7.1 Unless the strike at Vulcan Steel Works ends shortly, it will soon curtail our construction program. Our present supply of reinforcing barmats will last until mid-October, by which time an alternative source of supply must be found. I have researched other suppliers, but have been unofficially warned by union representatives that any attempt to obtain steel elsewhere may result in a walkout at other plants.

Where should such an entry appear in your progress report? The best position would be at the end of the Facts (Problems) section, immediately before or possibly part of the Outcome.

- Number your paragraphs and subparagraphs if your report is lengthy or comprehensive (see the examples above); this helps you to refer to a specific part of a previous report:

> The possibility of a shortage of steel mentioned in para 7.1 of my September report was averted when the strike at Vulcan Steel Works ended on October 6.

- Maintain continuity between reports. If you introduce a problem that has not been resolved in one report then you must refer to it in your next report, even though no change may have occurred or it has been solved only a day later. You must never simply drop a problem because it no longer applies.
- If management expects you to include project cost information in your progress report, insert it in three places:

> In the **Summary** (comment briefly on how closely you are adhering to projected costs).
>
> In the **Progress** section (give more details of costs, and particularly cost implications of problems).
>
> In the **Outcome** section (indicate future cost trends).

Costs are usually closely linked with your adherence to schedule: the more you drop behind schedule, the more likely you will have to report a cost overrun.

PROJECT COMPLETION REPORT

A project completion report is similar to an occasional progress report in that it describes work that has been done (that is, progress achieved). Normally, however, a project completion report omits any reference to present or future work, because when a job is complete no work is currently being done or is planned. Thus the Summary-Background-Facts-Outcome arrangement shown in Figure 4-2 can be adhered to fairly closely, with the **Facts** compartment being relabelled **Project Highlights.** Sometimes an additional compartment called **Exceptions** is inserted immediately after the Project Highlights to draw attention to deviations from the original plan for the project. These five compartments are outlined in Figure 4-7.

The project completion report written by Jack Binscarth at the end of his analysis of oil samples for Cantor Petroleums is shown below with the five writing compartments identified beside each part of the report. (Jack's progress report for this project is at Figure 4-5.) Note particularly that in a short report you can combine two, or sometimes more, writing compartments into a single paragraph. In Jack's project completion report, paragraph 1 contains both the **Summary** and the **Background,** and paragraph 2 contains both the **Project Highlights** and the **Exceptions.**

> To: F. Stokes
> From: J. Binscarth
> Date: October 25, 19xx
> Subject: Finalizing Cantor Petroleums' Project

Summary My analysis of oil samples for Cantor Petroleums was completed on October 24, eight days later than planned. The work was done at the Cheyenne Corner Refinery, as requested in Cantor Petroleums'

Background   purchase order No. 376188 dated September 4, 19xx, and was scheduled to start on September 11 and end on October 16. I was assigned to the project under our work order No. 2716.

Project Highlights   The work plan called for me to analyze 132 oil samples within the five-week period, but three problems caused me to overrun the schedule and complete four fewer analyses than specified. The delay was caused by a strike of refinery personnel and a faulty spectrophotometer that had to be sent out for repair and recalibration. The incomplete analyses

Exception   were caused by four contaminated samples that could not be replaced in less than six weeks.

Outcome   Russ Dienstadt, the refinery manager, agreed to a cost overrun and has corresponded with you separately on this subject. He also agreed that it would be uneconomical for me to return to analyze replacements for the four contaminated samples. When I delivered the 128 analyses to him on October 24, he accepted the project as being complete.

Jack

Figure 4-7. Writing plan for a project completion report.

PLAN FOR A PROJECT COMPLETION REPORT 1	
SUMMARY	A brief statement that the project or job is complete, plus a short description of the result(s).
BACKGROUND	The circumstances affecting the job, such as purpose, terms of reference, schedule, budget, and persons involved.
PROJECT HIGHLIGHTS	Major achievements, such as work accomplished, problems encountered (plus how they were resolved, and how they affected the project), targets met, and results obtained.
EXCEPTIONS	Variances from the project plan (if there are any), as work that either could not be completed or had to deviate from the plan; the reason for each exception, and its effect, is included.
OUTCOME	Normally a closing statement that identifies any follow-up action that has to be taken, such as invoicing a client or remedying an exception.

INSPECTION REPORT

An inspection can range from a quick check of a small building to assess its suitability as a temporary storage center, to a full-scale examination of an airline's aircraft, avionic equipment, repair facilities, and maintenance methods. In both cases the inspectors will report their findings in an inspection report. The building inspector's report will be brief: it will state that the building either is or is not suitable, and will give reasons. The airline inspector's report will be lengthy: it will describe in detail the condition of every aspect of the airline's operations and list every deficiency (condition that must be corrected). In both cases the inspectors' reports can follow the Summary-Background-Facts-Outcome arrangement.

Summary	The main result(s) of the inspection (very brief); what the reader most wants to know.
Background	Why the inspection was necessary; what was being inspected; who was involved; where and when the inspection took place.

Facts	What the inspection revealed (the details). There are two parts to the Facts:
	A. CONDITIONS FOUND. A description of:
	      Quality (condition) of an equipment
	      or a facility, or of work done.
	      Quantity of items examined, or of work
	      done.
	B. DEFICIENCIES. A list of:
	      Conditions that need to be corrected.
	      Work that needs to be done (or redone).

Outcome	A general statement of results, possibly with a recommendation.

Kevin Doherty's building inspection report in Figure 4-8 shows how these compartments helped him shape his report into a logical, easy-to-follow document. Note particularly how:

His **Summary** (1) tells the Production Manager the *one* thing he most wants to know: can they use the building?

Kevin has opened the **Conditions** section (3) with a summarizing general statement, and then supported it with facts.

He has presented the **Deficiencies** (3B) as a briefly stated list, which makes it easy to identify what has to be done.

The recommendation in his **Outcome** (4) supports his summary.

For a short inspection report like this, Kevin was correct in presenting all the Conditions first, and then listing all the Deficiencies. But for a long report that covers many items, such an arrangement could become cumbersome. For example, if Fran Hartley followed this sequence for an inspection at Remick Airlines, the organization of the Facts section would be like this:

MACRO ENGINEERING INC

From: Kevin Doherty

Date: November 5, 19xx

To: Hugh Smithson,
Production Manager

Subject: Inspection of Carter Building

The old Carter Building at the corner of River Avenue and 39th Street will make a suitable storage and assembly center for the Dennison contract. ①

Christine Lamont and I inspected the Carter Building on November 3, to assess its suitability for both storage and as a work area for 20 persons for 15 months. We were accompanied by Mr. Ken Wiens of Wilshire Properties. ②

We found the interior of the building to be spacious, to have good facilities, but to be unsightly. Our inspection showed that: ③

1. There are 4200 ft^2 of usable floor space (see attached building plan, supplied by Mr. Wiens); we need 3400 ft^2 for the project.

2. There are two offices, each 150 ft^2, and a large unimpeded space ideal for partitioning into a storage area and four work stations.

3. The building is structurally sound and dry, but it is very dirty and smells strongly (the previous tenant was a fertilizer distributor).

4. There are numerous power outlets, newly installed with heavy-duty circuits, and the building has excellent overhead lighting. ③A

5. Several wall surfaces are damaged and some contain obnoxious graffiti.

6. There is a new loading ramp on the north side of the building, suitable for semitrailers.

7. Washroom facilities are adequate for up to 30 persons, but one toilet and two washbasins are broken.

Before we rent the building, the rental agency will have to:

1. Clean it thoroughly.

2. Repair damaged walls, partitions, and toilet facilities. ③B

3. Redecorate the interior.

Ken Wiens said his firm would be willing to do all this.

I recommend we rent the Carter Building from Wilshire Properties, with the provision that the deficiencies listed above are corrected. ④

Kevin W.

enc

Figure 4-8. A short informal inspection report.

A. CONDITIONS FOUND:
　　1.　Electrical Shop
　　2.　Avionics Calibration Center
　　3.　Flammable Materials Storage

(etc . . .)

B. DEFICIENCIES:
　　1.　Electrical Shop
　　2.　Avionics Calibration Center
　　3.　Flammable Materials Storage

(etc . . .)

The more departments that are inspected, the longer the report would become, and the further apart each department's Conditions and Deficiencies sections would be.

　　To overcome this difficulty, Fran should treat each department as a *separate* inspection and reorganize the report so that for each department the Deficiencies section immediately follows the Conditions section. The organization of the whole report would then become:

SUMMARY
BACKGROUND
FACTS:

1. Electrical Shop:
　　A. Conditions Found
　　B. Deficiencies
2. Avionics Calibration Center:
　　A. Conditions Found
　　B. Deficiencies
3. Flammable Materials Storage:
　　A. Conditions Found
　　B. Deficiencies (etc. . .)

OUTCOME:

Conclusions
Recommendations (etc. . .)

　　Fran's inspection report now has a much more tightly knit, logical, coherent organization.

LABORATORY REPORT

There are two kinds of laboratory reports: those written in industry to document laboratory research or tests on materials or equipment, and those written in academic institutions to record laboratory tests performed by students. The former are generally known as test reports or laboratory reports: those written by students are simply called lab reports.

Industrial laboratory reports can describe a wide range of topics, from tests of a piece of metal to determine its tensile strength, through analysis of a sample of soil (a "drill core") to identify its composition, to checks of a microwave oven to assess whether it emits radiation. Academic lab reports can also describe many topics, but their purpose is different since they describe tests which usually are intended to help students learn something or prove a theory rather than produce a result for a client.

Laboratory reports generally conform to a standard pattern, although emphasis differs depending on the purpose of the report and how its results will be used. Readers of industrial laboratory or test reports are usually more interested in results ("Is the enclosed sample of steel safe to use for construction of microwave towers which will be exposed to temperatures as low as −40°C in a North Dakota winter?" a client may ask), than in how a test was carried out. Readers of academic lab reports are usually professors and instructors who are more likely to be interested in thoroughly documented details, from which they can assess the student report writer's understanding of the subject and what the test proved.

A laboratory report comprises several readily identifiable compartments, each usually preceded by a heading. These compartments are described briefly below.

Summary. A very brief statement of the purpose of the tests, the main findings, and what can be interpreted from them. (In short laboratory reports, the summary can be combined with the next compartment.)

Objective. A more detailed description of why the tests were performed, on whose authority they were conducted, and what they were expected to achieve or prove.

Equipment Set-up. A description of the test set-up, plus a list of equipment and materials used. A drawing of the test hook-up may be inserted here. (If a series of tests is being performed, with a different equipment set-up for each test, then a separate equipment description, materials list, and illustration should be inserted immediately before each test description.)

Test Method. A detailed, step-by-step explanation of the tests. In industrial laboratory reports the depth of explanation depends on the reader's needs: if a reader is nontechnical and likely to be interested only in results, then the test description can be condensed. For lab reports written at a college or university, however, students are expected to provide a thorough description here.

Test Results. Usually a brief statement of the test results or the findings evolving from the tests.

Analysis (also called **Interpretation**). A detailed discussion of the results or findings, their implications, and what can be interpreted from them. (The analysis section is particularly important in academic lab reports.)

Conclusions. A brief summing-up which shows how the test results, findings, and analysis meet the objective(s) established at the start of the report.

Attachments. These are pages of supporting data such as test measurements derived during the tests, or documentation such as specifications, procedures, instructions, and drawings, which would interrupt reading continuity if placed in the report narrative (in the Test Method section).

The compartments described here are those most likely to be used for either an industrial laboratory report or a college/university lab report. In practice, however, emphasis and labelling of the compartments probably will differ slightly, depending on the requirements of the organization employing the report writer or, in an academic setting, the professor or instructor who will evaluate the report.

Assignments

INCIDENT REPORTS

PROJECT 1: ACCIDENT AT A CONSTRUCTION SITE

You are an engineering technologist employed in the local branch of H. L. Winman and Associates. Currently you are supervising installation work at a remote construction project at Whirlpool Lake.

The day before you left for the construction site, your branch manager (Vern Rogers) called you into his office. "I'd like you to meet Harry Vincent," he said, and introduced you to a tall, gray-haired man. "Harry is with the Department of the Environment, and he wants you to take some air pollution readings while you're at Whirlpool Lake."

Mr. Vincent opened a wooden box about $14 \times 10 \times 10$ inches, with a leather shoulder strap attached to it. In the box, embedded in foam rubber, you could see a battery-powered instrument. "It's a Vancourt MK 7 Air Sampler," he explained, "and it's very delicate. Don't check it with your luggage when you fly to Whirlpool Lake. Always carry it with you."

For the next hour Mr. Vincent demonstrated how to use the air sampler, and made you practice with it until he was confident you could take the twice-daily measurements he wanted.

Now it is 10 days later and you have just finished taking the late-afternoon air sample measurements. You are standing on a small platform halfway up some construction framework at Whirlpool Lake, and are replacing the air sampler in its box.

Suddenly there is a shout from above, followed immediately by two sharp blows, one on your hardhat and the other on your forearm. You glimpse an 8-foot length of 4-inch square construction lumber tumble past you followed by the air sampler box, which has been knocked out of your hand. The box turns end over end until it crashes to the ground. When you retrieve it the box is misshapen and splintered and the air sampler inside it is twisted. Your hand also is throbbing badly and cannot grip anything. An examination at the medical center shows you have a fractured wrist, and now your forearm is encased in a cast. (Fortunately, it is not your writing hand.)

Part 1.   Write an incident report to Harry Vincent of the Department of the Environment. Tell him:

1. What has happened.
2. That you have shipped the damaged air sampler to him on Remick Airlines flight 673, for him to pick up at your city's airport (you enclose the airline's receipt with your report).
3. That if he wants you to take any more air pollution measurements he will have to send you another air sampler.

Harry Vincent's title is Regional Inspector and his address is: Department of the Environment, Suite 306, 2120 Longley Avenue of your city.

Part 2.   Write a memorandum-form incident report to Vern Rogers. You can mention that you were absent from the construction site for 24 hours, otherwise the incident has not affected your supervision work.

PROJECT 2: OUR SURVEYOR'S TRANSIT
HAS BEEN DAMAGED

You are the team leader of a four-person crew en route to an assignment at Minnowin Point in Alaska. On Sunday evening you take a cab to your local airport to catch Remick Airlines Flight 679 to Edmonton, Canada, the first stop on your route. As one of the crew's suitcases is lifted by an airline employee from the weigh scale to the conveyor belt, its shipping tag tangles with the strap of the case containing a surveyor's transit, which has been placed carelessly on the counter with its strap hanging over the edge. The moving suitcase pulls the transit to the floor, then drags it beside the conveyor belt and bumps it against obstacles along the way. You try to rescue the transit before it disappears through the swinging doors to the baggage room, but it crashes through them and disappears from view. When you retrieve the transit, its carrying case has been badly battered.

You consider that the transit cannot be used in its present condition and decide to leave it with a Remick Airlines' representative to be picked up by your company early in the week. You obtain a receipt for it and board the aircraft to Edmonton.

While in flight you write a report of the incident for your department supervisor, asking that the damaged transit be picked up and shipped to the manufacturer's service office. You also ask your supervisor to borrow or rent a replacement transit from the local manufacturer's representative, and to ship it on next Monday night's flight to Edmonton, for routing to you at Minnowin Point. You hand your report to the flight crew and ask them to send it by messenger to your office on their return flight to your home city.

Additional information you will need to write the report:

Type of Transit: Fennel "Tropl" 5¼ in. with optical plummet, Model A0150

Manufacturer: Otto Fennel Co., Kassel, West Germany

North American Service Office: 35 Kringle Street, White Plains, New York

Manufacturer's local representative: Engineers' Supply Company, 409 Cumberland Avenue of your city

The member of your crew who placed the transit carelessly on the counter: Doug Rickerson

Your department supervisor: J. D. Kamani.

PROJECT 3: INTERRUPTED TEMPERATURE TESTS

H. L. Winman and Associates has been carrying out a series of extreme cold and heat tests on electronic and mechanical switches for Terrapin Control Systems of Denver, Colorado. The tests have been running for four months and will last another two months. The schedule is tight because of initial problems with measuring equipment, which delayed the start by nine days and used up any spare time the project had available.

Currently, you are testing the switches for continuous periods of from 8 to 14 hours. The tests have two parts:

1. For the first 6 hours each day you increase or decrease temperature in 2°C increments until a predetermined high or low temperature is reached. At each 2° increment you test the switches and record how they perform.
2. For the remaining 2 to 8 hours, you bake or deep-freeze the switches at the preselected temperature. No monitoring is necessary during this period (although the switches are tested at room temperature the following day).

To avoid having a technician stay throughout part 2, which on some evenings runs as late as 12:30 a.m., you have installed electrical timers in the circuits of the oven and freezer chamber. At the end of part 2 each afternoon, the timers are set to switch off at the end of the prescribed bake and deep-freeze periods.

This morning when you remove batches 87H and 84C from the oven and freezer chambers you notice that, instead of being close to room temperature, the oven is still very hot and the freezer is still very cold. You check the electrical timers, but both are "off." Then you notice that the electric clock on the lab wall reads only 3:39; your wristwatch reads 8:47—a difference of 5 hours and 8 minutes. You telephone the local power utility:

"Was there a power cut last night?" you ask.

"Where do you live?" a voice replies.

"I'm calling from my office," you say, and quote the address.

"Yes, there was," the voice answers. "We had a transformer blowout at Penns Vale. It affected everyone in your area."

You ask when the power cut started and ended. The voice asks you to wait a minute.

"The transformer blew out at 9:18 last night," the voice eventually announces. "And we restored power to buildings in your area at 2:26 a.m."

You thank the voice, and consult your log for the previous day's tests:

You started part 1 at 9:55 a.m.

You started part 2 at 3:55 p.m., and set the timers to run for 8 hours (they were to switch off at 11:55 p.m.).

You consider what has happened:

The continuous bake and deep-freeze periods were interrupted part way through.

The oven temperature dropped, and the freezer temperature rose, for 5 hours and 23 minutes (but to what temperature?).

The power was restored and the oven temperature again increased, and the freezer temperature decreased (but to what temperature?).

The electric timers switched off at 5:03 a.m. (after their eight hours *total* running time).

You consider the implications of the power cut:

The batches have had uncontrolled, nonstandard testing and will have to be discarded.

Yesterday's tests will have to be run again (on two new batches).

The cost is:
 Labor: 14 hours (7 hours per batch) = $210.00.
 Materials: Two complete batches at $42.00 each.
 Time: One day extra to be added to the program schedule.

Write an occurrence report to your project coordinator (J. H. Grayson). Tell him what has happened, describe the implications, and possibly suggest what might be done to prevent a recurrence.

TRIP REPORTS AND PROGRESS REPORTS

PROJECT 4: ONE OF OUR CRATES
IS MISSING

You are employed by the local branch of H. L. Winman and Associates. When you arrive for work this morning, branch manager Vern Rogers calls you into his office.

"We've had a call from Hugh Smithson (he's production manager for Macro Engineering Inc in Phoenix)," Vern says. "Twelve cartons of special instruments they shipped three days ago were in a semitrailer which rolled near Verradon."

He pulls out a map and points to Verradon, which is 110 highway miles from your office. "The insurance company wants someone to look over the damage with one of their adjusters to confirm how much can be repaired or salvaged."

You drive to Verradon with Mike Shoemacher of Meredith Insurance Corporation. He takes you to a warehouse on the other side of town, where the smashed crates tell their own story of the violence of the accident. Very little of the delicate instruments could have survived such an impact.

You examine the crates one at a time. Broken glass, tangled wire, and chipped and splintered instrument cases are jumbled together. As you check each container, the adjuster notes the numbers in his book: 7, 4, 12, 11, 5, 3, 1, 9, 8 and 2 are totally beyond repair and obviously have no salvage value. Crate number 6 surprisingly is hardly marked: somehow its movements must have been cushioned. You open it carefully and check the instruments.

"This one seems okay," you say. "I think we could do something with it."

The adjuster adds up the totals. "Not very good for us," he says. "Ten out of eleven means a heavy claim."

"Twelve," you say. "There were twelve crates."

The adjuster checks his figures again while you count crates. There are 10 smashed ones and only one good one. And the smashed ones are not so smashed that one would be unrecognizable and so not be counted.

"One is missing," you say.

"Number 10," says Mike. And the two of you check the numbers against the crates.

Plainly, crate number 10 is missing. You and Mike check all the other items removed from the semitrailer, but number 10 is not there.

"Looks as if it has been stolen," Mike comments. "Probably before the accident. I don't think there would have been time afterward: the police were on the scene almost right away."

You telephone Vern Rogers and tell him what has happened.

"I'll telephone Hugh Smithson in Phoenix," he says. "In the meantime I want you to write me a report and send Hugh a copy by special delivery. And be sure to notify the local police."

Information you may need to write your report:

1. The shipping company was Merryhurst Express Lines of Phoenix, Arizona.
2. The waybill number was C7218.
3. Macro Engineering Inc's invoice number was R2261.
4. The 12 cartons were being shipped to Melwood Test Labs, Circular Route No. 1, Harmonsville.
5. Meredith Insurance Corporation's local address is 2800 Western Avenue, Room 414, in the same town as your office.
6. The semitrailer rolled at Berryman's Corner, two miles southeast of Verradon.
7. The crates are being held at Drayton Storage, 313 Crane Street, Verradon.
8. You report the missing crate to Verradon police at 2:25 p.m. on the day you checked the crates.

PROJECT 5: EXAMINING DEFECTIVE CABLE CONNECTORS

H. L.Winman and Associates is management consultant to Interstate Power Company, and currently is supervising the installation of parallel HV DC power transmission lines and a microwave transmitting system along a corridor between Weekaskasing Lake and Flint Narrows. The microwave transmission towers are located approximately 40 miles apart, and are numbered consecutively from No. 1 at Weekaskasing Lake to No. 17 at Flint Narrows. Each tower site has a small residential community and a maintenance crew.

The maintenance crew supervisor at tower site No. 6 reported a week ago today (work out the date) that she has found nine faulty cable connectors type MT-27 and has had to replace them. She telephoned your manager (Andy Rittman), asking whether the fault had been found elsewhere. "There are so many faulty connectors," she said, "I suspect there may be more at other sites."

Andy Rittman instructed you to fly to site No. 12, the site nearest the H. L. Winman office, to investigate whether there are any other faulty connectors. "Use company test procedure TP-33 to test all the connectors," he said, "and then write me a report on your return, so I can send a copy to all site maintenance supervisors."

You flew to site No. 12 two days ago and stayed there until this morning, when you flew back to your home city. The site maintenance crew supervisor was Don Sanderson, who asked you for a copy of your report. Here are details of what you found out:

1. There are 307 type MT-27 connectors on site, with 92 in stock and 215 installed along the lines and up the tower.
2. You tested 268 of the connectors.
3. You could not test the remaining 39 connectors because they were along part of the transmission line which was powered-up throughout your visit.
4. You placed each connector under tension using test procedure TP-33.

5. 233 of the connectors were OK.
6. 35 of the connectors proved to be faulty.
7. You identified the fault as a hairline crack, which became visible when a faulty connector was placed under tension.
8. You also noticed that, although the connectors looked similar, there seemed to be two kinds of connectors on site. One batch of connectors had the letters GLA on the base. The other had the letters MVK on the base.
9. Of the 268 connectors you checked, 196 were stamped MVK, and 72 were stamped GLA.
10. All the faulty connectors had the letters GLA stamped on the base. There were no faulty connectors with the letters MVK on the base.
11. You figured that the letters must identify either different manufacturers or different batches made by the same manufacturer.
12. You instructed the maintenance crew supervisor to replace all installed GLA connectors with MVK connectors. You also told him to place all the GLA connectors in a separate box and to mark them NOT TO BE INSTALLED.

Now you have returned to your home office and you are starting to write your trip report to Andy Rittman. Write the report, telling him of your trip and your findings.

PROJECT 6: INSTALLING VIDEOCONFERENCING EQUIPMENT

To improve intercompany communication, Macro Engineering Inc and H. L. Winman and Associates are installing videoconferencing equipment in the conference rooms at their two main business offices in Phoenix and Cleveland. The system is being installed by Westwind Video of Phoenix, and the Videocon Corporation of Montrose, Ohio. You have been appointed coordinator of the installation phase—you are based at the Cleveland office—and you are working under a tight deadline because Harvey Winman and Martin Dawes (in Cleveland) and Tina Mactiere and Wayne Robertson (in Phoenix) have set up their first videoconference for 1 p.m. (Eastern time) on the first day of next month. Today is the 18th of the current month.

The equipment is scheduled to be fully installed and ready for handover at both offices by the 25th, one week from today. The plan is to carry out system tests of the two installations between the 26th and the 30th, so that the videoconferencing system will be fully operational two days before the first videoconference.

A telephone call to both the contractor and Tina Mactiere in Phoenix quickly establishes that the western contractor is now two days ahead of schedule and has predicted that his installation work will be all wrapped up on the 22nd. Cleveland, however, is another story. Today the contractor is several days behind schedule. When you visit Dave Millies, Videocon's project manager, he says there have been problems.

"There was a seven-day delay in delivery of two of the video cameras," he explains. "They were shipped to Dayton, Ohio, in error, and it was four days before we were able to locate them. And then the in-desk control panel was damaged in transit. My technicians tried to repair it but after three days they realized it was a hopeless situation. So I had to order a replacement from the manufacturer."

"Is it here yet?" you inquire.

"Oh, no!" Dave Millies continues. "It won't be here for another week."

"How long will it take to install?" you ask.

"If my technicians are back on the job, two days," he replies. "You see, we're involved in a labor dispute right now, and no one is working. But it should be resolved in a week. I'm pretty sure it will."

You check the conference room and determine that one of the three video cameras has been installed, about 70% of the special lighting is in place, and 90% of the cabling has been done.

You figure that, with the delayed equipment and absent workers, the earliest the installation will be complete is the 27th or 28th.

Write a memo-form progress report to Martin Dawes. Inform him of the progress, problems, and delays, and predict a new installation completion date. Warn him that there will be only a very brief period for system tests and there is a possibility the first videoconference will have to be delayed.

PROJECT 7: REPORTING INSTALLATION OF ELECTRONIC EQUIPMENT

As an engineering technologist with Macro Engineering Inc you are often responsible for installing equipment at clients' offices. Currently you are working at Freedom Lake Narrows, and Judi Tarnicki has been assigned to assist you. Here are some background details pertaining to the project:

- You and Judi flew to Freedom Lake Narrows two weeks ago today (work out the date).
- The job is identified as Macro Engineering project M207 and the installation specification was dated one month ago today.
- The job requires you to: (1) install a Mantronics DS8 telecommunications link, a TR16 remote control system, and associated test equipment and hardware; and (2) test the system with the Lake Lawlong station to the south and Long Sutton station to the north.
- The job is expected to take four weeks, so that you should finish two weeks from today.
- As part of the contract you are required to submit a progress report at the midpoint of the project (today). It is to be sent to your project coordinator, Donna Williams, at Macro Engineering Inc.
- The installation specification you are working to is Macro Engineering specification MEI27-2; the test specification is MEI27-8.
- You work a five-day week (Monday to Friday).

Here are some of the events that have occurred during your two weeks at Freedom Lake Narrows:

1. During the first week you installed, with no significant problems to affect your progress:

    Transmitter-receiver TX/RX 106.
    The microwave antenna dish (on top of the tower).
    The cabling between the TX–RX 106 and the dish.

2. During the second week you ran into problems:

    (a) On Monday Judi dropped a crate onto one foot and had to be flown to the Freedomville hospital for treatment. The result: two broken toes. Judi was off work for three days and, although now back at work, she is moving about more slowly than usual. You estimate this has set your schedule back by two days.

    (b) Yesterday, when you opened the crate containing remote sensing unit RS201, you found that some parts were damaged. It looked as though the crate had been dropped or knocked about during shipment. You telephoned Macro Engineering for replacement parts, but production engineer Roly Thoms said it would be better to ship you a complete unit. Delivery is promised for one week from today.

3. You turned your attention to installing interequipment cabling and the test equipment, which will take you three more days. You will then install the AU9 voice communication station and the automatic switchover relays.

Providing the replacement RS201 arrives as promised, you will install it during the first two days of week 4. This will be followed by equipment calibration and testing, which you expect will take 6½ days, and client acceptance (two days).

You predict that because of the delays the system will not be ready for handover until three weeks from today. You could cut three days off the delay by working two weekends, but this would increase contract costs by approximately $760. You need approval to work overtime.

Now write your progress report to Donna Williams.

INSPECTION REPORTS

PROJECT 8: CHECKING AN EQUIPMENT INSTALLATION

You are an engineering technician with H. L. Winman and Associates and have been assigned to supervise the construction of a communications link along the HV DC power transmission line between Weekaskasing Lake and Flint Narrows. The communications link will have several relay stations along its length, and these are being built by contractors. You are one of three inspectors responsible for coordinating the contractors and inspecting their work as each phase is completed.

All construction work is complete and has been inspected at relay station MP-03 (Lake Lawlong). Currently an electronics firm known as Specialty Electronics Installations (SEI) is installing the electronics and data communications gear. Two days ago the SEI installation supervisor sent a message to your office that the equipment would be ready for inspection today. So yesterday Les Phillips, the senior project coordinator, gave you a memo which requested you to fly to Lake Lawlong today to inspect the equipment installation. You are working on project No. 3606.

You arrive at Lake Lawlong by charter helicopter at 9:15 a.m. and are greeted by Dale Yewskiw, the installation supervisor.

"I'm the only one here," he announces as he introduces himself. "I have sent the other two technicians on to the next site. There's no point in keeping them here, now the job is finished."

You open inspection checklist E27-2, which you have brought with you, and start working through it step by step. You place a checkmark next to every satisfactory item, and a comment beside every unsatisfactory item. There are 182 items to check.

You start your inspection outside the building, climbing the narrow ladder up the tower to the microwave dish near the top. Everything is OK, except for item 18 (which says: "Check that the cable to the microwave dish is laced firmly to the tower supports, and that the spacing between cable lacings does not exceed 36 in."). The cable is firm but you find three places where the lacings are too far apart. They measure 41 in., 39 in., and 44 in.

"The outside installation is excellent," you say to Dale as the two of you enter the small building, "except for that one thing."

Inside the building you start at the control unit, first checking the equipment installation. Step 41 calls for you to check Monitor RX-7, which is to be mounted on the wall beside the equipment rack. It isn't there. Dale explains that the mounting brackets and shelf (forming assembly 6100) were not supplied with the equipment, so he has temporarily put the monitor unit above the processor, on the top shelf of the rack. "I've ordered a replacement assembly," he says, "but the equipment can work this way until the shelf and brackets arrive." You accept this temporary change.

Steps 58 to 73 call for you to check all electrical connections. For the following items you write:

62. Check cable to ML17 at rear of unit RJ4 is enclosed within a shielded sleeve.	Sleeve missing.
63. Check resistance between connections TP9 and TP20 is 66 Ohms ± 6 ohms.	87 ohms (21 ohms too high).
71. Check that no connections show evidence of cold-soldered joints.	ML23 on RJ4 and B06 on TM14 are poorly soldered.

At this point you switch on the equipment and start doing electrical and operational tests. In the checklist you write:

113.	Simulate a power outage and check that the standby diesel unit cuts in automatically. (Limit: 6 seconds maximum)	Three attempts: Standby diesel cut in ok, but only after: 1. 9 sec 2. 10.5 sec 3. 8.5 sec
114.	When a power outage is simulated, check that diagnostic tester DT6 generates a fault signal and transmits it to data output.	No signal at data output.
115.	If no signal at data output, check whether the fault is at the diagnostic tester or in the transmission link.	Diagnostic tester generating signals ok. Transmission line faulty. Fault traced to defective PCB 10671.
144.	Check the voice transmission circuit by establishing two-way conversation with Flint Narrows.	Flint narrows receiving voice transmissions ok. Transmissions from Flint Narrows are intermittent.

The remainder of the tests are satisfactory, so you sit down with Dale Yewskiw and discuss the deficiencies.

"While you were doing the tests I resoldered the connections at ML23 and B06," Dale says, "and installed a shielded sleeve on the cable on ML17."

You check his work and approve it as satisfactory, and then insert checkmarks against steps 62 and 71. "You know," you mutter, "I'm going to call up Flint Narrows again. TM14 is part of the voice transmission circuit. Perhaps you have corrected the voice transmission problem."

This time when you call up Flint Narrows the voice transmission is perfect in both directions. "That's OK," you say, as you place a checkmark against step 144.

The helicopter returns at 3:45 p.m. and you prepare to fly out. "Overall, you've done a pretty good installation," you say to Dale as you leave. "There are very few deficiencies."

"Thanks," he replies. "I'll clear them up within a week, except for repositioning Monitor Unit RX-7. There's a back-order for the mounting assembly."

Now you are back in your office, writing your inspection report to your project coordinator, Les Phillips. You decide to recommend accepting the installation when all deficiencies have been corrected (except for moving the RX-7 to its proper position). You feel the contractor can be released from that task, since it could be handled easily by the client's resident technician (who will be joining the site next week). Write your report as a memorandum.

5

Longer Informal
and Semiformal
Reports

The previous chapter discussed short reports that deal primarily with facts, reports in which the writer identifies the relevant details and presents them briefly and directly. This chapter describes longer reports that often deal with less tangible evidence; reports in which the writer analyzes a situation in depth before drawing a conclusion and, sometimes, making a recommendation. They may describe an investigation of a problem or unsatisfactory condition, an evaluation of alternatives to improve a situation, a study to determine the feasibility of taking certain action, or a proposal for making a change in methods or procedures. All are written in a fluent narrative style that is both persuasive and convincing; their writers have concepts or new ideas to present and they want their readers to understand their line of reasoning.

INVESTIGATION REPORT

The term "Investigation Report" covers any report in which the writer describes how he or she performed tests, examined data, or conducted an investigation using tangible evidence. Basically, the writer starts with known data and then analyzes and examines it so that the reader can see how the investigation was conducted and the final results were reached. The report may be issued as a letter, as in Figure 5-2, an interoffice memorandum, as in Figure 5-6, or as a semiformal report.

Sometimes an investigation report is written first as a memorandum from a project group leader to a department head. Later its essential details are rewritten as a letter report from the department head to a customer or client. The investigation report in Figure 5-2 illustrates the informal yet businesslike tone of such a letter.

Although they are not always readily identifiable, there are standard parts to a well-written investigation report that help shape the narrative and guide the reader to a full understanding of its topic. They are shown in Figure 5-1, which is an expanded version of the basic compartments described at the start of chapter 4. These parts are easy to recognize in long investigation reports, where headings act as signposts introducing each parcel of information. They are more difficult to identify in short reports which use a continuous narrative. In the investigation report in Figure 5-2, the parts are identified by circled numbers:

(1) This is the **Summary.**
(2) The **Background** is only one sentence; the reader knows the circumstances, so details are omitted.
(3) The **Discussion** starts here, with a very brief reference to the **Method.**
(4) These are the **Findings.**
(5) This is the first **Idea** and its **Evaluation** or **Analysis.**
(6) These are **Ideas** 2 and 3, each followed by an **Evaluation.**
(7) The **Outcome** draws **Conclusions** and makes a **Recommendation.** (There are no **Attachments.**)

Figure 5-1. Compartments for an investigation report.

SUMMARY	A brief statement of the situation or problem and what should be done about it.
INTRODUCTION	**Background** of the situation or problem
DISCUSSION	The **Facts** or **Investigation Details,** comprising:
	METHOD — How the investigation was tackled.
	FINDINGS — What the investigation revealed.
	IDEAS — Different ways the situation can be improved or the problem resolved.
	ANALYSIS — Evaluation of each idea.
CONCLUSIONS	The **Outcome**, or result of the investigation; a summing-up.
RECOMMENDATION	A positive statement advocating action.*
ATTACHMENTS	Evidence: detailed facts, figures, and statistics which support the Discussion.*
	*Included only when appropriate.

H L WINMAN AND ASSOCIATES

PROFESSIONAL CONSULTING ENGINEERS
475 RESTON AVENUE
CLEVELAND, OHIO 44104

October 21, 19xx

Mr. D.R. Carlisle
Technical Operations Manager
Television Station DCMO-TV
PO Box 890
Montrose, OH 45287

Dear Mr. Carlisle:

Reduction in Noise Level - Satellite Studio Control Room

We have investigated the high background noise reported in the control room of Station DCMO-TV's satellite studio at 21 Union Road, and have traced it to the building's air-conditioning equipment. The noise could be reduced ① to an acceptable level by soundproofing the air-conditioning ducts and blowers, and by carpeting the control room. ②

Our investigation was authorized by your letter of 28 August 19xx. Tests ③ conducted with a noise meter at various locations in the control room established that the average ambient noise level is 36 dB, with occasional peaks of 39 dB near the west wall. This is approximately 8 to 10 dB higher than the noise levels measured in the control room of your No. 1 studio on Westover Road.

The unusually high noise level is caused by the air-conditioning equipment, ④ which is located in an annex adjacent to the west wall of the control room. Air-conditioner rumble and blower fan noise are carried easily into the control room because the short air ducts permit little noise dissipation between the equipment and the work area. The flat hardboard surface of the west wall also acts as a sounding board, and amplifies the noise.

We have considered three methods that could be used to reduce the ambient noise to an acceptable level:

1. The most effective method would be to move the air-conditioning equipment to a remote location. This, however, would require major structural alterations which would make the approach impractical except ⑤ as a last resort.

2/...

Figure 5-2. A letter-form investigation report.

2. A noise reduction of approximately 6 dB could be effected by replacing the existing blower fan assembly with a Model TL-1 blower manufactured by the Quietaire Corporation of Detroit, and by lining the ducts with Agrafoam, a new soundproofing product developed by the automobile industry in West Germany.

3. A further reduction of 1.5 dB could be achieved by replacing the vinyl floor tiles with indoor/outdoor carpet, a practice that has been successful in Air Traffic Control centers. Mounting carpet on the whole of the control room's west wall, and possibly along the first five feet of the north wall, would also reduce the noise by an additional 1 dB.

Since relocation of the air-conditioning equipment would be unduly expensive, we recommend that methods 2 and 3 be adopted to reduce overall ambient noise by approximately 8.5 dB for a total cost of $8080.00. We suggest that the modifications be made progressively so that noise reductions can be measured on completion of each step. This will permit either or both of the final steps to be eliminated if earlier modifications achieve better results than anticipated. The four steps will be:

Modification	Anticipated Ambient Noise Reduction	Approximate Cost
(a) Replace blower fan assembly	2.5 dB	$2560.00
(b) Line ducts with Agrafoam	3.5 dB	$2600.00
(c) Install floor carpet	1.5 dB	$1580.00
(d) Install wall carpet	1.0 dB	$1340.00

We believe that these modifications will provide the quieter working environment desired for your operating staff.

Sincerely,

RW Gray

for Peter Bell, Head
Mechanical Engineering Department

RWG:bls

Every investigation report will differ, depending on the topic you are investigating and the results you obtain. Sometimes you will use all of the standard parts, sometimes you will use only some of them, and occasionally you will need to devise and insert additional parts of your own. Typical standard parts are described below.

Summary.   Start with a brief description of the whole report, stated in as few words as possible. This will give busy readers a quick understanding of the investigation from which they can learn the results and assess whether they should read the whole report.

Background.   Describe events that led up to the investigation, knowledge of which will help the reader place the report in the proper perspective. This part is often referred to as the Introduction.

Investigation Details.   Describe the investigation fully, carefully organizing the details so that readers can easily follow your line of reasoning. This is the Discussion. In a long report the parts should contain information similar to that described here, and appear roughly in this order:

Start with Known Facts. Introduce readers to all the technical data and tangible evidence available at the start of the investigation or used during the investigation.

Introduce Guiding Factors. These are the requirements or limitations that controlled the direction of the investigation. They may be as diverse as a major specification stipulating definite results that must be attained, or a minor limitation such as price, size, weight, complexity, or operating speed of a recommended prototype or modification.

Outline Investigation Method. Tell readers the planned approach for the investigation so that they can understand why certain steps were taken.

Describe Equipment Used. If tests were performed that called for special instruments, describe the test setup and, if possible, include a sketch of it. In long reports this may have to be done several times, to keep details of each test setup close to its description.

Narrate Investigation Steps. Many investigation reports lend themselves to a chronological presentation of information. Use a logical step-by-step description if the investigation has progressed naturally through a series of steps involving the introduction to a problem, development of a means to overcome it, application of corrective measures, and tabulation of results. But when an investigation has included an analysis of processes, equipment, or methods, divide the information into suitable subject areas (such as equipment, processes, application) and examine the advantages and disadvantages of each. For more information on the chronological and subject method of report development, refer to the "discussion" section in Chapter 6.

Discuss Test Results. Analyze the results of each test and discuss whether they satisfy the needs of the investigation. (It is not sufficient simply to tabulate the test results and assume that readers will infer their meaning and implications.) Such an analysis after a test or series of tests can demonstrate that a specific trend influenced or altered the course of the investigation.

Develop Ideas and Concepts. At some point in your narrative you may want to introduce ideas and concepts which have evolved as a result of your investigation. Either introduce them periodically throughout the report to demonstrate what you had in mind before taking the next investigative step, or group them together near the end of the Discussion. They should develop naturally from the flow of the report and should be persuasive if they have to influence your readers' acceptance of a new technique or an unusual method you want to recommend.

Analyze Ideas. If you have developed alternative ideas (methods) for resolving a problem, evaluate them to assess which is best, using factors such as effectiveness, cost, and simplicity as your evaluation criteria. (Ensure that your readers understand what criteria you are using, and why they are important, before they read your evaluation.) You may evaluate each idea as soon as you present it, as has been done in the investigation of noise in a television studio (Figure 5-2), or you may present all the ideas first and then evaluate all of them, as Figure 5-1 suggests.

Conclusions.   In a brief summing-up, draw the main conclusions that have evolved from your investigation.

Recommendation(s).   State what steps need to be taken next. Recommendations are optional and must develop naturally from your Discussion and Conclusions.

Evidence.   Attach detailed data such as calculations, cost analyses, specifications, drawings, and photographs that support the facts presented in the **investigation details** section but would interrupt reading continuity if included with it. These pages of evidence are numbered consecutively and named Attachments or, in a long report, Appendixes, and placed at the end of the report.

The process of an engineering investigation, and the report that evolves from it, can be illustrated as a flow diagram (see Figure 5-3).

As an example, suppose that Martin Dawes, as president of H. L. Winman and Associates, decides to purchase an executive jet aircraft because he travels a lot, but cannot decide which aircraft would be most useful. He might assign two of his engineers to investigate a selected range of aircraft and to

Figure 5-3. Basic flow diagram for an investigation report.

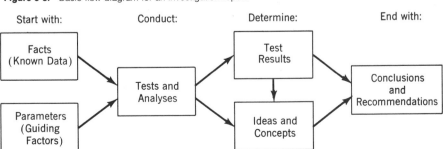

PROFESSIONAL CONSULTING ENGINEERS

INVESTIGATION REPORT

Report No.: __70/26__   File: __53-Civ-26__   Date: __March 20, 19xx__

Prepared for: __City of Montrose, Ohio__

Authority: __City of Montrose Letter HWY/69/38, November 7, 19xx__

Report Prepared by: _W. Rutichyslyn_ Approved by: __M. Downes.__

SUBJECT OR TITLE

INVESTIGATION OF STORMWATER DRAINAGE PROBLEM:

Proposed Interchange at Intersection of Highways 6 and 54

SUMMARY OF INVESTIGATION

The proposed interchange to be constructed at the intersection of Highways 6 and 54, on the northern perimeter of Montrose, incorporates an underpass which will depress part of Highway 54 and some of its approach roads below the average surface level of the surrounding area. A special method for draining the stormwater from the depressed roads will have to be developed.

Two methods were investigated that could contend with the anticipated peak runoff. The standard method of direct pumping would be feasible but would demand installation of four heavy duty pumps, plus enlargement of the three-quarter mile long drainage ditch between the interchange and Lake McKing. An alternative method of storage-pumping would allow the runoff to collect quickly in a deep storage pond which would be excavated beside the interchange; after each storm was over the pond would be pumped slowly into the existing drainage ditch to Lake McKing.

Although both methods would be equally effective, the storage-pumping method is recommended because it would be the most economical to construct. Construction cost of a storage-pumping stormwater drainage system would be $784,000, whereas that of a direct pumping system would be $967,000.

Figure 5-4. Title and summary page for a semiformal investigation report.

recommend the most suitable model. Starting with facts (specifications of the types of aircraft Martin Dawes is known to favor) and parameters (probable route mileage, maximum operating costs, maximum acceptable purchase price), they would analyze the different aircraft and determine which offers the most advantages under the proposed operating conditions. One of them might also conceive the idea of leasing an executive aircraft which could be rented from a local aviation company, complete with crew, on an as-needed basis. Armed with all this information the engineers would prepare a report of their findings, ending with a conclusion that identifies the most suitable aircraft to purchase but which also points out the economic advantage of leasing. The report they submit to Martin Dawes would be an investigation report that starts with known information but ends by developing an idea.

Some companies preface their investigation reports with a standard title and summary page similar to the H. L. Winman and Associates' design illustrated in Figure 5-4. This page saves a reader the trouble of searching for the summary and the report's identification details. Subsequent pages, which have not been included with the example, contain the report narrative, starting with the Background (Introduction). Reports written in this way tend to adopt a slightly more formal tone than memorandum and letter reports and are less likely to be written in the first person. Sometimes they are called Form Reports, although I prefer to identify them as Semiformal Investigation Reports.

EVALUATION REPORT

Where an investigation report starts with facts, an evaluation report starts with an idea or concept its author wants to develop (see Figure 5-5). The writer first establishes guidelines to keep the report within prescribed bounds, and then researches known data, conducts tests, and analyzes the concept to prove or disprove its viability. At the end of the evaluation the writer is able to draw the conclusion that the concept either is or is not feasible, or perhaps is feasible in a modified form.

Figure 5-5. Basic flow diagram for an evaluation report.

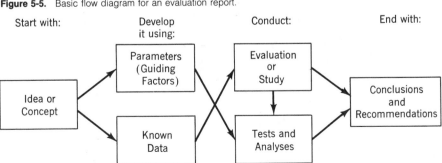

The Macro Engineering Inc report on training methods in Figure 5-6 is an evaluation report because it deals with ideas and concepts rather than with facts and events. The *idea* is management's belief that new training methods will increase production; the *known data* are the four training methods discussed by Fred Stokes; the *evaluation* is his assessment of each method, or combination of methods, in the light of the company's training needs. The standard parts that shape the narrative of this report are, however, the same as those for the investigation report.

The comments below identify the standard parts that Fred Stokes has used and discuss briefly how they have helped him fashion an effective report.

Paragraph 1 is a brief **summary** of the evaluation and main conclusions.

Rather than write a summary of the main points of the whole report, Fred homes in on the one point of immediate concern to management: How much is all this training going to cost? He has followed the first rule of summary writing: Identify what is most important to your readers *before you write*.

Paragraphs 2 and 3 provide **background information** and details of events that led up to the evaluation.

Fred knows that the person to whom the report is directed will already be aware of much of this information, but he also recognizes that there may be other readers to whom it may be new.

Paragraph 4 is a **guiding factor** that limits the extent of the evaluation.

Fred probably discussed these limitations with Tina Mactiere when he first established his terms of reference. No engineer or technician should attempt to write a report without first clearly establishing the objectives.

Paragraphs 5 through 9 comprise **investigation details.** Paragraph 5 outlines the **investigation method,** while paragraphs 6, 7, 8, and 9 each evaluate one of the training aids.

Fred's analysis is interesting and logical. He uses a three-step approach that makes the evaluation of each training method a small report in itself: (1) a brief **introduction** of the training method and how it is used; (2) a **discussion** of its application to industry and the implications of using it; and (3) a **conclusion** that sums up the value of the training method to Macro Engineering Inc.

Paragraphs 10 through 15 discuss **evaluation results.**

By first introducing the guiding factor that will most influence his selection, Fred is able to eliminate one training method quickly, identify another that has to be used, and establish that either of the two remaining methods can be used with it. His discussion is unquestionably logical, and follows naturally from the conclusions drawn at the end of the four analyses.

MACRO ENGINEERING INC

From: __F. Stokes, Chief Engineer__   Date: __November 20, 19xx__

To: __T. Mactiere, President__   Subject: __Updating our Training__
__Methods__

Summary

1. If we are to meet our production schedules with the semiskilled labor
    currently available, we will have to augment the informal on-the-job
    training method we currently use by introducing more comprehensive
    instruction, either in a classroom or by closed circuit television. A
    classroom would be less costly than a television system ($11,000 vs
    $36,000 to set up; $28,000 vs $31,000 annually to operate), but CCTV
    offers many peripheral advantages which make it the better choice.

Introduction

2. Over the past five years the scientific instruments we manufacture have
    become increasingly complex and have demanded highly sophisticated
    manufacturing techniques. At the same time our source of skilled labor
    has decreased, so that we have had to use semiskilled and unskilled
    labor for many positions on the assembly lines. This has resulted in
    a gradual slowdown of production, an increase in the number of manu-
    factured units rejected by Quality Control, and a consequent increase
    in manufacturing costs.

3. At the management meeting held on October 11, 19xx it was recognized
    that a better training program than we now have would do much to im-
    prove the manufacturing skills of our employees. I was requested to
    evaluate alternative training methods and to recommend a training
    technique that might increase productivity.

4. I have confined my report to an analysis of training methods that could
    be used in our plant. Development of a specific training program I
    will discuss in a separate report to be prepared after the training
    method has been selected.

Analysis of Industrial Training Methods

5. There are four basic training methods, ranging from inexpensive estab-
    lished methods to costly new techniques, that we could use to train our
    staff. The suitability of each is discussed generally here and is
    supported by a comparison chart attached as an appendix.

6. On-the-Job Training (OJT)

    6.1 This time-honored training method is used extensively in industry
         and is the method currently used on our production lines. It calls

Figure 5-6. An evaluation report.

for trainees to learn manipulative tasks in which a sequence of
movements has to be mastered. The trainees "learn by doing" under
the watchful eye of an instructor, who often is another fully
trained employee. When they have learned the task, they are given
a proficiency test to assess whether they can work on their own.

6.2 Every time we hire new employees, or transfer existing employees
from one production line to another, they are given OJT which continues until they are fully able to perform the new task. It is
an effective training method, but is slow and has hidden costs:
trainees are poor producers for a long time, and a higher-than-
normal proportion of their work is below standard and has to be
rejected.

6.3 To continue relying solely on OJT would be to perpetuate a training
method that has already been proved inadequate for our needs.

7. Classroom Instruction (C.I.)

7.1 The classroom situation is the form of training most of us know
best. Employees recognize this type of teaching and, even though
some may resent the schoolroom atmosphere it evokes, they usually
respond quite readily to it. Courses follow a basic pattern, with
the instructor setting objectives, preparing instructional material
to meet them, then presenting lectures. An end-of-course test
demonstrates whether the trainees have learned enough to be able
to perform their tasks under supervision.

7.2 To be effective, C.I. demands the proper atmosphere, a good instructor, and receptive trainees. This means providing space for
a properly equipped training room rather than just a corner of the
workshop; it means keeping an instructor who can teach specialist
subjects on the payroll, or drawing instructors from the supervisory line staff (this must be done carefully because not all
supervisors make good instructors); and it means making sufficient
employees available for training at the same time.

7.3 The cost of classroom instruction is high in relation to the degree
of training it provides. It is particularly suitable for teaching
theory, but of less value when purely manipulative skills must be
learned. Hence it must be supplemented by practical training
such as OJT.

8. Programmed Instruction (P.I.)

8.1 Programmed instruction is a relatively new technique that only
recently has been generally accepted as a useful training aid.
Because it permits students to work alone and at their own speed,
it is invaluable for teaching routine tasks to very small groups
or to individuals.

8.2 The type of programmed instruction most of us recognize is the
programmed textbook that teaches a small piece of information at
a time, tests the trainee on what he has just learned, provides

correction if the trainee still has not understood, and then proceeds to the next piece of information. Trainees progress quickly
or slowly through the program, depending on their previous knowledge
and ability to learn rapidly. They are tested periodically, and at
the end they are given a proficiency test. Except for monitoring
the trainees' progress and marking their final tests, an instructor
is not needed.

8.3 P.I. textbooks are ideal for teaching specialist subjects and new
techniques, but seldom can be used to teach manipulative skills.
Therefore, any P.I. manuals we produce would have to be supplemented by some verbal instruction and practical training. Preparation
of P.I. materials is also expensive and very slow -- too slow to
meet the rapid training needs of many of our production lines.

9. Closed Circuit Television (CCTV)

9.1 As an educational training aid, CCTV is rapidly gaining popularity
with departments of education across the country. Because its
initial cost is high, until recently it has been limited primarily
to colleges and schools. But there is no reason why the new, less
expensive systems should not be used in industry.

9.2 Training lessons showing manufacturing processes, assembly techniques, packaging methods, etc, could be prepared as TV "programs"
and stored on videotape ready for playback to trainees whenever
the need arises. We would be able to show programs repeatedly,
either to groups or individuals, at virtually no cost. We could
demonstrate the right and wrong ways of doing things, and enlarge
tiny objects for all to see simultaneously. And we could even
install monitors right on the production line for employees to
follow step-by-step as they assemble components. CCTV would give
us the ability to prepare and record instructional programs
quickly, so that teaching of new assembly methods would be able
to start very early in future projects.

9.3 The initial cost of setting up CCTV would be much greater than for
the other three methods, and operating costs would also be higher.
A staff member would have to be trained in TV production techniques, and a small room would be required for equipment storage
and playback. The advantages of CCTV should, however, outweigh
many of the objections to its high cost. The familiarity of TV,
combined with its uniqueness as an inplant teaching method, would
promote learning and employee acceptance of training courses. And
the image of a progressive company that it would convey would be
invaluable for building employee morale.

9.4 CCTV would not, however, be able to stand alone as a training
medium. We could use it to teach theory and to demonstrate manipulative skills, but it would not entirely replace the final step:
practice. This would still have to be provided by OJT.

Selection of a Training Method

10. The training method we select must teach both theory and manipulative skills, and also provide opportunity for practice. Since none of the training methods I have discussed meet all of these requirements, we will have to use a combination of methods that provides the best training for the least cost.

11. OJT must be retained because it is the only method which provides the practice that is essential before a trainee can become fully productive. But because OJT has so many hidden costs, it must be reduced to the minimum time possible by combining it with the most effective teaching method.

12. P.I. can be eliminated because of its high costs, slowness of preparation, and inability to teach complicated manipulative skills.

13. Either C.I. or CCTV could be combined successfully with OJT to provide a comprehensive training program. Both would significantly reduce the time required for OJT. The C.I./OJT combination would cost $45,000 the first year, and $34,000 per year thereafter. The CCTV/OJT combination would cost $73,000 the first year, and $37,000 per year thereafter. CCTV, however, offers many peripheral advantages that C.I. does not.

14. I consider that the advantages of the CCTV/OJT combination outweigh those of the C.I./OJT combination. The image that CCTV would convey and the enthusiasm it would spark are intangible factors that cannot be measured in dollars. CCTV equipment could also be used for activities such as sales seminars, management training, time and motion studies, and advertising, so that maximum value would be obtained from our investment.

15. Whichever plan is adopted, I will prepare specifications for purchasing the equipment, and will develop a detailed training implementation program.

Conclusions

16. Training methods can be improved at Macro Engineering Inc by combining the existing on-the-job training with either:

    16.1 classroom instruction, at moderate installation and operating cost, or

    16.2 a closed circuit television system, at high installation but moderate operating cost.

Recommendation

17. I recommend that we update our training methods by installing a CCTV system costing $36,000, and budget for an annual operating cost of $37,000.

F Stokes

APPENDIX: COMPARISON OF FOUR TRAINING METHODS

	OJT	C.I.	P.I.	CCTV
What is initial purchase and set up cost?	Nil	$11,000	$2,000	$36,000
What are annual operating costs? (including instructors salaries)	$6,000	$28,000	$64,000	$31,000
Can training method teach both theory and practice?	No; mostly practical skills	No; mostly theory	No; almost entirely theory	No; mostly theory, some practice
How quickly do trainees learn?	At moderate speed	At moderate speed	Depends on student & program	At moderate speed
How readily do trainees accept training method?	Readily	Usually readily	Varies; resisted by some	Usually readily
Can training be done on production line?	Yes	No; generally not	No; rarely	In part (direct assembly training)
How quickly can a specific training program be set up?	Rapidly	Fairly quickly	Slowly	Fairly quickly
Can students be taught independently or must they work in groups?	Mostly independently	Generally in groups	Only independently	Either
Can training method be used without an instructor?	Some of the time	No	Yes	Yes
Can equipment/facilities be used for other purposes?	No	Yes	No	Yes
Has training method any special requirements or limitations?	Interferes with production	Requires skilled instructor	Needs specialist writer; students must read well	Needs TV producer/instructor

Data source: File ENG/26/3107

In Paragraph 14 Fred outlines his preference and gives reasons for it, to prepare readers for the recommendation he will make. Although he knows management wants him to indicate his preference (see paragraph 3), he also recognizes that management may choose not to accept his recommendation (see paragraph 15).

Paragraphs 16 and 17 state the main **conclusions** and make a **recommendation.**

Fred ends his report by briefly outlining the two alternatives, and then stating clearly which he recommends. This section of a report is sometimes referred to as the **terminal summary.**

The **attachment** compares the methods in an easy-to-digest form.

Rather than support his short report with numerous pages of data, Fred has chosen to compare the training methods very generally in a single table and to refer readers who want more substantive data to the source of his information. (This can be done only when the source is readily available to readers.)

Note that Fred Stokes does more than simply describe each training method in the investigation details section. By discussing the advantages and disadvantages of each method, and then skillfully inserting a persuasive concluding statement at the end of each description, he conditions you to agree with his conclusions *before* you read them. This does not mean he can allow personal bias to creep into his work: his evaluation must be objective (for the good of the company), even though it may seem to be persuasive.

The first draft of Fred's report contained no headings: it was continuous narrative, just like an ordinary memorandum. He inserted them on the advice of Anna King, who told him: "In a multipage letter or memorandum, headings can help readers *see* how you have organized your thinking." The headings Fred uses are those in Figure 5-1. Paragraphs 5 through 15 of his report are his **discussion.** His use of paragraph numbers for an evaluation report is unusual, although common for field trip or inspection reports and in reports prepared for the military. (I have retained the paragraph numbers to illustrate how a paragraph-numbering system can be used for a longer report.)

The difference between an evaluation report and an investigation report can be seen more readily by comparing two similar situations that call for different types of reports. Earlier a situation was described in which Martin Dawes *has decided* to purchase an executive jet aircraft, and assigns two engineers to *investigate* which type of aircraft would be most suitable. Now suppose that Martin only *has an idea* that an executive jet might be useful to him and his company. He assigns an engineer to *evaluate* whether purchase of a private aircraft would be practicable and possibly to recommend what action should be taken.

This engineer's first step is to establish parameters: How often would an aircraft be used? How many persons would use it? What sort of budget allocation can be made available for its purchase and operation? What is the likely route length? The engineer then identifies the types of aircraft available, collects data on each (such as purchase price, operating costs, and performance), and looks at leasing as an alternative to purchase. Armed with all this data, he or she evaluates the advantages of having an executive jet under varying conditions, developing cost analyses for each situation. (In practice this would be done by computer, with the results possibly expressed as a series of benefit-cost ratios.) From this evaluation the engineer establishes whether the idea is economically feasible; if it is, whether it would be better to purchase or rent an aircraft; and which type of executive jet would be most suitable for the company's operations. The report submitted to Martin Dawes would be an evaluation report because it starts by outlining an idea, develops it by using known information, and ends by drawing definite conclusions based on an objective analysis of concepts, data, and parameters.

FEASIBILITY STUDY

A feasibility study is very similar to an evaluation report. It also starts with an idea or concept, and then develops and analyzes it to assess whether it is technically or economically feasible. The chief difference lies in the name and application of each document. An evaluation report is generally based on an idea that is originated and evaluated within the same company; hence it is nearly always informal. A feasibility study is normally prepared at a slightly higher level: the management of company A asks company B to conduct a feasibility study for it, because company A does not have staff experienced in a specific technical field. It is unlikely, for instance, that a company engaged solely in wholesale distribution of dry goods would have the capability to assess the feasibility of purchasing an executive jet. The company would seek advice from a firm of management consultants, who would submit their findings in a report called a feasibility study.

When a feasibility study is prepared for a client, the document that results may be either a letter or a formal report. A letter is used when the project is relatively small and there is no need to present the information formally. If Fred Stokes's memorandum report evaluating training methods (Figure 5-6) had been prepared for a client, it could have been called a feasibility study and written as a letter report. There would have been a few minor changes in narrative and format, but the basic information would have been the same. Alternatively, it could have been made more impressive if it had been prepared as a formal report, like the computer editing and elevator selection reports in Chapter 6. The margins of difference between the feasibility study and the evaluation report are so narrow that frequently personal preference dictates which label is applied.

TECHNICAL PROPOSAL

There are two kinds of technical proposals: those written on a personal level between individuals, and those prepared at company level and presented from one organization to another. The former are usually informal memorandums or letters; the latter are more often formal letters or reports.

Scientists, supervisors, technicians, engineers, and technologists write memorandums every day to their managers suggesting new ways of doing things. These memorandums become proposals if their writers fully develop their ideas. A memorandum that introduces a new idea without demonstrating fully how it can be applied is no more than a suggestion. A memorandum that introduces a new idea, discusses its advantages and disadvantages, demonstrates the effect it will have on present methods, calculates cost and time savings, and finally recommends what action needs to be taken, demonstrates that its originator has done a thorough research job before attempting to put pen to paper. Only then does a suggestion become a proposal.

A proposal follows the basic **Summary-Background-Facts-Outcome** arrangement suggested for reports:

1. A brief **Summary** of the situation or problem and how it can be resolved.
2. An **Introduction** that defines the situation and establishes why it is undesirable or is a problem.
3. A **Discussion** of how the situation or problem can be resolved, covering:
    • a description of the proposed solution,
    • what will have to be done to put it into effect,
    • what the proposed solution will achieve,
    • what advantages will accrue,
    • what disadvantages will be incurred, their likely effect, and how they can be resolved or ameliorated, and
    • what the cost will be.
4. A **Recommendation** for action, which should be a strong, positive statement or a direct request.

Anna King used this outline for the memorandum she wrote to Martin Dawes (see Figure 5-7) proposing that H. L. Winman and Associates consider converting to computer-based (on-line) writing and editing. Her proposal contains seven paragraphs:

• Paragraph 1 is her **Summary.**
• Paragraphs 2 and 3 are her **Introduction,** in which she describes the current unsatisfactory method.
• Paragraphs 4, 5, and 6 are her **Discussion,** in which she describes in detail what she wants to do and what the cost will be.
• Paragraph 7 (the last paragraph) is her **Recommendation** for action, in which she requests Martin Dawes's approval.

MEMORANDUM

To: Martin Dawes   Date: February 15, 19xx

From: Anna King   Subject: Proposal for stream-
                                      lining engineering
                                      writing and my editing

        The pen-and-paper method our engineers currently use
for writing their reports is outdated, cumbersome, and
time-consuming. To simplify and speed up report preparation
I propose introducing computer-based writing and editing
into our operations, and that we engage Macro Engineering
Inc to study the most effective and economical way to apply
this technology.

        The pen-and-paper system is cumbersome because it
creates unnecessarily repetitive work. When our engineer
authors finish their handwritten drafts they hand them in
for typing and wait one or two days for their typed drafts
to be returned. They then proofread the drafts, correct
errors and omissions, and make changes that invariably are
necessary when authors first see their words in type.   The
corrections are handed to the typist, who incorporates them
into a second typed draft.

        This is the draft that comes to me for editing. When I
have finished making my changes, and have obtained the
author's agreement and management's approval of the
document, the final report is typed. (If I have had to make
numerous changes, a third draft may be typed and checked
before the final version is submitted for management
approval.) This means that a minimum of three versions are
typed of each report, and with each typing the report has to
be completely proofread. Clearly, this is wasteful of both
the engineers' and typists' time.

        By writing directly into the computer our engineers
would be able to outline and type their reports right at
their video terminals. They would also be able to proofread
and correct their own work on the video screen, or on a
paper printout if they prefer. And when they feel their
reports are ready for editing, I would be able to call the
reports from their electronic files onto my video terminal,
type in my corrections and suggestions, and then return

1

Figure 5-7. A semiformal proposal, prepared at a computer terminal and printed on a
dot-matrix line printer.

their reports to them. Finally, when their reports are ready for printing in final form, they could be passed electronically to the word-processor typists for production formatting and esthetic refinements.

Although there are obvious advantages to computer writing and editing, unfortunately we cannot simply adopt the technology with our existing computer systems. The video terminals at most engineers' desks are part of the older Comron model 3000 mainframe computer which, although it is an excellent pure computer, has severe limitations as a word processor. Additionally, the mainframe computer is not compatible with the Vancourt 2200 word processors that I and the typists use. Consequently, some research needs to be done to identify what equipment purchases or system modifications are necessary, and what software will best achieve our objective.

I have talked to Vince Warchuk, the local branch manager of Vancourt Business Systems Inc (who supply our typists' word processing equipment), and to Tina Mactiere at Macro Engineering Inc, and both agree that my proposal is feasible. They suggest, however, that a study should be done beforehand to determine which of several alternative systems would be most economical to install and will have the best potential for upgrading as new technology becomes available. Tina Mactiere has offered to send one of her computer engineers here to study our requirements and prepare a report. The cost will be only $1100 if the study can be done during the week of March 10 to 14, because the engineer she plans to send will be coming to Montrose the previous week to design modifications for Multiple Industries' plant.

May I have your approval in principle to convert to computer-based writing and editing, and your authorization to spend $1100 for MEI to evaluate our needs?

Anna King

2

When there are alternative solutions for resolving a problem or unsatisfactory condition, the **Discussion** section of the proposal becomes more comprehensive because it has to describe and evaluate all the possibilities. The writing compartment is then expanded into four subsections:

The Objective: A definition of the parameters or criteria for an ideal (or good) solution.

The Proposed Solution: What it is, the result it would achieve, how it would be implemented, its cost, and its advantages and disadvantages.

Alternative Solution(s): For each: what it is, the result it would achieve, how it would be implemented, its cost, its advantages and disadvantages, and why it is not as effective as the proposed solution.

Evaluation: A brief tradeoff analysis, with reference to the **Objective,** comparing the effects of: (a) adopting the proposed solution, (b) adopting an alternative solution, and (c) taking no action.

Formal technical proposals are prepared by companies seeking to impress or convince another company, or the government, of their technical capability to perform a specific task. They are normally impressive documents prepared under extreme pressure, and call for techniques beyond the scope of most courses you are likely to encounter.

Assignments

PROJECT 1: SELECTING NEW TRUCKS AND AUTOMOBILES

Assume that you are employed by Federal Industries Inc (FII) and that your boss is Tom Westholm, who is Technical Services Manager. Today, Tom comes to you and says: "We have been having a lot of breakdowns among the company vehicles over the past year. I want you to investigate their condition and recommend what you think should be done. Write me a report, because I'll have to show it to the company directors if we are to spend money on new vehicles."

He adds that about $25,000 is available in the budget for purchasing new vehicles, and that you are to use that figure as a maximum.

The first department you talk to is the Service Department, which has two Nissan vans and a half-ton flatbed truck. Department Manager Wilf Friesen says one of the vans is five years old and the other is three years old. The older van has accumulated 170,000 miles on the odometer and the newer van has 115,000 miles. The older van is pretty rusty and breaks down frequently.

The Ford half-ton truck is eight years old. Its odometer is not working but Wilf says it travels some 18,000 miles per year, which means it has probably accumulated 144,000 miles. It has been a particularly troublesome vehicle, with repairs amounting to $840 last year and $790 the year before (these figures are for repairs only; maintenance costs were extra). Wilf says: "That's one truck we have got to get rid of; it's always off the road or breaking down on the road."

The next department you visit is Shipping and Receiving, where Stores Manager Janet Toshak describes the department's vehicles to you. "We've got a three-quarter ton Dodge flatbed truck which is in pretty good shape. It's three years old and it really doesn't get as much work as it should. You can tell that by looking at its odometer: only 41,000 miles in three years is pretty low mileage for a commercial vehicle."

"There's also a large three-ton Condor delivery van," she continues, "and it is quite a different story. Although it's old it doesn't give us much trouble. In nine years it has covered some 170,000 miles. The distance is low because it stands stationary a lot of the time while deliveries are being made. It costs us about $900 a year in maintenance and repairs, and I can live with that. A replacement van would be terribly expensive."

The third department manager you talk to is Rennie Cartier in Marketing. His technical sales representatives have three automobiles they use while on the road.

"The best one is a three-year-old Olds Cutlass," Rennie says. "It has 55,000 miles on it and gives us little trouble."

"The others are older," he continues, "and really have had their time. They are two eight-year-old Chevy Novas, one with 260,000 miles and the other with 274,000 miles. The Nova was one of the best cars GM put out, and I guess that's why we've kept them for so long. But now the interiors are getting pretty beat up and the exteriors are becoming rusted, particularly around the rear end."

From your investigation you realize there are four vehicles which need to be replaced:

> One Nissan van (Service Dept)
> One Ford half-ton truck (Service Dept)
> Two Chevy Novas (Sales Department)

You phone around and get some prices for replacements:

Nissan van	$8,000
Ford half-ton truck	$8,700
Chevy Nova equivalent	$20,000
(2 required, $10,000 each)	
TOTAL	$36,700

You phone around again and find what trade-in values are:

Nissan van	$1500
Ford half-ton truck	$ 850
Chevy Novas	$2200
($1100 each)	
TOTAL	$4550

Now you calculate the total cost of purchasing the new vehicles and deduct the trade-in values from the purchase prices. You find that the cost will be at least $7000 more than the budget figure quoted by Tom Weston when he assigned the project to you. So you have to find a cheaper way of meeting the company's requirements.

Then an idea occurs to you. You remember that Janet Toshak said that her department's three-quarter-ton Dodge truck is not used as much as it could be (that's why it's still in good condition). Suppose you could arrange for Janet to share the truck with the Technical Service Department? You talk to Janet and to Wilf Friesen, and they agree it could be done.

Now you recalculate the costs of buying three vehicles (one Nissan van and two Chevy Nova replacements), and allow a direct sale value of $600 for the half-ton truck (it will not be traded in on a new half-ton truck). This time you find the total cost will be slightly less than the budgeted total cost for buying new vehicles.

With this information in front of you, you prepare your investigation report for Tom Westholm. To it you attach specifications for the three vehicles you recommend should be purchased.

PROJECT 2: DEVELOPING A PRODUCTION LINE TRAINING SYSTEM

You are an engineering technician and you work for Magnum Electronics, a local firm that manufactures, sells, and services general electronic and telecommunications equipment. The production lines are partly computerized and automatic, to reduce manufacturing costs. But there also is intricate manual assembly work of small components, and this work keeps changing as products and market demands change.

Three weeks ago today (work out the exact date), production manager Bill Thorsteinson came to you and said: "We've got a problem and I want you to look into it."

Bill explained that training assembly workers takes a long time and is becoming too costly. "The assemblers either take too long to get trained," he says, "or are slow producers for too long a time, and this is increasing the cost of our products to the point where we are almost becoming uncompetitive."

At the moment all assembly workers are trained right on the job. When they start as new employees, or if they are to be taught how to assemble or solder a new product, they are told to sit at the bench and a supervisor shows them what to do. They work slowly, with the supervisor standing beside them at first and then stopping by frequently to check how they are getting on. Their initial attempts to assemble a component are done very slowly and the components often have to be either thrown away or stripped apart and reassembled.

"What we need is a proper training program," Bill says. "We need to take new employees into a separate room and teach them how to do the job before they go onto the production line. Then when they do start assembling they will be good producers immediately. I want you to investigate alternative ways of doing this training."

Bill takes you down the corridor to room 207. "This is to be the training room," he announces. "You can use it any way you want. I want you to find the most effective way to carry out training, and to do it quickly. When you have got all the information you want, and have done any tests you need, write me an investigation report."

The first step you take is to set up a small training bench in room 207 and equip it with the necessary tools for three employees (the maximum number you plan to train at any one time as part of your tests). You then plan to teach them how to assemble circuit board L284.

You arrange to have three new employees brought into the room at the same time and to be trained by you rather than trained on the assembly line. You spend one day teaching them how to do the work and then set them to do practice assembly work on their own. At the end of the training session you tabulate the time it took for each employee to be trained to the point where she or he could go directly onto the assembly line and immediately become a good assembler. The results are:

> Janet Deslauriers was fully trained in 3½ days; Wendy Williamson required 5½ days before you could send her onto the production line; and Frank Tarnicki took 4½ days. (Your teaching day at the beginning is included in each training time.)
> The total training time was 13½ days, plus your time of 5½ days.

Then you have an idea. Maybe it would be faster, you think, to use videotape to train the employees. So you arrange to rent three half-inch Nortone VCRs, a small, handheld Piedmont videocamera, and three Hakuri 12-inch monitors. You place a VCR beside each work station and a monitor above the work station in room 207.

From the production line you obtain an experienced assembler and videotape her as she assembles circuit board L284. Her name is Tami Wishart and you ask her to assemble the board slowly and carefully, keeping her hands out of the camera's view, so that you will be able to show a very clear picture to the new employees.

Now you bring in three new employees and this time spend only half of the first day (3½ hours) instructing them on what they have to do. You do this instruction partly yourself and partly by showing them the videotape several times. Then you tell them to sit at their work stations and to view the assembly on the TV screens above their individual work areas as they do their own assembling. You explain that they can stop the motion at any time, or repeat the whole sequence at any time.

Once again you measure how long it takes each employee to reach the point where she or he can be sent to the production line as a fully trained assembler. The times you measure are:

> Carl Sanders was fully trained in only 2 days (including the half-day instruction time), Marvin Friedman took 3 days, and Tricia Sauls needed 2½ days to be fully trained. This means they were all trained in a total of 7½ days.
>
> Your labor amounted to 3½ days (half a day to make the videotape, and 3 days to do the training).
>
> Tami Wishart's time for being the "model" whose work you videotaped amounted to half a day.

You compare the two methods of doing the training and calculate that using the VCRs and TVs is a much more effective method because it saves considerably more training time.

As well as calculating the time saved, you also calculate what it would cost to purchase a complete system:

3 Hakuri color monitors ($310 each)	$ 930.00
1 Piedmont color camera	$ 810.00
3 Nortone J244 VCRs, model P600 ($420 each)	$1260.00
10 Videocassettes ($15 each)	$ 150.00
TOTAL	$3150.00

You tell Bill Thorsteinson that you have examined two methods, that both are successful, but that one takes less time than the other and thus is more economical. He tells you to describe and present the costs of both methods in your report. He says you are to assume that you train 100 employees a year, and that they are paid $55 per day while they are being trained. Tami Wishart is paid $65 per day. When calculating your own time, he says you are to use a daily rate of $80.

You are to suggest these alternatives: (1) that classroom training be done every three weeks, with an "intake" of six new employees each time, 17 times a year; and (2) TV training be done every 10 days, with three new employees each time, 33 times a year. From all these figures you calculate whether or not it would be practicable to buy the TV equipment or cheaper to use normal classroom instruction.

Write your report to your manager, describing the results of your investigation.

PROJECT 3: NEW SPACE FOR THE DRAFTING DEPARTMENT

You are the senior draftsperson for Morton Consultants Inc, a firm that occupies the second and third floors of the Hartland Building at 200 Broadway Avenue of your city. Yesterday the Chief Engineer (David Carter) called you into his office and assigned you a drafting department relocation project.

"The area on the third floor which is now occupied by the Drafting Department is to become the location of the Marketing Department," he explained. "This means the Drafting Department will have to move in about six weeks. From what I can find out, there are three areas your department can move into, and they are shown on these sketches."

Mr. Carter shows you three sketches (see Figure 5-8) and describes them briefly to you.

"Area A is an L-shaped air-conditioned area of 1564 square feet in the basement. It's in the center of the building and has no windows. Rent would be $0.30 a square foot per month.

"Area B is a nearly square area of 1800 square feet on the third floor, close to where the Drafting Department is now, and adjacent to the Engineering Department (which is useful, for direct communication between the engineers and draftspersons). It has windows on the west and south sides, which make it a trifle warm in the summer, even with the air conditioning on, but on the other hand it has the best view in the building. From the windows you would have a magnificent view of the river and Memorial Park. And on warm days at lunchtime you could watch the sunbathers in the park from right up here!

"The cleanest area (from the point of view of having no obstructions such as pillars) is area C, an 80 ft long room of 1600 square feet along the north wall of the second floor. It remains cool in the summer because it gets little sun. It has the best windows and natural light in the building and is a very quiet area. Of course, it's not on the same floor as the Engineering Department, which would be a little inconvenient. And the rental cost for areas B and C would be $0.40 a square foot per month."

Mr. Carter also says that, as it will be a complete move for the Drafting Department, he feels it would be a good opportunity to completely re-equip the department with new drawing tables and stools, and possibly even new drawing cabinets. He instructs you to assess the three areas and to select one of them based on cost and overall suitability, and to identify the type of drawing tables that should be purchased. He will want a complete breakdown of costs plus a total price for buying new furnishings and for renting the space on an annual basis. He also adds that:

1. The blueprint machine will have to be replaced. The present one is a "Smooth-set" model 220B which is totally out of date (it is at least 20 years old). It possibly has a trade-in value of $150.
2. There are now eight draftspersons in the department, plus the Chief Draftsper-

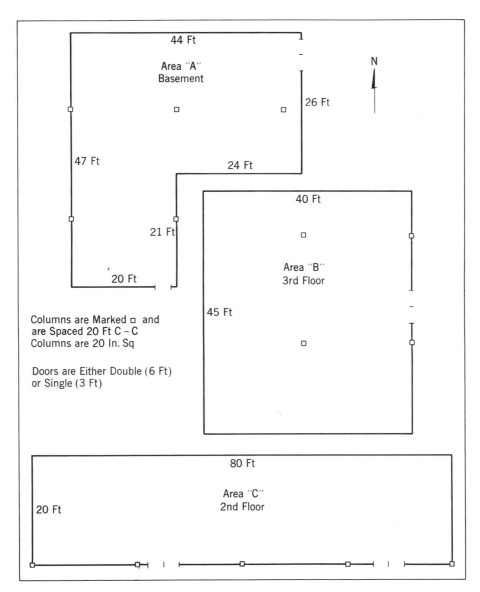

Figure 5-8. Proposed areas for drafting department.

son. You should allow for another two draftspersons who likely will be hired next year.

3. There are six four-drawer filing cabinets for drawings and they are nearly full. They can be piled three-high. At least one new cabinet will have to be purchased and should be included in your price.

You ask if a budget limitation has been established. "Not yet," he replies. "We're depending on your recommendations. Just the same, I imagine about

$8000 should be your maximum for equipment purchases, and $6000 to $6500 a year for rent."

To undertake this project and write your report you will have to do some research into the different types of drawing tables available and their costs and dimensions. You will also need to draw the available areas to scale and try plotting the positions of the tables, blueprint machine, and drawing cabinets on them (remember that the blueprinter has to be exhausted through an outside wall). Remember, too, that your report should make a recommendation as well as present all the information you have.

PROJECT 4: A VISIT TO WAKELING PROCESSORS INC.

Wakeling Processors Inc has its head office in Cleveland, Ohio, and numerous manufacturing plants throughout the country. Oliver R. Wakeling, president and general manager, calls in H. L. Winman and Associates to help him resolve a management problem with technical implications. The problem concerns the Wakeling Processors Inc plant at Weston, 95 miles from your city, and centers around the operation of the power house. This subsidiary is engaged in vegetable oil processing, for which the power house has to produce large supplies of hot water and a moderate supply of compressed air.

Mr. Wakeling is concerned because there has been a continuous upward increase in power house costs at the Weston plant over the past two years. Fuel consumption has risen by 18% and numerous breakdowns have occurred which have interfered with production, with a consequent increase in costs. Yet management visits have revealed little that could be attributed to poor operation; in fact, the power house has always been immaculate.

Mr. Wakeling believes that for him to make another inspection will not produce the answers he requires. He calls on Martin Dawes, at the head office in Cleveland, and outlines his problem. Martin suggests sending someone at the technician or technologist level who will be able to talk on technical subjects to all the people in the power house and at the same time evaluate production problems.

He telephones Vern Rogers, branch manager of the H. L. Winman and Associates office in your city, who in turn suggests assigning you to the project. They arrange a "conference call" between yourself, Martin, Vern, and Mr. Wakeling, during which Mr. Wakeling gives you a brief outline of the situation:

> The present chief engineer at the Weston Plant is David Skyla, and he is to retire in three months. Management has to decide whether to promote Barry Bishop, the existing senior shift engineer, or to bring in a new chief engineer from outside the plant. On paper, Bishop is ideal for the job: He has worked in the plant for 15 years (he is now 36) and always under Skyla, so his knowledge of the power house and its operating conditions cannot be challenged. Yet the rising costs indicate that all is not correct in the plant, and management wants to be sure that the new chief engineer does not perpetuate the present conditions.

Mr. Wakeling says he will advise Skyla and Bishop that he has engaged H. L. Winman and Associates to study the hot water and power generating system at their plant and that they should expect you. You visit the plant one week later.

During your talks to plant staff and tours of the powerhouse you make the following notes:

1. Housekeeping excellent—whole place shines (but is this only surface polish for impression of visitors?).

2. Maintenance logs are inadequately kept—need to be done more often. Need more detail. Equipment files not up to date and not properly filed.

3. Boiler cleaning badly neglected. Firm instructions re boiler cleaning need to be issued by head office.

4. Flow meters are of doubtful accuracy. May be overreading. Not serviced for three years. Manufacturer's service department should be contacted (these are Vancourt meters). Manufacturer needs to be called in to do a complete check and then recalibrate meters.

5. Overreading of meters could give false flow figures—make plant seem to produce more steam than is actually produced.

6. Good housekeeping obviously achieved by neglecting maintenance. Incorrectly placed emphasis probably caused by frequent visits from company president, who likes to bring in important visitors and impress them. Skyla likes reflected glory (so does Bishop).

7. Shift engineers are responsible for maintenance of pumps and vacuum equipment. Not enough time given over to this. They seem to prefer straight replacement of whole units on failure rather than preventive maintenance. Costly method! Obviously more breakdowns: they wait for a failure before taking action. A preventive maintenance plan is needed.

8. Bishop seems O.K. Genial type; obviously knows his power house. Proud of it! But seems to resist change. Definitely resents suggestions. Does he lack all-round knowledge? Is he limited only to what goes on in his plant? Is he afraid of new ideas because he doesn't understand them? Young staff hinted at this: too loyal to say it outright, but I felt they were restive, hampered by his insistence that they use old techniques that are known to work but are slow. Nothing concrete was said—I just "felt" it.

9. Skyla's done a good job training Bishop. Made him a carbon copy. Skyla doesn't do much now. Bishop runs the show, and has for over a year. He *expects* to get the job when Skyla retires. It'll be a real blow to him if he doesn't! Wakeling might even lose a good company man.

10. Discussed RAMSORT 2300 power panel with staff. Young engineers had read about it in "Plant Maintenance"—eager to have one installed (I described the one I'd seen at Abotinam Pulp and Paper). But Skyla and Bishop knew nothing about it—weren't interested. Obviously not keeping up to date with technical magazines.

When you return to head office you inform Vern Rogers verbally of your findings. He asks you to write an evaluation report for Mr. Wakeling, and suggests that you address both the technical problems and the personnel difficulties within the one report.

6

Formal Reports

Rarely written for single reader!

formal in tone impersonal in style

Semiformal report (found in other texts) - formal in tone but omits some of the elements (parts)

A formal report requires more careful preparation than the informal and semi-formal reports described in the previous chapters. Because it will be distributed outside the originating company, its writer must consider the impression it will convey of the entire company. Harvey Winman recognized long ago that a well-written, esthetically pleasing report can do much to convince prospective clients that H. L. Winman and Associates should handle their business, whereas a poorly written, badly-presented report can cause clients to question the company's capability. Harvey also knows that the initial impression conveyed by a report can influence a reader's readiness to plough through its heavy technical details.

The presentation aspect must convey the originating company's "image," suit the purpose of the report, and fit the subject it describes. For instance, a report by a chemical engineer evaluating the effects of diesel fumes on the interior of bus garages would most likely be typed on standard bond paper, and its cover, if it had one, would be simple and functional. At the other end of the scale, a report by a firm of consulting engineers selecting a college site for a major metropolis might be printed professionally and bound in an artistically designed book-type folder. But regardless of the appearance of a report, its internal arrangement will be basically the same.

Formal reports are made up of several standard parts, not all of which appear in every report. Each writer uses the parts that best suit the particular subject and the intended method of presentation. There are six major and several subsidiary parts on which to draw, as shown in the table on page 135. Opinions differ throughout industry as to which is the best arrangement of these parts. The two arrangements suggested in the section on The Full Report later in this chapter, and illustrated in the mini-reports in Figures 6-6 and 6-8, are those most frequently encountered. Knowledge of these parts and the two basic arrangements will help engineers and technicians adapt quickly to the variations in format preferred by their employers. Listed below are the parts as they appear in the most widely used format.

FORMAL REPORT: **(Major Parts** and *Subsidiary Parts)*

Cover or *Jacket*
Title Page
Summary
Table of Contents
Introduction
Discussion
Conclusions
Recommendations
References or *Bibliography*
Appendix(es)

[handwritten: Letter of transmittal – often (p. 168)]

[handwritten: Order - similar to book
cover -
title -
preface (summary, letter (ic abstract)
of trans.)
Table of contents]

A formal report is often accompanied by a *Cover Letter* or *Letter of Transmittal,* and may also contain other small parts such as a *List of Illustrations* (which follows the *Table of Contents*), a *Distribution List,* and an *Acknowledgments* paragraph.

[handwritten: appendix
bibliography]

MAJOR PARTS

Because the six major parts form the central structure of every formal report, technical writers must understand their purpose and function if they are to use them effectively.

SUMMARY

The summary is a brief synopsis which tells readers quickly what a report is all about. Normally it appears immediately after the title page, where it can be found easily. It identifies the purpose and most important features of the report, states the main conclusion, and sometimes make a recommendation. It does this in as few words as possible, condensing the narrative of the report to a handful of succinct sentences. It also has to be written so interestingly—so enthusiastically—that it encourages readers to read further.

The summary is considered by many to be the most important part of a report and the most difficult to write. It has to be informative, yet brief. It has to attract the reader's attention, but must be written in simple, nontechnical terms. It has to be directed to the executive reader, yet it must be readily understood by almost any reader.

Generally, the first person in an organization who sees a report is a senior executive, who may have time to read only the summary. If the executive's interest is aroused, he or she will pass the report down to the technical staff to read in detail. But if the summary is unconvincing, the executive is more likely to put the report aside to read "later;" if this happens, the report may never be read.

Because a summary is so important, always write it last, after the remainder of your report has been written. Only then will you be fully aware of the report's highlights, main conclusions, and recommendations. To write the summary first can prove difficult and frustrating, because you will not yet have hammered out many of the finer points. You need the knowledge of a soundly developed report firmly fixed in your mind if you are to fashion an effective summary.

For a summary to be interesting, it must be informative; if it is to be informative, it must tell a story. It should have a beginning, in which it states why the project was carried out and the report written; a middle, in which it highlights the most important features of the whole report; and an end, in which it reaches a conclusion and possibly makes a recommendation. The example below illustrates how the interest is maintained in an informative summary.

Informative Summary

A specimen of steel was tested to determine whether a job lot owned by Northern Railways could be used as structural members for a short-span bridge to be built at Peele Bay in northern Alaska. The sample proved to be G40.12 structural steel, which is a good steel for general construction but subject to brittle failure at very low temperatures.

Although the steel could be used for the bridge, we consider that there is too narrow a safety margin between the −51°C temperature at which failure can occur, and the −47°C minimum temperature occasionally recorded at Peele Bay. A safer choice would be G40.8C structural steel, which has a minimum failure temperature of −62°C.

Other informative summaries preface the two formal reports at the end of this chapter, and the semiformal report in Figure 5-4 of Chapter 5.

Some writers prefer to write a topical summary for reports that describe history or events, or that do not draw conclusions or make recommendations. As its name implies, a topical summary simply describes the topics covered in the report without attempting to draw inferences or captivate the reader's interest.

Topical Summary

Construction of Alaska's Minnowin Point Generating Station was initiated in 1978, and first power from the 1340 MW plant is scheduled for 1984. A general description of the structures and problems peculiar to construction of this large development in an arctic climate is presented. The river diversion program, permafrost foundation conditions, and major equipments are described. The latter include the 16 propeller turbines, among the largest yet installed, each rated at 160,000 horsepower.

Because they are less results-oriented, topical summaries are *not* recommended for most formal reports. In a formal report the summary should have a page to itself, be centered on the page, and be prefaced by the word "Summary" (or, sometimes, "Abstract"). If it is very short, it may be indented equally on both sides to form a roughly square block of information.

INTRODUCTION

Some report writers confuse the summary with the introduction. There is no need for this if they remember that the summary provides a *brief synopsis of the whole report,* whereas the introduction *introduces the subject* to readers so that they may read the remainder of the report more intelligently.

The introduction begins the major narrative of the report by preparing readers for the discussion that follows. It orients them to the purpose and scope of the report and provides sufficient background information to place them mentally in the picture before they tangle with technical data. A well-written introduction contains just enough detail to lead readers quickly into the major narrative.

The length of an introduction and its depth of detail depend mostly on the reader's knowledge of the topic. A writer who knows that the ultimate reader is technically knowledgeable will want to use technical terminology, but at the same time the writer has to cater to the executive reader who is probably only partly technical. To overcome this disparity in reading levels, skillful report writers often write the summary in a lay person's language and the introduction, conclusions, and recommendations in semitechnical language. They thus permit the semitechnical executives to gain a reasonably comprehensive understanding of the report without devoting time and attention to the technical details contained in the discussion.

Most introductions contain three parts: **purpose, scope,** and **background information.** Frequently, the parts overlap, and occasionally one of them may be omitted simply because there is no reason for its inclusion. The introduction normally is a straightforward narrative of one or more consecutive paragraphs; only rarely is it divided into distinct sections preceded by headings. It always starts on a new page (normally identified as page 1 of the report) and is preceded by the report's full title. The title is followed by the single word "Introduction," which can be either a center heading or a side heading as shown in Figure 6-1.

The **purpose** explains why the project was carried out and the report is being written. It may indicate that the project has been authorized to investigate a problem and recommend a solution, or it may describe a new concept or method of work improvement that the report writer believes should be brought to the reader's attention.

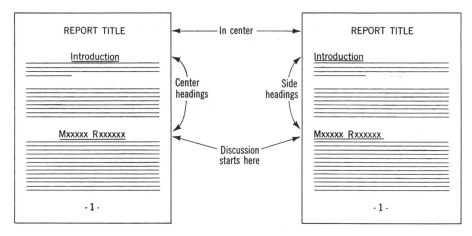

Figure 6-1. Headings used at the start of the introduction and discussion.

The **scope** defines the parameters of the report. It describes the ground covered by the report and outlines the method of investigation used in the project. If there are limiting factors, it identifies them. For example, if 18 methods for improving packaging are investigated in a project but only four are discussed in the report, the scope indicates what factors (such as cost, delivery time, and availability of space) limited the selection. Sometimes the scope may include a short glossary of terms that need to be defined before the reader starts to read the discussion.

Background information comprises facts readers must know if they are to fully understand the discussion that follows. The facts may include descriptions of conditions or events that caused the project to be authorized, previous investigations, and reports on the same or a closely related subject. In a highly technical report, or when a significant time lapse has occurred between it and previous reports, background information may also include a theory review and references to other documents. Generally, a lengthy theory review is placed in the appendix and only a brief summary of the theory is inserted in the introduction; similarly, a long list of documents is placed at the end of the report and the list is referred to in the introduction.

The introductions shown here, plus those forming part of the two sample reports at the end of this chapter, are representative of the many ways a writer can introduce a topic.

Introduction 1

Northern Railways plans to build a short-span bridge one half mile north of Lake Peele in northern Alaska and has a job lot of steel the company wants to use for constructing the bridge. In letter NR-70/LM dated March 20, 19XX, Mr.   *Background*

David L. Harkness, Northern Area Manager, requested H. L. Winman and Associates to test a sample of this steel to determine its properties and to assess its suitability for use as structural members for a bridge in a very low temperature environment.

Purpose

The test performed was the Charpy Impact Test. Two series of tests were run, one parallel to the grain of the test specimen and the other transverse to the grain, at 10° increments from +22°C down to −50°C.

Scope

Introduction 2

H. L. Winman and Associates was commissioned by Mr. D. C. Scorobin, President and General Manager of Auto-Marts Inc. of Dallas, Texas, to select a Cleveland site for the first of a proposed chain of drive-in groceries to be built in the northeastern United States. This area was chosen since it represents an average community in which to assess customer acceptance of this revolutionary form of service.

Purpose

Aside from exterior facade and foundation details, Auto-Marts are built to a standard 80 ft by 30 ft pattern with an order window at one end of the 80 ft wall, and a delivery window at the other. Auto-Marts carry only a limited selection of household goods, but boast 90-second service from the time an order is placed to its delivery at the other end of the building. For this reason, Auto-Marts attract people hurrying home from work, the impulse buyer, and the late-night shopper, rather than the selective shopper. The store depends more on the volume of customers than on the volume of goods sold to each individual.

Background

The chief consideration in selecting a site must therefore be a location on the homeward-bound side of a main trunk road into a large residential area. The site must have quick and easy entry onto and exit from this road, even during peak rush hour traffic. The residential area should be occupied mainly by upper and middle class people who own cars and have money to spend. And there should be little competition from walk-in grocery stores.

Scope

DISCUSSION

The discussion, which normally is the longest part of a report, presents as much evidence (facts, arguments, details, data, and results of tests) as readers will need to understand the subject. The writer must develop this evidence in an organized, logical manner to avoid confusing readers, and must present it imaginatively to retain the readers' interest.

In the discussion, a writer such as Dan Skinner (see Chapter 1) describes fully what he set out to do, how he went about it, what he actually did, and what he found out as the result of his efforts. He may consider the writing of these facts a chore and simply record them dutifully but unenthusiastically. Or he can respond to the challenge by organizing the information so interestingly that his readers feel almost as though they are reading a short story.

There is no reason why the discussion should not have a plot. It can have a beginning, in which Dan describes a particular problem and outlines some background events; a middle, which tells how he tackled the problem; dramatic effect, which he can use to hold readers' interest by letting them anticipate his successes, as well as share his disappointments when a path of investigation results in failure; a climax, in which readers are permitted to share his pleasure in a description of how the problem was eventually resolved; and a denouement (a literary word for final outcome), in which Dan ties up the loose ends of information and evaluates the final results. Storytelling and dramatization can be such important factors in holding reader interest that no report writer can afford to overlook them.

There are three ways you can build the discussion section of a report:

By *chronological development*—in which you present information in the order that events occurred.

By *subject development*—in which you arrange information by subjects, grouped in a predetermined order.

By *concept development*—in which you present the information as a series of ideas which reveal imaginatively and coherently how you reasoned your way to a logical conclusion.

Reports using the chronological or subject method offer less room for imaginative development than those using the concept method, mainly because they depend on a straightforward presentation of information. The concept method can be very persuasive. Identifying and describing your ideas and thought processes helps your readers organize their thoughts along the same lines.

As a report writer, you must decide early in the planning stages which method you intend to use, basing your choice on which is most suitable for the evidence you have to present. Use the following notes as a guide.

1. Chronological Development. A discussion which uses tbe chronological method of development is simple to organize and write. It requires minimal planning. You simply arrange the major topics in the order they occurred, eliminating irrelevant topics as you go along. It is useful for very short reports, for laboratory reports showing changes in a specimen, for progress reports showing cumulative effects or describing advances made by a project group, and for reports of investigations that cover a long time and require visits to many locations to collect evidence.

But the simplicity of the chronological method is offset by some major disadvantages. Because it reports events sequentially, it tends to give emphasis to each event regardless of importance, and this may cause readers to lose interest. If you read a report of five astronauts' third day in orbit, you do not want to read about every event in exact order. It may be chronologically true to report that they rose at 7:15, breakfasted at 7:55, sighted the second stage of their rocket at 9:23, carried out metabolism tests from 9:40 to 10:50, extinguished a cabin fire at 11:02, passed directly over Houston at 11:43, and so on, until they retired for the night. But it can make deadly dull reading. Even the exciting moments of a cabin fire lose impact when they are sandwiched between routine occurrences.

When using chronological development, you must still manipulate events if you are to hold your readers' attention. You must emphasize the most interesting items by positioning them where they will be noticed, and deemphasize less important details. Note how this has been done in the following passage, which groups the previous events in descending order of interest and importance:

> The highlight of the astronauts' third day in orbit was a cabin fire at 11:02. Rapid action on their part brought the fire, which was caused by a short circuit behind panel C, under control in 38 seconds. Their work for the day consisted mainly of metabolism tests and. . . .
> They sighted the first stage of their rocket on three separate occasions, first at 9:23, then at . . . , and passed directly over Houston at 11:43. Their meals were similar to those of the previous day.

This narrative uses *chronological* order to describe the events on a day-to-day basis, and a modified *subject* order to describe the events for each day.

Over a five-year period H. L. Winman and Associates has been investigating the effects of salt on concrete pavement. Technical editor Anna King has suggested that the final report should have chronological development, because the investigation recorded the extent of concrete erosion at specific intervals. She wanted the engineering technician writing the report to describe how the erosion increased annually in direct relation to the amount of salt used to melt snow each year, and for the final conclusion to demonstrate the cumulative effect that salt had on the concrete.

2. Subject Development. If the previous investigation had been broadened to include tests on different types of concrete pavements, or if both pure salt and various mixtures of salt and sand had been used, then the emphasis would have shifted to an analysis of erosion on different surfaces or caused by various salt/sand mixtures, rather than a direct description of the cumulative effects of pure salt. For this type of report Anna King would have suggested arranging the topics in *subject* order.

The subject order could be based on different concentrations of salt and sand. The technician would first analyze the effects of a 100% concentration of salt, then continue with salt/sand ratios of 90/10, 80/20, 70/30, and so on, describing the results obtained with each mixture. Alternatively, the technician could select the different types of pavement as the subjects, arranging them in a specific order and describing the effects of different salt/sand concentrations on each surface.

The head of the drafting department, Barry Brewster, has been investigating blueprint machines and he plans to recommend the most suitable model for installation in his department. He writes his report using the subject method. First, he tells readers that he intends to evaluate the speed, economy of operation, and usefulness of each blueprinter, and then he emphasizes the most suitable machine by discussing it either first or last. If he chooses to describe it first, he can state immediately that it is the best blueprinter, and say why. He can then discuss the remaining models in descending order of suitability, comparing each to the selected machine to show why it is less suitable. If he prefers to describe the best machine last, he can discuss the others in ascending order of suitability, stating why he has rejected each one before he starts describing the next.

Also suitable for subject development is an analysis of insulating materials being conducted by laboratory technician Rhonda Moore. In her report, she groups materials having similar basic properties or falling within a specific price bracket. Within the groups she describes the test results for each material and draws a conclusion as to its relative suitability for a specified purpose. Like Barry, she describes the materials in ascending or descending order of suitability.

The subject method of development permits Barry and Rhonda to either hide their personal preferences until almost the end of their reports or, if they prefer, let their preferences show all the way through. They have the choice of using an impersonal, objective approach, or a personal, subjective approach similar to that for the concept method. These alternative approaches are illustrated in Figure 6-2.

Rhonda prefers to use the primarily objective approach. She discusses each insulating material (the subject) without using any words that might make her readers think she believes the material to be a good or bad choice:

> Material A has an R-17 rating. It is a dense, brown, tightly packed fibrous substance enclosed in a continuous, one-meter wide envelope, which is supplied in 10-, 20-, and 50-meter rolls. To cover a standard 4 × 3 meter area would cost $108.00. The fire resistance of Material A is. . . .

Barry prefers the more colorful subjective approach: he wants to persuade his readers to accept or reject each blueprinter (subject) as they read about it:

> The Image-copy is the only machine that meets all our technical requirements. It has a high speed of 25 feet per minute, an extended continuous-roll feature with automatic cutter, and both manual and automatic stacking capabilities. Although its purchase price of $5700.00 is $700.00 above budget, its high reliability (reported by users such as Multiple Industries and Apex Architectural Services)

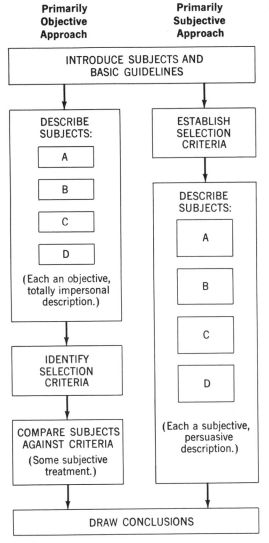

Figure 6-2. Alternative methods for a comparative analysis.

would probably cause less equipment downtime and consequent production frustrations than other models. Specific features are. . . .

Barry must establish the criteria or parameters which most affect his choice, and must convince his readers why they are important, *before* he describes each blueprinter; if he does not, readers will not readily accept his subjective statements. Rhonda has only to introduce the insulating materials, and possibly outline how they were selected, before she describes them. She need not identify her selection criteria until almost the end of the discussion, after she has objectively described each material but before she subjectively compares their main features.

Whichever approach you use for a comparative analysis, your readers should find that the discussion leads comfortably and naturally into the conclusions you eventually draw. The conclusions should never contain any surprises.

3. Concept Development. By far the most interesting reports are those using the concept method of development. True, they need to be organized more carefully than reports using either of the previous methods, but they give the writer a tremendous opportunity to devise an imaginative arrangement of the topic—which is the best way to hold a reader's attention.

They can also be very persuasive. Because the report is organized in the order in which you reasoned your way through the investigation, your readers will understand the subject much more quickly. They will readily appreciate the difficulties you encountered and frequently draw the correct conclusion even before reading it. This helps readers to feel they are personally involved in the project.

You can apply this approach to your reports by thinking of each project as a logical but forceful procession of ideas that form a total concept. If you are personally convinced that the results of your investigation are valid, and remember to explain in your report *how* you reached the results and *why* they are valid, then you will probably be using the concept method properly.

Always try to anticipate reader reaction. If you are presenting a concept (an idea, plan, method, or proposal) which readers are likely to accept, then use a straightforward four-step approach:

1. Describe your concept in a brief overview statement.
2. Discuss how and why your concept is valid; offer strong arguments in its favor, starting with the most important and working down to the least important.
3. Introduce negative aspects, and discuss how and why each can be overcome or is of limited importance.
4. Close with a restatement of your concept, its validity, and its usefulness.

But if you are presenting a controversial concept, or need to overcome reader bias, then modify your approach. Try to overcome objections by carefully establishing a strong case for your concept *before* you discuss it in detail.

Special Projects Department Head Andy Rittman used this approach in a report he prepared for Mark Dobrin, owner/manager of a company making extruded plastic and metal parts for the defense industry. Manufacturing costs had risen steeply over the past two years, and Mark's prices had rapidly become uncompetitive. Mark thought he should replace some of his older, less efficient equipment, so he asked H. L. Winman and Associates to evaluate his needs.

Andy Rittman quickly realized that if Mark was to avoid going completely out of business he would have to replace far more of his equipment than he expected, and would have to introduce microprocessor-controlled methods of production and inventory control. Because the cost would be high, he would have to lease rather than buy the equipment.

Here, Andy had a problem. Mark was as old-fashioned as some of his extruders and shapers, and throughout his life he had steadfastly refused to acquire anything that he did not own outright. He was unlikely to change now.

In his report, Andy used a carefully reasoned argument to prove to Mark that he needed a lot of new equipment and that the only feasible way he could acquire it would be to lease it. Throughout, Andy wrote objectively but sincerely of his findings, hoping that the logic of his argument would swing Mark around to accepting his recommendation. Very briefly, here is the step-by-step approach Andy used:

> He opened with a summary which told Mark that to avoid bankruptcy he would have to invest in a lot of expensive equipment and make extensive changes in his operating methods.
>
> Andy then produced financial projections to prove his opening statement, and discussed the productivity and profitability necessary for Mark to remain in business.
>
> He discussed why Mark's equipment and methods were inefficient, introduced the changes Mark would have to make, established why each change was necessary, and demonstrated how each would improve productivity. (Andy referred Mark to an appendix for specific equipment descriptions, justifications, and costs.)
>
> Andy introduced two sets of cost figures: one for making the minimum changes necessary for Mark's business to survive, and the second for more comprehensive changes that would ensure a sound operating basis for the future. He commented that both would require capital purchases likely to be beyond the financial resources of Mark's business.
>
> He outlined alternative financing methods available to Mark, the implications and limitations of each, and the financial effect each would have on Mark's business. (Although he introduced leasing, Andy made no attempt to persuade Mark that he would have to lease; he let the figures speak for themselves.)

Andy concluded his discussion by summarizing the main points he had made: new equipment *must* be acquired; to buy even the minimum equipment was beyond Mark's financial resources; and, of the various financing methods available, leasing was the most feasible. But not until his recommendation did Andy come right out and suggest that Mark should make comprehensive changes and lease the new equipment. By then, Andy had become so involved with Mark's predicament that he wrote strongly and sincerely in favor of his recommendation.

Even though the concept method challenges a writer to fashion interesting reports, it is not always the best reporting medium. For instance, the concept method could possibly have been used for the blueprinter report mentioned earlier, but it is doubtful whether the topic would have warranted full analysis of the author's ideas. When a topic is fairly clear-cut, there is no need to lead the reader through a lengthy "this is how I thought it out" discussion. Reserve the concept method for topics that are controversial, difficult to understand, or likely to meet reader resistance.

Whichever method you use, avoid cluttering the discussion with detailed supporting information. Unless tables, graphs, illustrations, photographs, statistics, and test results are essential for reader understanding while the report is being read, banish them to the appendix. But always refer to them in the discussion, like this:

> The test results attached at Appendix C show that aircraft on a bearing of 265°T experienced considerably weaker reception than aircraft on any other bearing. This was attributed to. . . .

ref. to graphs, etc.

Readers interested only in *results* would consider that this statement tells them enough and would continue reading the report. Readers interested in knowing *how the results were obtained*, who want to see the overall picture, will turn to Appendix C to find out how you went about performing the tests. They probably will not do so immediately, but will return to it when they have finished the report or reached a suitable stopping place. In neither case are readers slowed down by supporting information which, though relevant, is not immediately essential to an understanding of the report.

Unless the discussion is very short, it will consist of a series of sections each containing information on a major topic. Insert an informative heading at the beginning of each section to indicate the section's contents. If the section is

Sections — w/ subtopics if necessary

Figure 6-3. Headings show subordination of ideas.

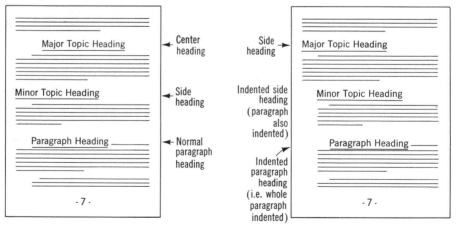

long or naturally subdivides into subtopics, also use subordinate headings. Major topics normally demand center or side headings, while minor topics need only side or paragraph headings, as shown in Figure 6-3 (also see Figure 11-4 in Chapter 11). These headings eventually form the table of contents for the report.

After each heading, start the section with an overview statement to describe what the section is about and suggest what conclusion will be drawn from it. Overview statements are miniature summaries which direct a reader's attention to the point you want to make. If the section is short, the overview statement may be a single sentence: if the section is long, it will probably be a short paragraph.

At the end of each major section insert a concluding statement that summarizes the result of the discussion within that section. From these section conclusions you can later draw your main conclusions.

CONCLUSIONS

Conclusions briefly state the major inferences that can be drawn from the discussion. They must be based entirely on previously stated information and *never* surprise readers by introducing new material or evidence to support your argument. If there is more than one conclusion, state the main conclusion first and follow it with the remaining conclusions in decreasing order of importance. This is shown in the two examples below, which present the same conclusions in both narrative and tabular form.

Narrative Conclusion

The operational life of QK 3801 magnetrons can be increased 30% by installing modification XL-1 on our radar transmitters. This will save $8300 over the next 12 months, but will require us to increase magnetron run-in time from 4 to 8 hours.

Tabular Conclusion

Installation of modification XL-1 on our radar transmitters will result in:
1. A 30% increase in the operational life of QK 3801 magnetrons.
2. An $8300 saving over the next 12 months.
3. A change in operating procedures to increase magnetron run-in time from 4 to 8 hours.

Whether you use narrative or tabular conclusions depends on personal preference. Both are acceptable, although the advantage of the tabular arrangement is that its separate conclusions are easier to identify. Perhaps the best rule is to use narrative style when there are very few conclusions and to tabulate them when there are several.

Because conclusions are *opinions* (based on the evidence presented in the discussion), they must never tell the reader what to do. This must *always* be left to the recommendations.

RECOMMENDATIONS

Recommendations appear in a report when the discussion and conclusions indicate that further work needs to be done, or if a specific solution to a problem has evolved from the writer's investigations. Write them in strong, definite terms to convince readers that the course of action you advocate is valid. Use the first person and active verbs:

> I recommend that we build a five-station prototype of the Microvar system and test it operationally.

Unfortunately, in many industries, and often in government circles, personal involvement is severely frowned upon. (No one seems to know whether this stems from natural reticence supposedly displayed by technical writers, or from fear that a faulty recommendation might be traced to a specific individual.) As a result, many reports fail to contain really strong recommendations because their writers hesitate to say "I recommend . . . ," and so use the indefinite "It is recommended. . . ." Surely a technical person who has researched a subject well enough to advocate definite action can present recommendations more convincingly than this! You need not always use the personal "I" (most reports prepared for clients use the plural "we," to indicate that they represent the company's viewpoint). For example:

> We recommend building a five-station prototype of the Microvar system. We also recommend that you:
>
> 1. Install the prototype in Railton High School.
> 2. Commission a physics teacher experienced in writing programmed instruction manuals to write the first programs.
> 3. Test the system operationally for three months.

Because recommendations must be based solidly on the evidence presented in the discussion and conclusions, they can *never* introduce new evidence or new ideas.

APPENDIX

Related data not necessary to an immediate understanding of the discussion should be placed further back in the report, in the appendix. The data can vary from a complicated table of electrical test results to a simple photograph of a blown resistor. The appendix is a suitable place for manufacturer's specifications, graphs, comparative data, drawings, sketches, excerpts from other reports or books, and correspondence. There is no limit to what can be placed in the appendix, providing it is relevant and reference to it is made in the discussion.

The term "appendix" applies to only one set of data. In practice there can be many sets, each of which is called an appendix and assigned a distinguishing letter, such as A, B, or C. The sets of data are collectively referred to as either "appendixes" or "appendices"; the former term is preferred in the United States.

The importance of an appendix has no bearing on its position in the report. Whichever set of data is mentioned first in the discussion becomes Appendix A, the next set becomes Appendix B, and so on. Each appendix is considered a separate document complete in itself and is page-numbered separately, with its front page labeled 1. (Not every organization follows this practice. Although most companies now number appendix pages separately, some continue to number them consecutively as an integral part of the report.) Examples of appendixes appear at the end of Report 1 of this chapter.

SUBSIDIARY PARTS

In addition to the six major parts of a report, there are several additional parts that perform more routine functions. Although I refer to these as "subsidiary," they nevertheless contribute much to the effectiveness of a report. A writer must never assume that these seemingly peripheral details will look after themselves. Because they directly affect the image conveyed of both the writer and the company, they must be prepared no less carefully than the rest of the report.

COVER

Almost every major formal report has a cover. It may be made of cardboard printed in multiple colors and bound with a dressy plastic binding, or it may be only a light cover of colored fiber material stapled on the left. In either case it has one obvious purpose: to inform the reader of the main topic of the report. Its more subtle purposes are to convey an immediate image of the type of company that has produced the report, and to reflect the type of subject that the report describes. This "matching" of subject matter and company image plays an important part in setting the right tone for a report.

The cover should contain few words. The report title is an obvious choice, since a good title will invite a prospective reader to open the report and discover what it is about. The words must also stand out clearly but tastefully on the page. Printing a title like a newspaper banner headline in two-inch high letters may attract attention, but will have a negative effect on the reader's opinion of the author's technical competence. A few words neatly arranged and well separated from other printed matter will help to present the correct image. The only other words should be the originating company's name and perhaps the name of the author, if he or she is well known enough to lend weight to the technical level of the report and advance the company's image.

Choice of an informative title is particularly important. It should be short, yet imply that the report has a worthwhile story to tell. Compare the vague title below with the more informative version written beneath it.

Original vague title

RADOME LEAKAGE

Revised informative title

POROSITY OF FIBERGLAS CAUSES RADOME LEAKS

Technical officers at remote radar sites would glance at the first title with only a muttered, "Looks like somebody else has the same problem we've got." But the second title would encourage them to stop and read the report. They would recognize that someone may have found the cause of (and, hopefully, a remedy for) a trouble spot that has bothered them for some time.

Note how the following titles each *tell* something about the reports they precede:

REDUCING AMBIENT NOISE IN AIR TRAFFIC
CONTROL CENTERS

EFFECTS OF DIESEL EXHAUST ON LATEX PAINTS

SALT EROSION OF CONCRETE PAVEMENTS

These titles may not attract every potential reader who comes across them, but they will certainly gain the attention of anyone interested in the topics they describe.

TITLE PAGE

The title page normally carries four main pieces of information: the report title (the same title that appears on the cover); the name of the person, company, or organization to whom the report is directed; the name of the company originating the report (sometimes with the author's name); and the date the report was completed. This page may also contain the authority or contract under which the report has been prepared, a report number, a security classification such as CONFIDENTIAL or SECRET, and a copy number (important reports given only limited distribution are sometimes assigned copy numbers to control and document their issue). All this information must be tastefully arranged on the page, as has been done in the first sample report later in this chapter.

TABLE OF CONTENTS

All but very short reports contain a Table of Contents (T of C). The T of C not only lists the report's contents, but also shows how the report has been arranged. Just as a prospective book buyer will scan the contents page to find out what a book contains and whether it will be of interest, potential readers will scan a report's T of C to assess how the author has organized the work. They will be searching for a clue to the author's competence to arrange technical information logically. If they sense that the report writer knows his or her business both as a technical person and as a writer, then they will delve into the report.

A satisfactory arrangement for a T of C is shown on page 171. (Note that the single word CONTENTS often is preferred to "Table of Contents.") This page also contains a list of appendixes attached to the report, with each identified by its full title. In long reports, a list of illustrations with their page numbers may also be helpful. If used, it should be inserted between the T of C and the list of appendixes.

REFERENCES (ENDNOTES), BIBLIOGRAPHY, AND FOOTNOTES

A report writer who refers to another document, such as a textbook, journal article, report, or correspondence, or to other persons' data or even a conversation, must identify the source of this information in the report. To avoid cluttering the report narrative with extensive cross-references, they are placed in a storage area at the end of the report or at the foot of the page. This storage area is known as a List of References, a Bibliography, or Footnotes. Specifically:

A **List of References** is the most convenient and popular way to list source documents. The references are typed as a sequentially numbered list at the end of the narrative sections of the report (usually immediately ahead of the appendix). Such numbered references are frequently referred to as *Endnotes.* For a short example, see page 180.

A **Bibliography** is simply an alphabetical listing of the documents used to research and conduct a project. The documents are listed in alphabetical order of authors' surnames, and the list is placed at the end of the report narrative. A bibliography may list many more documents than are referred to in the report. An example appears on page 156.

Footnotes are like *Endnotes* but are printed at the foot of the page on which the particular reference appears. Footnotes are becoming less popular because they are awkward to arrange on the page and, since they are visible, distract reading continuity. Turn to page 157 for an example.

Most industries and engineers prefer to use a List of References for their technical reports.

Preparing a List of References (Endnotes). References should contain certain information, arranged in this order:

(a) Author's name (or authors' names).
(b) Title of document (article, book, paper, report).
(c) Identification details, such as:

For a book: city and state (or country) of publication, publisher's name, and year of publication.

For a magazine article or technical paper: name of magazine or journal, volume and issue number, and date of issue.

For a report: report number, name and location of issuing organization, and date of issue.

For correspondence: name and location of issuing organization; name and location of receiving organization; and the letter's date.

For a conversation or speech: name and location of speaker's organization; name, identification, and location of listener(s); and the date.

(d) The page number (if applicable) on which the referenced item appears or starts.

For example, if your first reference is to an illustration on page 74 of a book by Laurinda K. Wicherly, your first entry in your list of references would contain:

(a) Author's name (in natural order: first name and/or initials, and then surname)
(b) Book title (always underlined)
(c) City of publication, publisher's name, and year of publication (all within parentheses)
(d) Page number (the first page of the referenced pages)

And it would look like this:

1. Laurinda K. Wicherly, <u>Fiberoptic Modes of Communication</u>, (New York: The Moderate Press Inc, 1985), p. 74.

If a book has two authors, both are named:

2. David B. Shaver and John D. Williams, . . .

But if there are three or more authors, only the first-named author need be listed:

3. Donald R. Kavanagh and others, . . .

If a book is a second or subsequent edition (as this book is), the edition number is entered immediately after the book title:

4. Ron S. Blicq, <u>Technically-Write!</u> 3rd ed. (Englewood Cliffs, NJ: Prentice-Hall, Inc., 1986).

Some books contain sections written by several authors, each of whom is named within the book, with the whole book edited by another person. If your reference is to the whole book, then identify it by the editor's name and insert the abbreviation "ed." immediately after the name:

5. Robert M. Woelfle, ed., <u>A Guide for Better Technical Presentations</u> (New York: IEEE Press, 1975).

But if your reference is to a particular section of the book, identify it by the specific author, enclose the section title in quotation marks, underline the book title, and then name the editor:

6. J. M. Lufkin, "The Slide Talk: A Tutorial Drama in One Act" in <u>A Guide for Better Technical Presenations</u>, Robert M. Woelfle, ed. (New York: IEEE Press, 1975), p. 14.

Similarly, if you are referring to an article in a magazine or journal, and it is your seventh reference, you would list the article like this:

(a) Author(s)'s name(s) (in natural order)
(b) Title of article (always in quotation marks, and *not* underlined)
(c) Title of journal or magazine (underlined)
(d) Volume and issue numbers (shown as two numbers separated by a colon)
(e) Journal or magazine issue date
(f) Page on which article or excerpt starts (optional entry)

7. William L. Everitt, "The Engineer: A Perennial Student," <u>IEEE Spectrum</u>, 14:12, December 1977, p. 21.

When a magazine article, technical paper, or report is published as one of several documents bound into a volume, then it is listed within quotation marks, and only the title of the volume is underlined, as has been done here. But if the article, technical paper, or report is published *separately* (not in a magazine or journal), its title is underlined and the quotation marks are omitted:

8. Derek A. Lloyd, <u>Effective Communication and Its Importance in Management Consulting.</u> Report No. 61, Smyrna Development Corporation, Atlanta, Georgia, February 18, 1984).

If a report or magazine article does not show an author's name, then the first entry for that item will be the title of the report or article:

9. "Continuing Education: Challenge of the 80's" in Engineering Technologist, 8:5, May 1978, p. 113.

For a letter or memorandum, the entry should be like this:

10. Christine Lamont, Macro Engineering Inc, Phoenix, Arizona. Letter to Marlene Hudson, No. 7 Architectural Group, Dallas, Texas, December 12, 1985.

And for a conversation or speech:

11. Elwood R. Phillips, Lakeside Power and Light Company, Montrose, Ohio, in conversation with Anna King, H. L. Winman and Associates, Cleveland, Ohio, March 16, 1986.
12. Emily K. Schlesinger, Media, Pennsylvania, speaking to the PCC84 Communication Conference, Atlantic City, New Jersey, October 11, 1984.

Every entry in a list of references must have a corresponding reference to it in the Discussion section of your report. At an appropriate place in the narrative you should insert a superscript (raised) number to identify the particular reference. It should look like this:

Earlier tests[3] showed that speeds higher than 2680 rpm were impractical. *(Alternatively, the superscript could go here.)* ⎯⎯⎯⎯⎯⎯⎯→

If your report is being prepared on a microprocessor, computer, or electronic typewriter that cannot type superscript numbers, you may state the reference number in parentheses:

Earlier tests (3) showed that speeds . . .

If you refer to the same document several times, your list of references need show full details for that document only the first time you refer to it. Subsequent references can be shown in a shortened form containing only the author's name (or authors' names) and the page number. For example, if the first reference you make is to an item on page 48 of the particular book described below, the entry in the list of references would be:

1. Wayne D. Barrett, Management in a Technical Domain (San Francisco, Calif.: Martin-Baisley Books, 1983), p. 48.

Now suppose that your second and third sources are other documents but for your fourth source you again refer to *Management in a Technical Domain,* this time quoting from page 159. Now you need list only the author's surname and the new page number:

> 4. Barrett, p. 159.

And you would do the same for each future reference to the same document, simply changing the page number each time. You can even make repeated references to several different documents by the same author by simply inserting the year of publication for the particular document between the author's name and the page number:

> 9. Barrett, 1983, p. 159.

Preparing a Bibliography.   A bibliography lists not only the documents to which you make direct reference, but also many other documents which deal with the topic. The major differences between the elements of the list of references and those of the bibliography are:

1. Bibliography entries are *not* numbered.
2. The name of the first-named author for each entry is reversed, so that the author's surname becomes the first word in the entry. (If there is a second-named author, his or her name is *not* reversed.)
3. The *first* line of each bibliography entry is extended about five typewriter spaces to the left of all other lines (see Figure 6-4).
4. The entries are organized in alphabetical order of first-named authors.
5. Punctuation of individual entries is significantly different, with each entry being divided into three compartments separated by periods: (1) author identification; (2) title of book or specific article; and (3) publishing details. (How the periods are positioned is shown in Figure 6-4.)
6. Page numbers usually are omitted, since generally the bibliography refers to the whole document. (Reference to a specific page can be made within the narrative of the report.)

Because a bibliography is not numbered, you cannot cross-refer directly to it simply by inserting a superscript number in the report narrative, as can be done with a list of references. The most common method is to insert a parenthetical reference in the narrative, and in it include the author's name (or authors' names) and the page number:

> Although the tests conducted in Alaska (Faversham, p. 261) showed only moderate decomposition . . .

The full descriptive listing for Faversham's book or report is carried in the bibliography.

BIBLIOGRAPHY

Armstrong, Karyn B. "A Zipcode-Oriented Filing System."
Proceedings of the Sixteenth National Marketing
Conference. The Marketers' Association of America,
Seattle, Wash., June 22, 1985.

Blicq, Ron S. Technically-Write!, 3rd ed. Englewood Cliffs,
N.J.: Prentice-Hall, Inc., 1986.

"Continuing Education: Challenge of the 80's." Engineering
Technologist, 8:5, May 1978.

Everitt, William L. "The Engineer: A Perennial Student."
IEEE Spectrum, 14:12, December 1977.

Lamont, Christine, Macro Engineering Inc, Phoenix, Arizona.
Letter to Marlene Hudson, No. 7 Architectural Group,
Dallas, Texas, December 12, 1985.

Lloyd, Derek A. Effective Communication and Its Importance
to Management Consulting. Report No. 61, Smyrna
Development Corporation, Atlanta, Georgia, February 18,
1984.

Lufkin, J. M. "The Slide Talk: A Tutorial Drama in One Act."
A Guide for Better Technical Presentations, ed. Robert
M. Woelfle. New York: IEEE Press, 1975.

Phillips, Elwood R. Lakeside Power and Light Company, Montrose,
Ohio. Conversation with Anna King, H. L. Winman and
Associates, Cleveland, Ohio, March 16, 1986.

Shaver, David B., and John D. Williams. Numerical Control
for Grain-Handling Equipment. St. Paul, Minn: Mid-
western Publishers Inc., 1983.

Wicherly, Laurinda K. Fiberoptic Modes of Communication.
New York: The Moderate Press, Inc., 1985.

Figure 6-4. A typical bibliography.

If several publications by the same author are listed in the bibliography, then the date of the particular publication is included as a parenthetical reference to identify which document is being mentioned:

> The most significant tests were those conducted 22 miles south of Old Crow, N.W.T. (Crosby, 1981, p. 17), which showed that . . .

The alternative (but less preferred) method is to use footnotes in combination with the bibliography, by inserting superscript numbers into the report narrative and placing footnotes at the bottom of the page. If you use this combination, keep the footnotes short because the documents are described fully in the bibliography. The footnotes should look like this:

> [1] Faversham, p. 261.
> [2] Crosby, 1981, p. 17.

Figure 6-4 shows a bibliography formed from some of the publications listed earlier as examples of references (endnotes).

Preparing Footnotes. Although footnotes once were common, they are now the least-preferred method of source referencing, particularly for technical and business reports. You will still see them used, however, in scholarly works and some scientific papers.

Typical footnotes appear at the foot of this page. The information they contain is the same as that for a list of references. The major differences are:

> The first line of each footnote entry is indented slightly and the footnote number is raised slightly, as shown in the example at the foot of this page.[1]
>
> The footnotes may be numbered consecutively on a page-by-page basis, with each page starting afresh with footnote 1, or they may be numbered consecutively throughout the whole report.
>
> A footnote can be an explanatory note,[2] rather than a reference to another document, although use of footnotes for this purpose is not recommended.
>
> If a document is referenced more than once, it is listed in full only the first time. For subsequent references, only the author's name (or authors' names) and the relevant page number are listed.[3]

The most effective way to refer readers to source material is to insert superscript numbers in the report narrative and to print a numbered list of references at the end of the report. Although footnotes or a bibliography may be used in place of a list of references, they have disadvantages which make them less suitable for engineering and industrial reports.

Source referencing is an essential part of a report. The entries must be

---

[1] Norman Berstein, *The Engineers' Notebook* (New York: The Alternative Press Inc., 1986), p. 174.
[2] In typeset documents, footnotes normally are printed in a smaller typeface.
[3] Berstein, p. 106.

complete and accurate, so that readers can identify, and locate or order, every document you list. For that reason, always transcribe numbers, names, titles, dates, and other details extremely carefully from the original document.

DISTRIBUTION LIST

A report sent to several readers will often contain a distribution list that identifies all the persons who are to receive it. The position of the distribution list depends on company practice, the preference of the author, and the length of the list. Probably the most suitable position is immediately after the bibliography, although a short distribution list may appear at the foot of the T of C page. A long list may have a page to itself and be inserted as a separate appendix.

ACKNOWLEDGMENTS

Acknowledgments are inserted whenever a report author wishes to acknowledge special help received during the investigation or study which is the topic of the report. Acknowledgments should be brief, simple, and limited only to those persons who have made a significant contribution. They may be inserted on a page by themselves after the summary, after the T of C, as part of the introduction, or immediately after the discussion.

COVER LETTER

The cover letter is a brief courtesy letter that identifies the report and states briefly why it is being forwarded to the addressee:

Dear Mr. Merrywell:

We enclose our report No. 8-23, "Selecting New Elevators for the Merrywell Building," which has been prepared in response to your letter WDR/71/007 dated April 27, 19xx.

If you would like us to submit a design for the enlarged elevator shaft, or to manage the installation project on your behalf, we shall be glad to be of service.

Sincerely,

Barry V. Kingsley
H. L. WINMAN AND ASSOCIATES

This cover letter accompanied formal report 2 (see pages 186 to 190) when it was delivered to Mr. David Merrywell, H. L. Winman and Associates' client.

A cover letter may be prepared as either a letter or an interoffice memorandum, and is attached to the outside front of the report. A letter is used for reports distributed outside the originating company, an interoffice memorandum for reports confined to internal distribution.

LETTER OF TRANSMITTAL

Occasionally you will find a letter of transmittal inserted as a foreword or preface to a report. Normally written and signed by top management, it reflects company policy and sometimes management's interpretation of the report's findings. Because it repeats much of the information contained in the introduction, conclusions, and recommendations, it sometimes replaces the summary.

The letter of transmittal can also serve a useful purpose when it is used with a major technical proposal, which is a form of technical report in which a company describes how it can successfully tackle a proposed task at an economical price for the government or another company. In this case it is used chiefly to discuss financial and legal implications, such as specifications and statements that may be open to more than one interpretation.

An alternative to the letter of transmittal is the **executive summary,** which is an analytical summary of the purpose of the report, its main findings and conclusions, and the author's recommendation. Unlike the normal report summary prepared for all readers, the executive summary presents detailed information on aspects of particular concern to senior executives and often will introduce financial implications. It may be prepared as a letter, as in Figure 6-9, or as a more formal page titled "Executive Summary."

A letter of transmittal or executive summary is bound inside a report, behind the cover or title page.

THE FULL REPORT

THE MAIN PARTS

This description of the full formal report covers only the ten parts that contribute directly to the report itself. The distribution list, cover letter, letter of transmittal, and acknowledgments are omitted because they do not contribute directly to the technical content. Other parts that sometimes may be omitted are the cover and T of C (if the report is very short), the recommendations (if the writer has none to make), and the endnotes and appendix (if there are no references or supporting data).

TRADITIONAL ARRANGEMENT

(Conclusions and Recommendations *after* Discussion)

These ten main parts, complete with capsule descriptions, are listed in Figure 6-5. The order in which they are presented is common to most formal technical and business reports and is the format illustrated in the mini-report in Figure 6-6. Note that with this arrangement there is a logical flow of informa-

THE MAIN PARTS OF A FORMAL REPORT

Cover:	Jacket of report; contains title of report and name of originating company; its quality and use of color reflect company "image."
Title Page:	First page of report; contains title of report, name of addressee or recipient, author's name and company, date, and sometimes a report number.
Summary:	An abridged version of whole report, written in nontechnical terms; *very* short and informative; normally describes salient features of report, draws a main conclusion, and makes a recommendation; always written last, after remainder of report has been written.
Table of Contents:	Shows contents and arrangement of report; always includes a list of appendixes and, sometimes, a list of illustrations.
Introduction:	Prepares reader for discussion to come; indicates purpose and scope of report, and provides background information so that reader can read discussion intelligently.
Discussion:	A narrative that provides all the details, evidence, and data needed by the reader to understand what the author was trying to do, what he or she actually did and found out, and what he or she thinks should be done next.
Conclusions:	A summary of the major conclusions or milestones reached in the discussion; conclusions are only opinions so can never advocate action.
Recommendations:	If the discussion and conclusions suggest that specific action needs to be taken, the recommendations state categorically what must be done.
References:	A list of reference documents which were used to conduct the project and which the author considers will be useful to the reader; contains sufficient information for the reader to correctly identify and order the documents.
Appendix(es):	A "storage" area at the back of the report that contains supporting data (such as charts, tables, photographs, specifications, and test results) which rightly belong in the discussion but, if included with it, would disrupt and clutter the major narrative.

Figure 6-5. Capsule descriptions of formal report parts.

Use as pattern.

FORMAL REPORT – ARRANGEMENT OF PAGES

PRELIMINARY PAGES

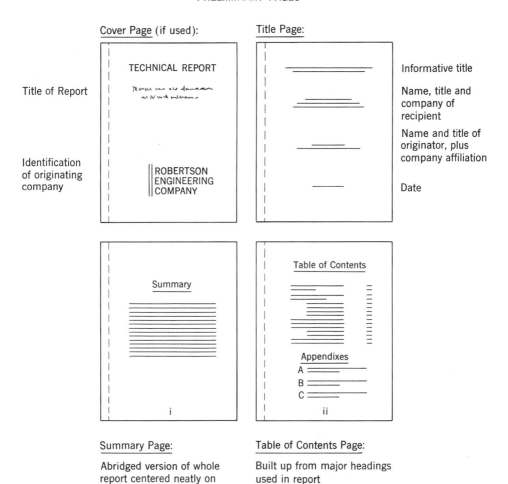

Cover Page (if used):

Title of Report

Identification of originating company

TECHNICAL REPORT

ROBERTSON ENGINEERING COMPANY

i

Title Page:

Informative title

Name, title and company of recipient

Name and title of originator, plus company affiliation

Date

Summary

Table of Contents

Appendixes
A
B
C

ii

Summary Page:

Abridged version of whole report centered neatly on page

Always has page to itself

Table of Contents Page:

Built up from major headings used in report

NOTE: Preliminary pages are identified by lower case roman numerals

Figure 6-6. Formal report—traditional arrangement.

Use this as the pattern.

FORMAL REPORT — ARRANGEMENT OF PAGES
(Conclusions and Recommendation *following* Discussion)

Introduction, Discussion, Conclusions, Recommendation and References

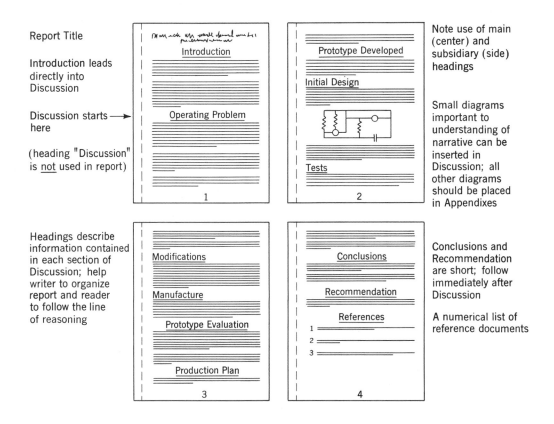

Report Title

Introduction leads directly into Discussion

Discussion starts here

(heading "Discussion" is <u>not</u> used in report)

Introduction

Operating Problem

1

Prototype Developed

Initial Design

Tests

2

Note use of main (center) and subsidiary (side) headings

Small diagrams important to understanding of narrative can be inserted in Discussion; all other diagrams should be placed in Appendixes

Headings describe information contained in each section of Discussion; help writer to organize report and reader to follow the line of reasoning

Modifications

Manufacture

Prototype Evaluation

Production Plan

3

Conclusions

Recommendation

References
1
2
3

4

Conclusions and Recommendation are short; follow immediately after Discussion

A numerical list of reference documents

NOTE: Page numbers are arabic numerals either at foot of page as shown here or at top right-hand corner

Left-hand margin is wider to allow for binding edge

FORMAL REPORT – ARRANGEMENT OF PAGES

SUPPORTING DATA (Appendixes)

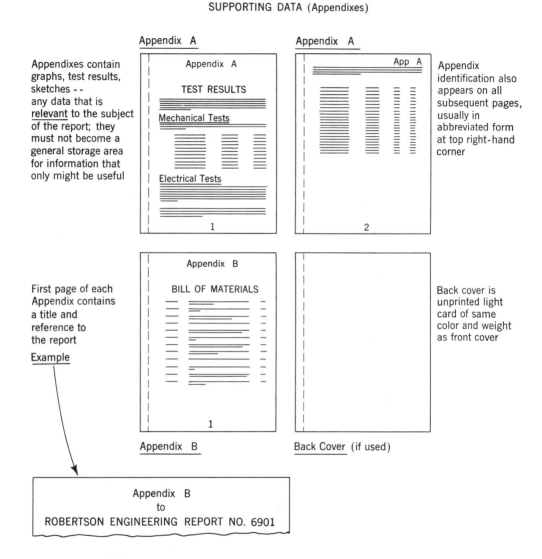

Appendixes contain graphs, test results, sketches -- any data that is <u>relevant</u> to the subject of the report; they must not become a general storage area for information that only might be useful

Appendix identification also appears on all subsequent pages, usually in abbreviated form at top right-hand corner

First page of each Appendix contains a title and reference to the report

<u>Example</u>

Back cover is unprinted light card of same color and weight as front cover

Appendix A

Appendix A

Appendix A

TEST RESULTS

Mechanical Tests

Electrical Tests

1

App A

2

Appendix B

BILL OF MATERIALS

1

Appendix B

Back Cover (if used)

Appendix B
to
ROBERTSON ENGINEERING REPORT NO. 6901

NOTE: Appendixes are inserted in the order that they are referenced in the report

Each Appendix is considered a complete document in itself, thus is numbered separately using arabic numerals

tion: the **introduction** leads into the **discussion,** from which the writer draws **conclusions** and makes **recommendations** (the two latter parts sometimes are referred to jointly as the terminal summary). The sample formal report, "Evaluation of Word Processors for H. L. Winman and Associates's Engineering Department" (Report 1), follows the traditional arrangement of report parts.

ALTERNATIVE ARRANGEMENT

(Conclusions and Recommendations *before* Discussion)

In recent years, more and more report writers have altered the organization of their reports to serve the needs of their readers. This alternative arrangement brings the conclusions and recommendations forward, positioning them immediately after the introduction so that executive readers do not have to leaf through the report to find the terminal section (the report's outcome). The advantages of this "pyramid style" are immediately evident: Busy readers have only to read the initial pages to learn the main points contained in the report, and the writer can help them along by gradually increasing the technical content of the report, catering to semitechnical executive readers up to the end of the recommendations, and to fully technical readers in the discussion and appendix. Although the natural flow of information that occurs in the traditional arrangement is disrupted, Figure 6-7 shows there is now a reader-oriented flow, with the three compartments of information containing progressively more technical details.

This gradually increasing development of the topic in three separate stages is similar to newspaper technique. If you read any well-written news item, you will find that the first paragraph or two contain a capsule description of the whole story. The next three or four paragraphs contain a slightly more detailed description, and then the newspaper invites you to continue reading on a subsequent page. The final eight or nine paragraphs repeat the same story, but this

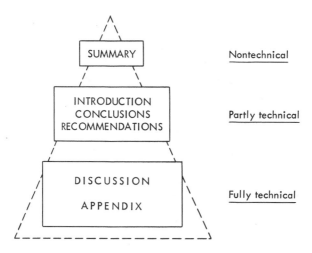

Figure 6-7. The formal report arranged "pyramid style."

time with more names, more peripheral information, and more details. Newspapers cater to the busy reader who may not have time to read more than the opening synopsis, and also to the leisurely reader who wants to read all the available information.

A mini-report layout of this alternative arrangement appears in Figure 6-8. This figure contains only the central page of the layout (containing the introduction, conclusions, recommendations, and discussion). The remaining two pages, with the layout for the preliminary information and appendixes, are

Figure 6-8. Formal report—alternative arrangement.

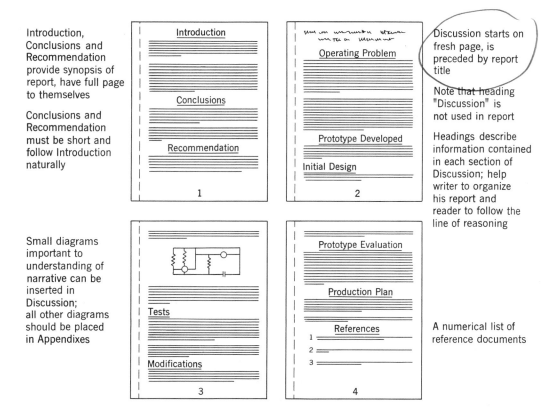

FORMAL REPORT – ARRANGEMENT OF PAGES
(Conclusions and Recommendation <u>before</u> Discussion)
(ALTERNATIVE ARRANGEMENT)

Introduction, Conclusions, Recommendation, Discussion & References

Introduction, Conclusions and Recommendation provide synopsis of report, have full page to themselves

Conclusions and Recommendation must be short and follow Introduction naturally

Small diagrams important to understanding of narrative can be inserted in Discussion; all other diagrams should be placed in Appendixes

Introduction

Conclusions

Recommendation

1

Operating Problem

Prototype Developed

Initial Design

2

Tests

Modifications

3

Prototype Evaluation

Production Plan

References
1
2
3

4

Discussion starts on fresh page, is preceded by report title

Note that heading "Discussion" is not used in report

Headings describe information contained in each section of Discussion; help writer to organize his report and reader to follow the line of reasoning

A numerical list of reference documents

NOTE: Page numbers are arabic numerals either at foot of page as shown here or at top center

Left-hand margin is wider to allow for binding edge

the same for both arrangements and can be obtained from the first and third pages of Figure 6-6. An example of the alternative arrangement of report parts can be seen in the technical report, "Selecting New Elevators for the Merrywell Building," Report No. 2 in this chapter.

TYPICAL FORMAL REPORTS

The excerpts from two formal reports that follow are typical of the work expected of graduate engineers, technologists, and technicians. They comprise:

Report 1: Evaluation of Word Processors for H. L. Winman and Associates Engineering Department.	The writer of this report has placed the conclusions and recommendations after the discussion, following the traditional arrangement of report parts. The complete report is presented here.
Report 2: Selecting New Elevators for the Merrywell Building.	In this report the writer has adopted the alternative arrangement of report parts, placing the introduction, conclusions, and recommendations as a single unit ahead of the discussion. (Only the summary, introduction, conclusions, recommendations, and first two pages of the discussion are shown here.)

On the pages preceding each report I have inserted a few comments, plus some of the data that initiated each project and resulted in the report being written. Both reports have been typed single-spaced. Double-spacing is more common, although some companies prefer to single-space their reports and leave two lines between main paragraphs, as in these two examples.

Formal Report 1
Evaluation of Word Processors
for H. L. Winman and Associates
Engineering Department

This report was prepared by a computer systems analyst at Macro Engineering Inc, in response to Anna King's proposal shown in Figure 5-7 of Chapter 5. Its author—Ken Nishimura—typed and edited the report on a computer, using the computer as a word processor, and had the finished report typed automatically by a letter-quality line printer. The word-processing program automatically inserted page numbers and assembled and printed the Contents page. To understand the circumstances leading up to the report, first read Anna King's proposal in Figure 5-7 (see pages 123 and 124).

COMMENTS ON THE REPORT

The report contains all the major parts described earlier in this chapter. The comments below identify how Ken has used these parts to develop a convincing case to present to his client. Although he seems to be identifying one particular remedy for H. L. Winman and Associates to adopt, in reality he has left the door open for the company to select an alternative method if the executives want to spend more money than they had originally planned.

- The **Letter of Transmittal** (Figure 6-9), which Ken drafted for his boss's signature, is an executive summary that quickly highlights the main points of the report and draws attention to significant aspects.
- Because he recognizes his report will be read by many managers and supervisors at H. L. Winman and Associates, not all of whom will be familiar with computer terminology, Ken has refrained from using "jargon" terms such as RAM, ROM, kilobyte, and megabyte.
- The first paragraph of the **Introduction** contains *background* details, the second sentence of the same paragraph contains the report's *purpose,* and paragraphs 2 and 3 (including the four subparagraphs) outline the report's *scope.*
- The background information on pages 1 and 2 helps familiarize readers with the situation. Ken has assumed that Martin Dawes and Anna King will not be his only readers.
- The **Discussion** generally has a chronological arrangement, as is evident from examining the **Table of Contents.** Within the Discussion, however, there is a central section (report pages 2 to 4) which presents and discusses the four proposed methods in a subject arrangement.
- Rather than leap swiftly into a full description of the four methods (report page 3), Ken first provides a capsule description of each method and then defines in general terms the main criteria he will use to evaluate it. "Setting the scene" like this helps readers follow an author's line of reasoning much more easily.
- The table in report Figure 2 is a simple means for illustrating the report's main findings. The differences between each method cannot be deduced easily from the narrative descriptions, but can be visualized readily from this illustration. The criteria listed on page 7 and the comparative analysis in Appendix A show how Ken arrived at the points score he has assigned to each factor.
- The **Conclusions** highlight only the *practicable* alternatives available to H. L. Winman and Associates, while the **Recommendation** identifies which alternative the report writer favors. In this way Ken leaves the final choice to his client.
- The **Appendixes** contain details which support the statements Ken has made throughout the Discussion. If the cost figures in Appendix B, for example, had been inserted in the narrative sections of the report, they would have interrupted the natural flow of information.

March 26, 19xx

MACRO ENGINEERING INC

600 Deepdale Drive
Phoenix, AZ
85007

Mr. Martin Dawes, President
H. L. Winman and Associates
475 Reston Avenue
Cleveland, Ohio 44104

Dear Martin:

In the enclosed report, Ken Nishimura shows that the simplest way
to make computer writing and editing available to H. L. Winman and
Associates' engineers would be to purchase additional Vancourt 2200
computers and assign them solely for engineering use. This would
leave your present computer systems intact and create the least
disruption. The price would be $28,250, which is $1750 less than
the cost you have budgeted.

A more comprehensive alternative, but also considerably more expensive, would be to replace all of your computers with an integrated
MEI "04" system. Long range, this is probably the better choice
because of its greater adaptability and increased compatibility
with other systems. However, its price of $48,300 (which includes
a 35% intercompany discount and the revenue from selling your present computers) is $18,300 more than you have budgeted. Consequently
it is Ken's second choice.

Please call me or Ken Nishimura if you have any questions regarding
the report.

Regards,

Tina Mactiere

Tina R. Mactiere, President

KTN:rb
enc

Figure 6-9. An executive summary accompanying a formal report.

MACRO ENGINEERING INC

EVALUATION OF WORD PROCESSORS
FOR H. L. WINMAN AND ASSOCIATES
ENGINEERING   DEPARTMENT

Prepared for:

Mr. Martin Dawes, President
H. L. Winman and Associates

Prepared by:

Ken T. Nishimura
Senior Systems Analyst
Macro Engineering Inc

Report No. C509

March 26, 19xx

Report 1.

SUMMARY

A study has been conducted at H. L. Winman and Associates
to determine what methods could be introduced to enable
engineers to write and edit reports at their computer
terminals. The existing Comron computer system would be
inefficient as a word processor, even with enhanced
software.

The simplest and most effective alternative is to purchase
additional Vancourt word processing terminals and install
them at strategic locations throughout the engineering
department. The cost would be $28,250, which is $1750
less than budgeted for the project.

i

Report 1.

CONTENTS

—Illustrations inserted here—

APPENDIXES

ii

Report 1.

171

EVALUATION OF WORD PROCESSORS
FOR H. L. WINMAN AND ASSOCIATES' ENGINEERING DEPARTMENT

INTRODUCTION

(handwritten: Background)

In a memorandum dated February 15, 19xx, technical editor Anna King proposed introducing computer-based writing and editing at H. L. Winman and Associates, in place of the much slower and time consuming pencil-and-paper method currently used by the company's engineers.[1] Subsequently, Macro Engineering Inc was commissioned to evaluate H. L. Winman and Associates' needs and to identify the most effective and economical way to use computer technology in a report-writing environment.[2]

(handwritten: Purpose)

The following factors were established by H. L. Winman and Associates as requirements to be taken into account:

(handwritten: scope)

1. The cost for purchasing hardware and software should not exceed $30,000.

2. Existing hardware should be used as much as possible.

3. The system should be easy to learn and use, so that minimum training will be necessary.

4. The system must be IBM compatible, be able to communicate with other computer systems, and have built-in potential for upgrading as new computer technology is introduced.

The study was to be completed by March 14, and the report submitted by March 28, 19xx.

EXISTING COMPUTERS AT H. L. WINMAN AND ASSOCIATES

(handwritten: Chronological w/ subject ord)

Currently H. L. Winman and Associates has two main computer systems, one used purely as a computer and one used for word processing. The computer is the Comron model 3000, which was installed eight years ago and is used solely by the engineering department. It comprises a central processing unit (CPU), 58 video terminals, and two dot-matrix line printers. The word-processing system comprises 10 Vancourt Business Systems Inc model 2200 video terminals, nine operated by secretaries and word-processor typists in the office services department, and one used by the company's technical editor. They are tied in to a Vancourt 2000 central processing unit and a letter-quality line printer. There are also three stand-alone Kenta 405 personal computers and a dot-matrix printer shared by the materiel control and personnel departments. The three separate systems are illustrated in figure 1.

1

Report 1.

ENGINEERING	OFFICE SERVICES	MATERIEL CONTROL
Comron 3000 Mainframe Computer	Vancourt 2000/2200 Word Processor	Kenta "405" Personal Computers
58 Terminals	10 Terminals	3 Terminals

Fig. 1.   The three existing computer systems.

As the objective is to achieve across-the-company word processing capability at H. L. Winman and Associates' head office, full communication is essential between the Comron and Vancourt computer systems.   (Communication between the three Kenta personal computers and the two larger systems is less important.) Intersystem communication is hampered, however, because the two systems in their present configurations are neither electronically nor functionally compatible.

Compatibility is difficult to achieve because the two systems operate under different premises:

subject arrangement

The Vancourt 2000/2200 is a "dedicated" system designed entirely for word processing.   Its keyboards have 12 special function keys which enable operators to perform complex word-handling maneuvers rapidly and with only one or two key strokes for each command.   Its software is comprehensive and is built into the system, ready for immediate access. The Vancourt 2000/2200 can also function as a computer, but requires additional software (i.e. computer programs) to fulfill this function.

The Comron 3000 is designed primarily to handle complex computer problems. Its keyboards have far fewer function keys than the Vancourt 2200 terminals, so that word processing commands have to be assigned to normal keys which have to be operated in combination to achieve special maneuvers.   As each command requires three or four key strokes, the word processing functions are difficult to learn, hard to remember, and slow to execute.   Word processing software can be purchased, but as a separate program that occupies large chunks of the computer's available memory. A recently developed modification kit, comprising both hardware and software, is reputed to enhance the Comron computer's word processing capability and system compatibility.[3]

2

Report 1.

METHODS FOR ACHIEVING FULL COMPUTER WRITING CAPABILITY

Although upgrading the Comron computer software initially seemed to be the simplest way to achieve computer writing and editing among H. L. Winman and Associates' engineers, I felt I should consider other alternatives that conceivably might offer better potential. The four methods I evaluated were:

1. Upgrading the Comron 3000 computer so that the 58 Comron terminals are compatible with the Vancourt computer and can be used for word processing.

2. Replacing the Comron 3000 computer system with Vancourt 2200 video terminals, and upgrading the Vancourt systems' computing capabilities.

3. Replacing the three existing computer systems (Comron, Vancourt, and Kenta) with a single, integrated system with full computing and word processing capabilities.

4. Leaving all three existing systems intact, but purchasing additional Vancourt 2200 video terminals for use by the engineering department.

In my evaluation I considered eight factors affecting each method's suitability and effectiveness, and compared the results in the table in appendix A. Four factors I considered to be of major importance:

* The method's total cost.

* The method's ability to meet H. L. Winman and Associates' current and future word processing/computing needs.

* Compatibility between computer systems, both within H. L. Winman and Associates and between the company's computers and external computers.

* The training that engineers and secretaries would need before they would become thoroughly familiar with the computer system and software programs.

A brief description of the four proposed computer systems, including comments on how successfully each meets these factors, is provided below.

Upgrading the Comron Computer for Word Processing

To upgrade the Comron 3000 computer's hardware will require inserting additional circuit boards and memory into the system's central processing unit (CPU), purchasing an interface to enable the Comron and Vancourt computer

3

Report 1.

174

systems to communicate with each other, and purchasing two letter-quality line printers, one for the engineering department and one for the secretaries and typists in the office services department. Software also will have to be upgraded by purchasing the Comwrite 2 word processing program. The equipment, software, and costs are listed in appendix B.

Cost. The cost to upgrade the Comron computer for word processing will be $10,615, which is $19,385 less than the $30,000 budget.

System Capability. The Comron 3000W, as it is identified by the manufacturer in its word processing configuration, will be able to perform full text processing without downgrading the system's current computing capabilities. Capability for further upgrading will remain high from a computing standpoint but, because the Comron 3300 keyboards have only five special function keys, the possibility for total word processing enhancement will be limited.

Compatibility. The Comron 3000W will be fully compatible with the Vancourt 2000 computer and line printers, but will not be compatible with the three Kenta personal computers. Because the Comron computer is not IBM compatible, its ability to communicate with external computers (i.e. computers and word processors at other locations) also will be limited.

Simplicity. Operators using the Comron 3300 keyboards and Comwrite 2 word processing software will have to memorize special commands to achieve full word processing capability. I estimate engineers will require a minimum of eight hours training before they will be able to type and edit their reports with reasonable confidence. Word processor typists and secretaries also will need to be trained on the system, because the key strokes and commands for the Comwrite 2 program differ substantially from those they are accustomed to using. Their training will likely take six hours per person.

Replacing the Comron Computer with Vancourt Terminals

Installing Vancourt 2200 terminals throughout the engineering department will mean substantially increasing the memory of the Vancourt 2000 CPU, purchasing 48 Vancourt 2200 video terminals and two line printers for engineering use, and selling the Comron computer system. The existing word processing software program will continue to be used, but a "Vancalc" computer program will have to be purchased.

Cost. The cost of 58 Vancourt video terminals, 2 line printers, and Vancourt software will be $81,900. This will be partly offset by selling the Comron computer system for an estimated $9200, resulting in a net cost of $72,700. This is more than double the proposed maximum cost for the word processing system.

4

Report 1.

System Capability.   Every terminal in the engineering department will have full word processing capabilities equal to those currently available from the 10 Vancourt 2200 terminals in the office services department. Capability for future upgrading of the word processing function will be high because the Vancourt 2000/2200 system is designed for modification to incorporate new technology as it evolves. However, although Vancourt's Vancalc software program is rated good by its users,[4] it lacks some of the more complex features currently available from Comron's built-in software and has only limited potential for future enhancement.

Compatibility.   There will be full internal compatibility, not only between the Vancourt video terminals in the engineering and office services departments but also between them and the Kenta computers in the materiel control department. As the Vancourt 2000 system is IBM compatible, it will also be able to communicate with IBM compatible computers at other locations throughout North America.

Simplicity.   Previous experience with the Vancourt 2000 word processing system and the Vancourt 2200 terminals has demonstrated the system's simplicity from both the learner's and the user's viewpoint. Engineers already familiar with the Comron 3000 computer will need approximately three hours apiece to familiarize themselves with both the word processing and Vancalc program software. Typists who are already using the word processing software will require about half an hour apiece to learn how to transfer files and identify writers' changes.

Replacing All Computers with a Single, Integrated System

Installing a single, integrated system will mean selling all existing computers and replacing them with 71 new video terminals, a CPU, and four line printers (two letter-quality and two dot-matrix). The Macro Engineering Inc MEI 04 series has been chosen as a representative integrated system because a 35% discount can be obtained by purchasing the system from an H. L. Winman and Associates' affiliate. Purchasing an integrated system from a single supplier also offers the opportunity to buy a limited number of portable microprocessors as part of the mix. The proposed system will comprise: an MEI 804 CPU, 68 MEI model 404 desktop video terminals, 3 model 504 portable computers, and 4 model 704 line printers.

Cost.   Purchase price for the integrated system will be $71,950 (see appendix B for a cost breakdown), which will be offset by an approximate sale value of $23,650 for H. L. Winman and Associates' existing systems. The total cost will be $48,300, which is $18,300 more than the proposed budget.

System Capability.   The MEI 04 integrated system with built-in "Macrotype" software has both word processing and computing capabilities fully equal to those provided by the existing Vancourt and Comron computers

5

combined. The mainframe CPU also has space for installing system improvements as they become available.

Compatibility. The question of internal compatibility will not arise because H. L. Winman and Associates will have a single, integrated system. And, as the MEI 04 system is fully IBM compatible, communication with external computers also will be readily achieved.

Simplicity. The integrated software and 15 special function keys of the MEI 04 system provide rapid cursor movement and text manipulation for word processing, and a complete range of computer operating modes including extended graphics. I estimate engineers will need approximately five hours training time, $1\frac{1}{2}$ hours for familiarization with the computer's operating modes, and $3\frac{1}{2}$ hours for learning the word processing functions. Typists will need $2\frac{1}{2}$ hours each to learn the special key functions and document transfer techniques.

Purchasing Additional Vancourt 2200 Terminals

For this option H. L. Winman and Associates' three existing computers will be left intact, but additional components for the Vancourt computer system will be purchased and installed in the engineering department. These will be used solely for report writing and editing, and will be connected to the existing CPU. The purchase will comprise: 10 Vancourt 2200 video terminals, 3 Vancourt 2200P portable computers, a Vancourt 2800 line printer, and an enhanced CPU memory.

Cost. The purchase cost for the Vancourt components will be $28,250, which is $1750 less than the allotted budget for system upgrading.

System Capability. The engineering department's Vancourt video terminals will provide full word processing capability for up to 13 engineers at any one time, and will readily transfer reports to and from the typists' terminals in the office services department. The portable computers will also enable up to three engineers at a time to do their report writing at remote locations or at home. Capacity for future system upgrading will be high, in line with Vancourt's policy for continual system improvement. The Comron computer system will retain its existing capabilities and capacity for upgrading.

Compatibility. System compatibility will remain unchanged: there will be no internal compatibility between the Comron and Vancourt computer systems. (The need for compatibility between the Comron and Vancourt computers will be lessened, however, because engineers will have access to Vancourt terminals specifically designated for keyboard writing.) Because the Vancourt system is IBM compatible, it will provide access

6

Report 1.

to external IBM compatible systems, although compatibility between the Comron computer and external systems will remain limited.

Simplicity. Assuming that engineers using the Vancourt terminals will already be familiar with the Comron terminals, I estimate they will require three hours apiece to become thoroughly familiar with the special function keys and word processing software. Typists will require an average half hour each to learn how to transfer files and incorporate the engineering writers' corrections.

COMPARING THE ALTERNATIVES

To determine which of the four proposed systems meet most of H. L. Winman and Associates' requirements, I prepared a comparative analysis table (see figure 2) which rates each requirement according to the following criteria:

        4 - Meets all the requirements.
        3 - Almost meets the requirements; has minor limitations.
        2 - Only partly meets the requirements; has moderate limitations.
        1 - Meets only a minimum of the requirements; has severe limitations.
        0 - Fails to meet the requirements.

I then established criteria for each requirement:

        Cost   -- $30,000 maximum.

        Capabilities   -- Full word processing and computing capabilities,
                           plus good capacity for expansion.

        Compatibility   -- Full compatibility with other in-house computers,
                           plus IBM compatibility for communicating with
                           computers at other locations.

        Simplicity   -- Training time not to exceed 352 hours, based on
                           80 engineers at 4 hours apiece and 16 secretaries/
                           word processor typists at 2 hours apiece.

Most emphasis was assigned to the capability of the selected system to perform at the highest level, as both a word processor and a computer, now and in the future.

Although simply upgrading the Comron computer for use as a word processor is by far the cheapest option, its poor showing in almost all other factors results in its achieving the lowest points score. Similarly, the particularly high cost and limited expansion capability of replacing the Comron computer system with an enhanced Vancourt system makes this method a low-score option.

7

Report 1.

RATING THE ALTERNATIVES

Method:	Upgrade Comron	Replace Comron; Install Vancourt	Replace All; Install MEI 04	Install Vancourt Word Proc for Eng Use
Cost:	4	0	1	4
Capability:				
As a word processor:	2	4	4	4
As a computer:	4	3	4	4
For expansion:	2	2	4	3
Compatibility:	2	4	4	4
Simplicity: (for training)	1	4	3	4
	—	—	—	—
Total Points:	15	17	20	23

Fig. 2. Comparison of Proposed Word Processing Systems

The remaining two options score equally highly if only their capability, compatibility, and simplicity of operation are considered. However, the higher cost of selecting the integrated MEI 04 computer system for installation throughout H. L. Winman and Associates' head office relegates it to the second-choice position.

The highest points score is achieved by the proposal to install additional Vancourt computer terminals in the engineering department, solely for use as dedicated word processors. The only factor in which this option does not fully meet H. L. Winman and Associates' requirements is its capacity for expansion, and even then it scores 3 out of 4 points. It is also the only high-scoring method to retain all of the company's existing computer hardware.

CONCLUSIONS

Computer-based writing and editing of engineering reports can be achieved effectively and economically at H. L. Winman and Associates. Two methods are particularly suitable:

1. Purchasing 10 additional Vancourt 2200 computer terminals, plus 3 Vancourt 2200P portable computers, and assigning them to the engineering department for use as dedicated word processors, would provide a simple method that is $1750 below the targeted budget of $30,000.

8

Report 1.

2. Replacing all the existing computers with an MEI 04 integrated computer system would provide a comprehensive, state-of-the-art system that is $18,300 above the established budget.

Simply upgrading the word processing capability of the existing Comron computer would be the least expensive method, at $10,615, but it would provide a more difficult system to learn and operate and would offer limited intersystem compatibility.

RECOMMENDATION

I recommend that H. L. Winman and Associates purchase 10 Vancourt model 2200 computer terminals, 3 Vancourt model 2200P portable computers, 1 model 2200L line printer, and an additional model 2200D disk drive, and install them in the engineering department for a total purchase price of $28,250.

one chosen method

REFERENCES

1. Anna King, H. L. Winman and Associates, Cleveland, Ohio. Memorandum to Martin Dawes, February 15, 19xx.

2. Martin Dawes, H. L. Winman and Associates, Cleveland, Ohio. Letter to Tina R. Mactiere, Macro Engineering Inc, Phoenix, Arizona, March 1, 19xx.

3. Victor Van Chornisse, "Comron Enhancement Offers Semi-dedicated WP System," Word Processing Systems News, 3:8, August, 19xx.

4. William G. Strang, Jr., The Word Processor and Computer Handbook (Chicago, Illinois: Best-byte Press Inc., 19xx), p. 231.

9

Report 1.

APPENDIX A

COMPARISON OF ALTERNATIVE METHODS FOR INTRODUCING WORD PROCESSING INTO THE ENGINEERING DEPARTMENT

Proposed System:	Upgrade Comron Computers for Word Processing	Replace Comron Computers with Vancourt WPs	Replace All Computers With MEI "04" System	Purchase 13 Vancourt WPs for Engineering
Cost	$10,615	$72,700	$48,300	$28,250
Capabilities as a Word Processor (WP)	Some limitations; few function keys	Comprehensive	Comprehensive	Comprehensive
Simplicity of Use (as a WP)	Complex; requires multiple key strokes	Simple, fast, easy to use	Simple, fast, moderately easy to use	Simple, fast, easy to use
Capabilities as a Computer	Comprehensive	Moderate; some limitations	Comprehensive	Not applicable (Comron retained)
Compatibility with Other Systems	Good in-house; limited externally; not IBM compatible	VG in-house and externally; IBM compatible	Excellent; IBM compatible	Fair/good in-house; VG externally; IBM compatible
Capability for Future Upgrading	Limited for WP; Excellent for computing	Excellent for WP; only fair for computing	Excellent for WP and computing	Excellent for WP; Comron computer retained: Exc.
Training Required for Word Processing	736 hours	248 hours	440 hours	248 hours

Report 1.

APPENDIX B

EQUIPMENT LISTS AND COST ANALYSES

1. UPGRADING COMRON 3000 COMPUTER FOR WORD PROCESSING

   Hardware Purchases

2 Finewriter 801 letter-quality printers, compatible with Vancourt and Comron computers, @ $2250 each	$ 4,500
1 Interface, to enable Comron and Vancourt computers to communicate	2,600
1 CPU memory package CP3006	2,830

   Software Purchases

Comron word processing program "Comwrite 2"	685
Total Cost:	$10,615

2. REPLACING COMRON 3000 COMPUTER SYSTEM WITH VANCOURT 2200C WORD PROCESSORS UPGRADED FOR COMPUTER OPERATION

   Hardware Purchases

58 Vancourt 2200C terminals @ $1300 each*	$75,400
1 Vancourt 2205 letter-quality line printer	2,200
1 Heavy duty dot-matrix line printer, model 444	1,600
1 CPU memory upgrade package VBS 2000M	1,950

   Software Purchases

Vancourt "Vancalc" computer program	750
Total Purchase Price:	$81,900

   Hardware Sales

58 Comron 3300 terminals @ $100 each	5,800
1 Comron CPU 3000	3,000
2 Comron dot matrix line printers @ $200 each	400
Total Sales:	$ 9,200
Total Cost:	$72,700

* Special price for quantity purchase
(more than 30 units)

1

Report 1.

3. REPLACING ALL COMPUTERS WITH MEI'S "04" COMPUTER SYSTEM

Hardware Purchases[**]

68 MEI 404 terminals @ $700 each	$47,600
3 MEI 504 portable computers @ $2050 each	6,150
1 MEI 804 CPU, including built-in software	13,200
2 MEI 704D dot-matrix line printers @ $800 each	1,600
2 MEI 704L letter-quality line printers @ $1700 each	3,400

Software Purchases - none necessary

Total Purchase Price:	$71,950

Hardware Sales (Estimated)

1 Complete Comron computer system (from item 2, above)	9,200
10 Vancourt 2200 terminals @ $550 each	5,500
1 Vancourt CPU 2000	3,200
1 Vancourt 2205 line printer	900
3 Kenta 405 personal computers @ $1500 each	4,500
1 Kenta 110 dot-matrix line printer	350
Total Sales:	$23,650
Total Cost:	$48,300

4. PURCHASING VANCOURT 2200 WORD PROCESSORS FOR
    THE ENGINEERING DEPARTMENT

Hardware Purchases

10 Vancourt 2200 video terminals @ $1600 each	$16,000
3 Vancourt 2200P portable computers @ $2700 each	8,100
1 CPU memory package VBS 2000M	1,950
1 Vancourt 2205L letter-quality line printer	2,200

Software Purchases - none necessary

Total Cost:	$28,250

** MEI equipment prices include 35% discount

2

Report 1.

Excerpts from
Formal Report 2
Selecting New Elevators
for the Merrywell Building

Before reading these excerpts, read the client's letter authorizing H. L. Winman and Associates to initiate an engineering investigation (see Figure 6-10). By comparing it with the conclusions and recommendations, you can assess how thoroughly Barry Kingsley (the report's author) has answered the client's requests.

COMMENTS ON REPORT

The summary is short and direct because it is written primarily for one reader: the President of Merrywell Enterprises Inc. It encourages him to read the report immediately, and to accept its recommendations, by offering the opportunity to save $50,000.

Although the background information contained in the first two paragraphs of the introduction seems to repeat details the client already knows, Barry recognizes he must satisfy the needs of other readers who may not be fully aware of the situation in the Merrywell Building. He then defines the purpose and scope of the investigation by stating the client's terms of reference in paragraph 3 of the introduction. (Note that he has copied them almost verbatim from Mr. Merrywell's letter.)

The conclusions present Barry's answers to Mr. Merrywell's four requests. Their order is different from that in paragraph 3 of the introduction because he has chosen to present the main conclusion first (in this case, the best combination of elevators that can be purchased within the stipulated budget), and to follow it with subsidiary conclusions in descending order of importance. Barry is aware that when using the alternative report format he must write conclusions that evolve naturally and logically from the introduction, *because his readers have not yet read the discussion.*

Barry uses the first person plural to open his recommendation because, although he alone is the report's author, he is representing H. L. Winman and Associates' views to the client.

The first two pages of the discussion have been included to show how, early in his report, Barry establishes criteria that will subsequently influence how he selects a combination of elevators that will best meet his client's needs. By carefully identifying the five criteria and describing why each is valid, he shows his readers the direction his report will take. (In later sections of his report—not included in the sample pages—he identifies various combinations of elevators that could be installed, and demonstrates which do or do not meet the criteria until he finally reaches an optimum configuration.)

MERRYWELL ENTERPRISES INC
617 Carswell Avenue
Montrose, Ohio 45287

File: WDR/71/007

April 27, 19xx

Mr. Ian Bailey, P.E.
Head, Civil Engineering
H. L. Winman and Associates
Professional Consulting Engineers
475 Reston Avenue
Cleveland, Ohio 44104

Dear Ian:

The elevators in the Merrywell Building are showing their age. Recently
we have experienced frequent breakdowns and, even when the elevators are
operating properly, it has become increasingly apparent that they do not
provide adequate service at the start of work, at noon, and at the end
of the working day. I have therefore decided to install a complete range
of new elevators, with work starting in mid-August.

Before I proceed further, I would like you to conduct an engineering
investigation for me. Specifically, I want you to evaluate the structural condition of my building, assess the elevator requirements of the
building's tenants, investigate the types of elevators available, and
recommend the best type or combination of elevators that can be purchased
and installed within a proposed budget of $500,000.

Please use this letter as your authority to proceed with the investigation. I would appreciate receiving your report by the end of June.

Regards,

David P. Merrywell, President
Merrywell Enterprises Inc

DPM:tk

Figure 6-10. A letter authorizing an investigation.

H L WINMAN AND ASSOCIATES

(Sometimes called abstract)

A condensed version of the actual report

SUMMARY

The elevators in the 71-year old Merrywell Building
are to be replaced. The new elevators must not only
improve the present unsatisfactory elevator service,
but must do so within a purchase and installation
budget of $500,000.

Of the many types and combinations of elevators considered, the most satisfactory proved to be four
8 ft by 7 ft (2.45 x 2.13 m) deluxe passenger elevators manufactured by the YoYo Elevator Company, one
of which will double as a freight elevator during
non-peak traffic times. This combination will provide the fast, efficient service requested by the
building's tenants for a total price of $450,000,
which will be 10% less than the budgeted price.

i

Report 2: Summary page.

186

SELECTING NEW ELEVATORS FOR THE MERRYWELL BUILDING

INTRODUCTION

When in 1964 Merrywell Enterprises Inc purchased the Wescon property in
Montrose, Ohio, they renamed it "The Merrywell Building" and renovated the
entire exterior and part of the interior. The building's two manually
operated passenger elevators and a freight elevator were left intact, although
it was recognized that eventually they would have to be replaced.

Recently the elevators have been showing their age. There have been frequent
breakdowns and passengers have become increasingly dissatisfied with the inadequate service provided at peak traffic hours.

In a letter dated April 27, 19xx to H.L.Winman and Associates, the President
of Merrywell Enterprises Inc stated his company's intention to purchase new
elevators. He authorized us to evaluate the structural condition of the
building, to assess the elevator requirements of the building's occupants,
to investigate the types of elevators available, and to recommend the best
type or combination of elevators that can be purchased and installed within
the proposed budget of $500,000.

CONCLUSIONS

The best combination of elevators that can be installed in the Merrywell
Building will be four deluxe 8 ft by 7 ft (2.45 x 2.13 m) passenger models,
one of which will serve as a dual-purpose passenger/freight elevator. This
selection will provide the fast, efficient service desired by the building's
tenants, and will be able to contend with any foreseeable increase in traffic.
Its price at $450,000 will be 10% below the proposed budget.

Installation of special elevators requested by some tenants, such as a full-
size freight elevator and a small but speedy executive elevator, would be

1

Report 2: Introduction and Conclusions.

feasible but costly. A freight elevator would restrict passenger-carrying capability, while an executive elevator would elevate the total price to at least 20% above the proposed budget.

The quality and basic prices of elevators built by the major manufacturers are similar. The YoYo Elevator Company has the most attractive quantity price structure and provides the best maintenance service.

The building is structurally sound, although it will require some minor modifications before the new elevators can be installed.

RECOMMENDATIONS

We recommend that four Model C deluxe 8 ft by 7 ft (2.45 x 2.13 m) passenger elevators manufactured by the YoYo Elevator Company be installed in the Merrywell Building. We further recommend that one of these elevators be programmed to provide express passenger service to the top four floors during peak traffic hours, and to serve as a freight elevator at other times.

Report 2: Conclusions (continued) and Recommendations.

SELECTING NEW ELEVATORS FOR THE MERRYWELL BUILDING

Evaluating Building Condition

We have evaluated the condition of the Merrywell Building and find it to be structurally sound. The underpinning done in 1958 by the previous owner was completely successful and there still are no cracks or signs of further settling. Some additional shoring will be required at the head of the elevator shaft immediately above the 9th floor, but this will be routine work that the elevator manufacturer would expect to do in an old building.

The existing elevator shaft is only 24 feet wide by 8 feet deep (7.35 x 2.45 m), which is unlikely to be large enough for the new elevators. We have therefore investigated relocating the elevators to a different part of the building, or enlarging the existing shaft. Relocation, though possible, would entail major structural alterations and would be very expensive. Enlarging the elevator shaft could be done economically by removing a staircase that runs up the center of the building immediately east of the shaft. This staircase is used very little and its removal would not conflict with fire regulations since there are also fire staircases inside the east and west walls of the building. Removal of the staircase will widen the elevator shaft by 11 feet (3.35 m) which will provide just sufficient space for the new elevators.

Establishing Tenants' Needs

To establish the elevator requirements of the building's tenants we asked a senior executive of each company to answer the questionnaire attached as Appendix A. When we had correlated the answers to all the questionnaires, we identified five significant factors that would have to be considered before selecting the new elevators. (There were also several minor suggestions that we did not include in our analysis, either because they were impractical or because they would have been too costly to incorporate.) The five major factors were:

> Every tenant stated that the new elevators must eliminate the lengthy waits that now occur. We carried out a survey at peak travel times and established that passengers waited for their elevators for as much as 70 seconds. Since passengers start becoming impatient after 32 seconds[1], we estimated that at least three, and probably four, faster passenger elevators would have to be installed to contend with peak-hour traffic.

3

Report 2: Start of Discussion.

Although all tenants occasionally carry light freight up to their offices, only Rad-Art Graphics and Design Consultants Inc considered that a freight elevator was essential. However, both agreed that a separate freight elevator would not be necessary if one of the new passenger elevators was large enough to carry their displays. They initially quoted 9 feet (2.75 m) as the minimum width they would require, but later conceded that with minor modifications they could reduce the length of their displays to 7 feet 6 inches (2.3 m). All tenants agreed that if a passenger elevator doubled as a freight elevator they would restrict freight movements to non-peak travel times.

The three companies occupying the top four floors of the building requested that one elevator be classified as an express elevator serving only the ground floor and floors 6, 7, 8, and 9. Because these companies represent more than 50% of the building's occupants, we consider that their request is justified.

Three companies expressed a preference for deluxe elevators. Rothesay Mutual Insurance Company, Design Consultants Inc, and Rad-Art Graphics all stated that they had to create an impression of business solidarity in the eyes of their customers, and they felt that deluxe elevators would help to convey this image.

The management of Rothesay Mutual Insurance Company and Vulcan Oil and Fuel Corporation requested that a small key-operated executive elevator be included in our selection for the sole use of top executives of the major companies in the building. We asked other companies to express their views but received only marginal interest. The consensus seemed to be that an executive elevator would have only limited use and the privilege could easily be abused. However, we retained the idea for further evaluation, even though we recognized that an executive elevator would prove costly in terms of passenger usage[2].

We decided that the first two of these factors are requirements that must be implemented, while the latter three are preferences that should be incorporated if at all feasible. The controlling influence would be the budget allocation of $500,000 stipulated by the landlord, Merrywell Enterprises Inc. In decreasing order of importance, the requirements are:

1. Passenger waiting time must be no longer than 32 seconds.
2. At least one elevator must be able to accept freight up to 7 ft 6 in. long (2.3 m).
3. An express elevator should serve the top four floors.
4. The elevators should be deluxe models.
5. A small private elevator should be provided for company executives.

Report 2: Second page of Discussion.

Assignments

The four formal report assignments that follow require varying levels of technical knowledge and personal involvement in the decision making. Projects 1, 2, and 4 are multipart projects that require several short memos and reports to be written before the formal report is tackled.

PROJECT 1: A TRAFFIC STUDY FOR THE TOWN OF TARNAPIN

You are a recently graduated engineering technician employed by the local branch of H. L. Winman and Associates. Today is July 7, and this morning branch manager Vern Rogers calls you into his office and assigns a small project to you.

Part 1: Learning the Background Details. "I want you to take on this project for the resort town of Tarnapin," Vern says. "Tarnapin is a lakeside town in Tarnapin State Park, in (a neighboring state), just 15 miles over the state border. I have received a letter from the president of the Tarnapin Business Owners' Association—her name is Wendy Kibarre—and she has asked us to carry out a traffic movement study."

Vern hands you Wendy Kibarre's letter and suggests you read it before he describes the project any further. This is the letter:

Dear Mr. Rogers:

The Tarnapin Business Owners are concerned because the State Department of Highways and Traffic Control is planning to reroute Highway 271 so that it passes north of the resort town of Tarnapin. This, they feel, will significantly reduce business activity.

We believe that the rerouting plan has been based on an inadequate traffic assessment, and are asking H. L. Winman and Associates to conduct a two-stage study on behalf of the Tarnapin Business Owners' Association. Specifically, we are requesting a representative of H. L. Winman and Associates to:

1. Meet with me to gain initial information from which to assess the depth of study required.

2. Prepare a proposal outlining your study plan and the total cost.

3. On our acceptance of your proposal, conduct an in-depth study of traffic movements.

4. Prepare a report of your findings, complete with recommendations.
I look forward to meeting your representative.

Sincerely,

Wendy L. Kibarre
President, Tarnapin Business
Owners' Association

"What I suspect is happening," Vern suggests, "is that the Highways
Department wants to take the traffic away from the town because the highway
passes right along the lakeshore. I've been there, and if I remember correctly
Highway 271 becomes Lakeshore Drive as it passes through the town. People
wanting access to the beach along the edge of Lake Tarnapin have to cross
Lakeshore Drive."

Vern suggests you drive to Tarnapin on July 10. "Get all the information
you can," he says, "and then write me a trip report on your return. I would also
like to see your proposal before you submit it to Wendy Kibarre."

You drive to Tarnapin on the morning of July 10 and find that Wendy
Kibarre owns the Cottage Crafts Gift Shop at 119 Lakeshore Drive. You introduce yourself and Wendy takes you into her small office at the back of the gift
shop.

"The problem is a little more complex than I said in my letter to Mr.
Rogers," Wendy announces as she hands you the map in Figure 6-11. "The
dotted line traces the route for the highway bypass that the Department of
Highways and Traffic Control is proposing. You will notice that the proposed
highway passes through the tiny community of Aspen Heights. Unfortunately,
some of the Tarnapin business owners feel there has been some collusion between the Director of State Highways and Traffic Control and the principal
property owner in Aspen Heights."

"In what way?" you ask.

"The principal property owner in Aspen Heights is Vincent Warnock.
He also owns and runs the service station and general store. Vincent is married
to Julia Chendra, whose brother Jason Chendra is Director of State Highways
and Traffic Control. Quite likely the connection is a coincidence and there is no
conflict of interest, but nevertheless some of the Tarnapin business owners are a
bit suspicious."

You ask Wendy what the inadequacies were in the traffic study conducted previously, which she had mentioned in her letter to Vern Rogers.

"Oh, it's just that there seems to be too little information on which to
base such a major change," she replies, and pulls the map toward her and points
to two locations along Lakeshore Drive. "The measurements were taken at the
points I have marked as 'B' and 'D'," she says. "They simply took a traffic count
at each location, and then broke the total down into different types of vehicles.

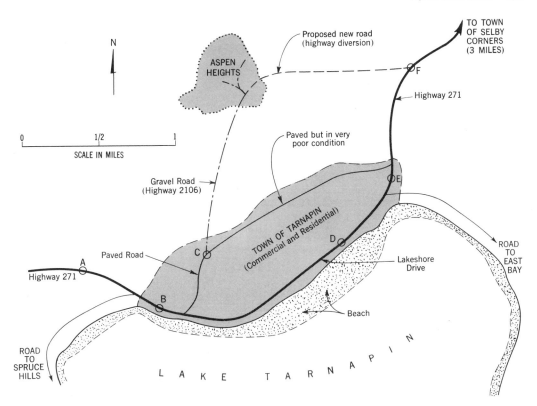

Figure 6-11. Map of resort town of Tarnapin.

No attempt was made to see if the traffic was as heavy on the approaches to Tarnapin, along Highway 271."

Before you leave, Wendy gives you a copy of the report prepared by the State Department of Highways and Traffic Control, for you to use as your starting point. You then drive around the small town and along the approach roads, and decide that for your study you would also want to measure traffic at points 'A,' 'C,' and 'E' so that you can assess traffic along the length of Highway 271 and along the road behind the town which, presumably, some people already use to bypass the built-up area. Then you drive back to your office.

Part 2. Writing a Trip Report. The next day you read the report Wendy Kibarre gave you. Titled "Report on Traffic Density: Tarnapin State Park," it is dated September 18, 19xx (last year). You note that its authors are Connie T. Warhaft and Ken A. Haaken, that the report bears identification number 83/20, and that it was issued by the State Department of Highways and

Traffic Control. There are three parts of the report which you mark with a yellow marker, considering them of importance to your project:

From Page 7:

Traffic movements measured at two points on Highway 271 within the town of Tarnapin in Tarnapin State Park show that through traffic, particularly at weekends, is singularly heavy along Lakeshore Drive, which passes through the resort and skirts the beach. Traffic counts show traffic movements of 4310 and 4285 vehicles per day on weekdays, and 7792 and 7621 vehicles per day on weekends. Assuming that most traffic movements occur between 7 a.m. and 11 p.m., this is equivalent to 140 movements per hour on weekdays and 440 movements per hour on weekends. This, we believe, is too heavy a concentration on a lakeshore road crossed regularly by families desiring access to the beach.

From Page 9:

A new highway built west from a point on highway 271 three miles south of Selby Corner to Aspen Heights, and then south to Tarnapin along existing gravel Highway 2106, would reroute through traffic away from Lakeshore Drive. We estimate that this action alone will reduce traffic along Lakeshore Drive by 70%.

From Page 12:

APPENDIX B

Traffic Count by Vehicle Types

A breakdown of traffic, by specific types of vehicles, for a one-hour period measured between 11:00 a.m. and 12 noon on Saturday, August 12, 19xx, shows the following distribution:

Vehicle type	Point B	Point D
No. of vehicles	533	528
Distribution:		
Automobiles	73%	71%
Recreational vehicles	11%	15%
Commercial vehicles	13%	11%
Two-wheeled vehicles	3%	3%

You feel that the study done by the State Department of Highways and Traffic Control was inadequate, that additional traffic counts need to be done at other points on the highway, and that a check of drivers' destinations should be made to give the traffic and vehicle counts more meaning. You decide to write a proposal to Wendy Kibarre, in which you will outline the scope of the project you will undertake for the Tarnapin Business Owners' Association and what the cost will be.

Write a trip report addressed to your branch manager (Vern Rogers), in which you describe the results of your trip and outline what you plan to do.

Part 3: Writing a Proposal.   After he has read your trip report, Vern Rogers discusses your plans and then you outline what you will write in your proposal to Wendy Kibarre. You tell Vern that you will measure traffic at five points, four along the highway and one on the road that skirts the small town. You also say that you will take a survey of vehicle drivers to determine their destinations, so that you can predict the routes they will take through the town. In particular, you want to assess whether they are going to drive right through or will be stopping in town to shop or go to the lakeshore. The points where you will be conducting your traffic survey are identified as A through E on the map.

You work out the time required as follows:

Technicians to be on site: 2 people, 4 days each
Technician to analyze results: 2 days
Technician to write report: 2 days
Draftsperson to prepare map: 2 days
Technician to check drawing and typing: 6 hours

Vern says to use a basic rate of $15.00 per hour for technician and draftsperson salaries. "And count a working day as 7½ hours," he adds.

You calculate travel and accommodation expenses to be $360.00.

"Calculate your total costs," Vern says, "and then add 30% to the total for H. L. Winman and Associates' fee. Just present the total dollar cost to your client; don't display how you reached the total."

Armed with this information you prepare a proposal for carrying out the study. Address it to Wendy Kibarre, as Chairperson of the Tarnapin Business Owners' Association. Prepare your proposal as a business letter and in it describe what you propose to do and what the total cost will be. Remember that the members of the Business Owners' Association will want to know that you are going to provide a more comprehensive survey than they have already received from the Department of Highways and Traffic Control, and they must be reasonably convinced that you will be providing more definitive information. They will be looking for value for the money you are asking them to spend.

Part 4: Carrying Out the Investigation.   You and Frank Wizniak drive to Tarnapin on the evening of Wednesday, August 8, 19xx (this year). You are the senior technician/team leader, and Frank is your assistant technician. You meet Wendy Kibarre and explain that from Thursday to Saturday you will be taking automatic and visual traffic counts, and will also be carrying out a destination study by stopping vehicles and asking each driver to tell you his or her destination.

First, late on Wednesday evening, you set up five automatic counters, one at each of the locations identified on the map as points A through E. At each location you tack down a rubber-covered cable from one side of the road to the other, and connect it to a Vancourt model 2120 counter. Your plan is to count traffic over two 24-hour periods on Thursday August 9 and Saturday August 11. (You are treating Monday through Thursday as light traffic days, and Friday through Sunday as heavy traffic days.)

As Saturday is the heaviest traffic day, you plan to count various vehicle types passing each of these points for one-hour periods between 11 a.m. and 5 p.m., and treat them as being typical distributions. Both you and Frank will carry out the traffic counts, one sitting on either side of the road at each point, and then average your results.

The third test will be to question drivers at locations B and D to determine their immediate destinations. You also will do this twice, on Thursday and Saturday. Since it would be impractical to stop all traffic, you plan to take a random sample of 200 vehicles travelling in each direction.

Your Thursday count of vehicles, which you read at midnight from the automatic counters, provides the following statistics:

```
Point A - 1425 vehicles
Point B - 4456 vehicles
Point C -  620 vehicles
Point D - 4341 vehicles
Point E - 1926 vehicles
```

Your Saturday visual count of vehicles at the five measurment points shows the following distribution (these figures are the averages of your and Frank's counts):

Location:	A	B	C	D	E
Automobiles:	101	444	40	405	182
Recreation Vehicles:	9	53	3	79	39
Commercial Vehicles:	18	70	17	56	29
Two-wheeled Vehicles:	3	18	1	22	10
Start of Measurement:	11:00	12:15	4:00	1:30	2:45

The Saturday count of vehicles crossing the automatic recording point shows a signifcant increase over the Thursday count:

```
Point A - 1968 vehicles
Point B - 7880 vehicles
Point C -  734 vehicles
Point D - 7736 vehicles
Point E - 3844 vehicles
```

The drivers you stop and interview at points B and D give the following answers to your question: "What is your destination right now?"

POINT B	Thursday	Saturday
WESTBOUND		
Highway 271	53	68
Spruce Hills	147	132
EASTBOUND		
Tarnapin	165	144
Highway 271	27	34
East Bay	8	22
POINT D	*Thursday*	*Saturday*
WESTBOUND		
Tarnapin	155	141
Highway 271	26	37
Spruce Hills	19	22
EASTBOUND		
Highway 271	94	126
East Bay	106	74

Part 5: Writing the Formal Project Report.   When you return to your office on Monday morning, August 13, you start using the data you have acquired to form a traffic pattern. From this data you determine whether or not the recommendations of the previous report were valid, and whether traffic should be rerouted. If rerouting is important, then there are three possibilities:

1. Building the highway recommended in the State Department of Highways and Traffic Control report. The highway would travel from point F on the map, west through Aspen Heights, and then south to point C. The cost would be $4,700,000 (this cost figure was quoted in the previous report).
2. Rebuilding the road from point E to point C, at an estimated cost of $780,000. The road is currently in bad shape and needs extensive work. You received a rough estimate of the repair cost from Ozzie's Road Construction Service of Selby Corner.
3. Building a new stretch of highway from point C to point A, at a total cost of $1,650,000. You also received this rough estimate from Ozzie's. The two quotations are in a letter dated August 14, 19xx (this year), which is signed by Ozzie Brehaut, owner-manager.

Now write the report, making sure you not only present all the facts but also provide a sound analysis of the results. Remember, too, that Wendy Kibarre and the members of the Tarnapin Business Owners' Association will be expecting you to draw conclusions and make recommendations.

You also have to include a list of references. One reference will be the

State Department of Highways and Traffic Control report provided by Wendy Kibarre. Another you examined during the study was issued by the Federal Parks Administration and describes traffic studies taken along the lakefront at Spruce Lake, Minnesota, in Spruce Woods National Park, during the summer of 1985. The report is titled "Traffic Study at SWNP, July 1985" and is authored by Dennis K. Walkicz and Andy G. Horoschuk. The report bears number 1985/22 and is dated September 23, 1985. You used the report as a guide for conducting your traffic measurements.

PROJECT 2: COPING WITH A NOISY NEIGHBOR

Part 1.   You are an engineering technologist in the Cleveland office of H. L. Winman and Associates. This morning Department Head Peter Bell calls you into his office.

"I've had a letter from the area manager of Mansask Insurance Corporation," he explains. "His staff are complaining it's too noisy in the office—that they get headaches from it and go home tired."

Peter hands you the letter, dated yesterday:

MANSASK INSURANCE CORPORATION
210 - 381 Fort Street
Montrose, Ohio 45287

Mr. P. Bell, P.E.
H. L. Winman and Associates
475 Reston Avenue
Cleveland, OH 44104

Dear Mr. Bell:

My staff complains that for the past three months the noise level in our office has been too high. They claim that it is affecting their work and causing fatigue.

Will you please look into this problem for me to see if their complaints are justified. If they are, will you suggest what can be done to remedy the problem, recommend the most suitable method, and include a cost estimate.

Yours truly,

David L. Wizowaty
Area Manager

"I also talked to him briefly on the phone," Peter continues. "He figures the staff are exaggerating a bit, that the noise isn't nearly as bad as they make out. Just the same, he's worried: staff turnover has been much higher recently, and it's costing him a fortune to train the replacements he has to hire."

Peter asks you to take on the project and assigns it project number C62.

At 4:00 p.m. the same day you visit Mansask Insurance Corporation and talk to David Wizowaty. You notice a background hum, which you consider to be caused by motors in the electric typewriters and accounting machines. You are still there when the office staff quits at 4:30. After they go, you notice you can still hear the hum, but at a lower level.

You walk around the office with Wizowaty, who plagues you with questions: "What do you think?" he asks. "Seems like the same noise level you get in any business office. Eh?"

It's apparent he is hoping for a good report from you, which he can use to prove to his staff that their complaints are imaginary.

"I can't tell without taking readings," you hedge. "Noise is a pretty tricky thing. What some people think is too noisy, others hardly notice."

But you do notice that the hum gets significantly louder near the east wall of the office. Then suddenly it stops; or, rather, dies away. The time is 4:45.

"What's on the other side of this wall?" you ask.

"Oh, that's Artistic Novelties," Wizowaty replies. "They distribute cheap imports—that sort of thing."

"Have you talked to them about the hum?"

"Yeah! I asked Frank Doherty about it—he's the manager next door. Pretty hostile, he was."

"And what time do they quit work?" you ask.

"Right now," Wizowaty replies. "You can always tell, because they switch their machines off."

You arrange with Wizowaty to take sound-level measurements one week from today. You want to find out how much of the noise in the insurance office is generated by normal office activity and how much by the machines next door.

You consider a visit to Artistic Novelties is essential, since you think that the machines next door offer the only real problem. You want to know the sound levels on both sides of the wall between the two companies, to assess the extent of soundproofing you probably will recommend.

Write to Frank Doherty, Manager of Artistic Novelties, to ask permission to carry out sound-level measurements one week from today. His business is at room 208, 381 Fort Street.

Part 2. It is now one week later. Taking a GenRad 1551-C Sound-Level Meter with you, you return to 381 Fort Street. You plan to measure sound levels at various locations in the Mansask Insurance Corporation offices under four circumstances:

When both businesses are empty.
When only Artistic Novelties is working (4:30–4:45).
When only Mansask Insurance is working (8:00–8:15).
When both businesses are working.

You also plan to take readings in Artistic Novelties' office.

As Mr. Doherty has not replied to your letter, yesterday afternoon you telephoned him to ask if you could come in today to take the measurements. He said he was "terribly busy" and that it's "damned inconvenient," but he somewhat reluctantly agreed.

You record the measurements you take (see table), and compare them to the general ratings for office noise, which you obtain from City of Montrose standard SL2020, dated January 13, 1986. The recommended sound levels for an office are:

Quiet office: 30–40 dB
Average office: 40–55 dB
Noisy office: 55–75 dB

You note that the sound level in the Mansask office increases as you move toward the dividing wall between the two offices (see Figure 6-12).

You also notice there seem to be two components of noise in Mansask's offices, some being transmitted through the air and some being transmitted through the structure (from Artistic Novelties' machines, through the floor). A hand placed on the walls or floor feels the vibration. Floors in both offices are tiled.

Before leaving Mansask you tell David Wizowaty that there seems to be a

Figure 6-12. Plan of Mansask Insurance Corporation's Office.

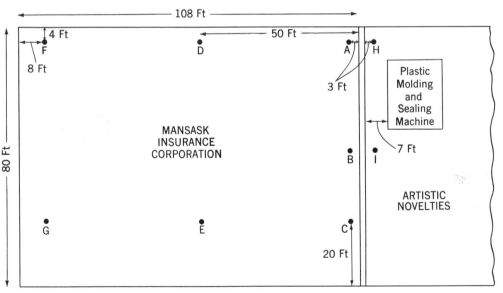

● = Points at which sound levels recorded

noise problem, but it can be corrected. You warn him, however, that it may prove expensive. He says he will have a job justifying the costs to his head office.

Appendix B

AVERAGE SOUND LEVELS SECOND FLOOR, 381 FORT STREET, MONTROSE, OHIO				
Mansask Insurance Corporation	Both Offices Working	Only Artistic Working	Only Mansask Working	No One* Working
A	74 dB	73 dB	48 dB	27 dB
B	71 dB	69 dB	51 dB	27 dB
C	66 dB	65 dB	50 dB	27 dB
D	64 dB	61 dB	52 dB	26 dB
E	63 dB	59 dB	51 dB	28 dB
F	59 dB	53 dB	49 dB	26 dB
G	58 dB	51 dB	48 dB	28 dB
Artistic Novelties				
H	86 dB	—	—	25 dB
I	83 dB	—	—	26 dB

*Mostly air conditioner noise.
Note: Measurements made with GR 1551-C Sound-Level Meter set to "A" scale.

On returning to your office you write a progress report for Peter Bell. (To do this, you may want to use some of the information in Part 3).

Part 3. You consider possible ways to reduce the sound levels in Mansask Insurance Corporation's offices:

1. Erect a false wall, insulated internally with Corrugon, from floor to ceiling on Mansask's side of the wall between the two companies.
2. Glue ¾ in. black cork panels on the Mansask side of the wall.
3. Install carpet throughout the Mansask office.
4. Mount Artistic Novelties' molding machine on Vib-o-Rub (insulating rubber that eliminates transmission of vibration from machine to building structure).

You recognize that remedies 1 and 2 are alternatives (they both deal with sound transmitted through the air). Remedies 3 and 4 also are alternatives (they both dampen vibrations and sound carried through the structure). Remedy 3 also quite effectively dampens internal office noise.

You consider the approximate costs:

Remedy 1—$6,800
Remedy 2—$1,400
Remedy 3—$13,800
Remedy 4—$800

You consider possible problems each remedy may present:

1. Corrugon is in short supply; delivery time would be a minimum of 3 months.
2. To some people, cork has an offensive smell; this can be partly corrected by treating the cork with polymethynol.
3. The carpet must be dense and have a good quality rubber underlay (included in the approximate cost above).
4. Depends on cooperation of Artistic Novelties' manager.

You calculate probable noise reductions for each method:

Remedy 1: 6 to 10 dB
Remedy 2: 4 to 7 dB
Remedy 3: 8 to 12 dB
Remedy 4: 3 to 5 dB

(These anticipated reductions apply to the Mansask office only, when both businesses are working.)

You consider which alternatives to recommend to Mansask Insurance Corporation and then write an investigation report describing your findings and suggested corrective measures. You also write Mr. Wizowaty a brief cover letter summarizing your findings. Prepare the letter for Peter Bell's signature.

NOTE: You prepare a *formal* report because David Wizowaty has mentioned he might have difficulty convincing his head office executives that they must authorize the cost of the remedial measures. And, because David Wizowaty knows little about noise and its effects, you decide to include some explanatory information. (You would also be wise to research and document such information at a learning resources center, to establish positive evidence for the statements you make in your report.)

PROJECT 3: BUILDING A TOOTHPICK TOWER

A local organization known as "The Association of Structural Engineers Inc" (ASEI) has offered an award to students of the course you are taking. The President of ASEI (Frederick K. Bartley, P. E.) has written to the head of your English/Communications Department to announce the award. His letter says (in part):

> . . . it is to be known as the ASEI Communication Award and it will be presented annually. It will be an engraved plaque displaying the award winner's name, the name of the educational institution, and the year. The winner will be the student who shows the most ingenuity in completing a project we assign and concurrently displays the greatest competence in technical communication and report writing.

You are to compete for this year's award, for which ASEI has published the following requirements:

You are to design, build, and test to destruction a structure that supports the maximum weight for the lowest possible price. Specifications for the structure are:

1. It is to be built solely of wooden toothpicks and glue. Only flat style toothpicks are to be used and they are to be referred to as "structural members." The type of glue to be used is at the student's discretion.
2. The top of the structure must be a level surface on which the load will be exerted. It may be either:
   (a) a horizontal $5 \times 3 \times \frac{1}{4}$ in. wooden platform, the top surface of which must be no less than 10 in. and no more than 12 in. above the base of the structure, or
   (b) a horizontal bridge which spans two supports placed 12 in. apart.
3. There must be no protrusions above the platform or bridge's upper surface, to permit a load to be placed or exerted on it.
4. The cost of the structure is to be calculated according to the number of structural members used. You are to assume that each structural member costs $60. (Short portions of structural members that are half or less the length of a full member may be assumed to cost $30.)
5. An economy vs. load factor is to be calculated in dollars per pound of maximum load supported. For example, if 200 structural members are used to build a structure that supports 18 pounds before it fails, the structure can be rated at $666.67 per pound of load. This figure is derived as follows:
   No. of members $\times$ cost per member $\div$ maximum load supported
   or: $200 \times \$60 \div 18 = \$666.67$/lb.
6. Since the objective is to build a structure which will withstand the highest possible load for the lowest possible price, the winning design will be that which achieves the lowest cost per pound of load.

At the end of the project you are to prepare a formal report that describes the project and the results achieved, and suggests what you could have done to obtain a better cost/load ratio.

Some of the topics you should consider when planning and writing your report are:

Why the project was undertaken.
Your terms of reference.
The different types of structure that could be used, and the advantages and disadvantages of each.
Why you chose a particular design for your structure.
The effect that striving to achieve an optimum strength/cost ratio had on your design decision.
Whether a designer should aim for high strength at high cost, or low strength at low cost, to achieve this optimum ratio.

The types of glue available, their influence on structure strength, and why you selected a particular type of glue.

How the structure was built.

How the structure was tested.

How and why the structure failed.

The results of the test, and how the cost/load factor calculations were made.

How the structure's strength could have been increased without an equivalent increase in cost.

Your conclusions on the success achieved by your structure, and some recommendations for improving it.

An unsuccessful toothpick tower.

Your instructor will inform you of any other requirements and where, when, and how your structure is to be tested. The photograph shows a previous (not very successful) design.

PROJECT 4: SHAVING THE PEAK

You are employed by the local branch of H. L. Winman and Associates and your branch manager is Vern Rogers. This morning Vern approaches you carrying a letter.

"I've had both a telephone call and this letter from Arnold Grabowski," he says, "He's plant manager for Broderick Metal Products Inc. in Crestview. He's looking for ways to cut electric power costs and wants us to make suggestions. Will you go over there and see what can be done?"

First you read the letter:

BRODERICK METAL PRODUCTS INC.
216 Garrick Crescent
Crestview

Dear Mr. Rogers:

As I mentioned to you on the telephone yesterday, I am asking your firm to conduct a small study into our operations to determine in what ways we can reduce our electric power consumption. The nature of our manufacturing cycle requires that we occasionally experience an unusually high power demand, and these peaks are setting the level of our monthly bills. I want you to identify areas in which we can economize on our overall power consumption, and so reduce the occasional high peaks.

Please use this letter as authorization for H. L. Winman and Associates to carry out the study.

Sincerely,

Arnold T. Grabowski
Plant Manager

Vern explains the billing procedure used by Crestview Power Company (CPC) for large commercial customers, manufacturers, and industrial facilities. (CPC is the local power utility.) "Commercial and industrial customers are billed in two ways," he says. "First they are charged for the total power consumption (in kilowatt hours) recorded during the month."

"Like a private homeowner?" you ask.

"Exactly," Vern replies. And then he says that commercial and industrial customers also pay an additional charge based on the peak demand (in kilowatts) that the customer draws during the month. "This can be a significant additional charge," he continues. "Suppose, for example, that a company's average demand during a month is 3000 kW (kilowatts), but just once during the month an emergency occurs which causes the company to draw 6000 kW for 28 minutes. The company would then be charged as though its demand had been 6000 kW for the whole month."

"That seems a little unfair," you say.

"Not really," Vern replies. "You have to remember that the power utility—CPC in this case—has to keep that reserve power *available at all times,* so that it can continually meet peak demands. And it costs the utility a lot of money to build and operate sufficient generating plants to meet such demands."

The charges for peak demand power are often greater than the charges for normal power consumption, Vern explains. Additionally, they increase exponentially as a series of "steps," so that the higher steps cost progressively more than the lower steps. And although locally the peak demand applies only to the month in which the demand occurs, in some states the peak demand can apply to the next three, six, or even 12 months. "So you can see why Mr. Grabowski is eager to reduce his plant's overall consumption."

Part 1: Learning the Background Details. You call on Mr. Grabowski two days later and he describes the Broderick Metal Company Inc. plant to you.

"Garson Broderick started the company in 1919, right after the first world war," Mr. Grabowski says. "He's dead now, but it's still a family business run by his daughter Mary and son Carl. There are 520 employees in the plant and we make all kinds of metal fixtures, cabinets, machine parts, tools, and so on, under contract to various other companies. We're also into fiberglass extrusions. Much of the equipment we use is pretty sophisticated, which almost makes us a high-tech company. And some of it draws a lot of power."

He explains that the company tries to schedule its manufacturing operations so that the equipments that draw the most power are not used at the same time. "We have a complex manufacturing plan that brings in each major equipment in rotation, or during a separate shift, and it works well most of the time. But once in a while—about every ten days or two weeks—we have to bring several major equipments into use at the same time. There is no other way to get around it if we are to meet our manufacturing contract dates." He explains that most contracts carry an expensive penalty clause for late delivery. "The penalty is usually costlier than incurring a peak load," he adds.

"What happens if Crestview Power Company has a power outage during a severe storm?" you ask.

"We have a standby generator for emergencies. It can deliver 120 kW, just enough to keep some of the primary manufacturing equipment going," he explains. "We shut down absolutely everything else. Fortunately it happens rarely, and even then for only a few minutes. We've yet to experience a serious outage."

You describe your plan to Mr. Grabowski. "I want to tour your whole plant, to determine where economies can be made. Will there be any difficulty getting into any of the shops?"

"No," he replies. "I'll give you a memo authorizing you to go everywhere, and requesting cooperation of the area supervisors."

"That will be great!" you say. And you ask him to provide some power consumption information. "I'll need two sets of figures: a list of everything that draws power, showing the average each equipment draws and the peak it can draw if pushed to the limit; and a short list showing the *average* power you draw at any time during a normal day, and the peaks you hit and end up being charged for as your monthly power demand."

Your tour of BMPI quickly shows that the plant is being run economically. In the manufacturing departments you notice no excess power being drawn unnecessarily and that the departments work on a load-sharing basis: machines are batched in threes and the operators check with each other so that only two machines operate simultaneously. (For example, they "stagger" their lunch hours and morning and afternoon breaks as part of their operating arrangements.)

In the office area you notice 1.5 kW electric heaters being used as "toe warmers" by employees whose working positions are along the north wall. (Some rearrangement of the working stations, away from the north wall, you feel would save 6 kW of power.) And some of the office machines seem to be left on unnecessarily when no one is working with them. (Possibly a small saving here, you think; about 2 kW.)

You examine the number of overhead lighting fixtures to identify possible further savings, but see immediately that considerable "pruning" has been done. In the office area every third fluorescent fixture has been disconnected, and in halls and passageways every second fixture has been disconnected. Only in the shipping and stores departments are all the fixtures still lit. You estimate some 30 fixtures could be disconnected for a saving of 3 to 4 kW.

Within the manufacturing departments some pruning has also been done, but further fixtures could be disconnected or some fluorescent tubes could be removed for a total saving of about 10 to 12 kW. This is particularly true of areas where machines are equipped with an operator's lamp.

"You run a pretty tight ship here," you announce to Mr. Grabowski when you return to his office. "There are very few places where I can suggest savings."

"We've been chipping away at cutting our power consumption," he admits. "But not nearly enough. That's why we had to call in H. L. Winman and Associates. The task has grown beyond our capability."

He hands you several sets of figures. The first is a list of all the equipments or operations that draw electric power (see Table 1). "As you can see," he says, "if all the equipment is operating in a normal way, our demand would be an average 656 kW. In practice we don't operate at that level, since only rarely is all equipment being used at once. Our normal average demand is about 520 kW."

"We have the potential to demand as much as 900 kW," he continues, "if every piece of equipment is operated at peak capacity. In practice we never reach

TABLE 1
List of all Equipment Drawing Electric Power
at Broderick Metal Products Inc.

	Power Consumption (kW)	
	Average	Maximum
Manufacturing Equipment		
Lathes and drills	46	70
Plastics extruders	66	96
Metal presses	90	140
Metal shapers	64	87
Paint drying chamber	60	92
Metal forge	20	38
Metal cutters	24	38
Packaging machines	18	25
Miscellaneous equipments	30	44
Total	418	630
Manufacturing Support Equipment		
Packing department	32	44
Test and inspection department	16	20
Total	48	64
Plant Support Facilities		
Office equipment	25	28
Heating*	12	12
Air conditioners†	28	28
Lighting	125	140
Total	190	208
PLANT TOTAL	656	902

* Auxiliary heating only; plant heated by natural gas
† Summer only

Table 2
NORMAL AND MAXIMUM POWER DEMANDS
AT BRODERICK METAL PRODUCTS INC.

Normal demand (average): 520 kW
Peak demand (average):   715 kW*

* Peak demand occurs approximately once
every 10 to 12 days.

that level, and never will. Our actual peak demands, which seem to occur roughly every 10 days, are much nearer 700 to 730 kW. I've shown the average regular demand and peak demand on this sheet." And he hands you the sheet (shown as Table 2.)

"All right," you say, "let's convert that to dollars and cents." From your pocket you take a folded sheet which you obtained from CPC before coming to BMPI. Titled CRESTVIEW POWER CO. ELECTRIC POWER DEMAND CHARGES, it contains a long list of the "steps" for each power demand level. The list is preceded by a short introductory paragraph:

> The monthly power demand charge levied on commercial and industrial customers is based on the peak demand drawn by the customer during each one-month billing period, and costed as shown below.

You identify the section of the list that applies to the power demand drawn by Broderick Metal Products Inc.:

Peak Demand (kW)	Cost/kW ($)
501–550	1.40
551–600	1.50
601–650	1.60
651–700	1.75
701–750	1.90
751–800	2.10

You take out your calculator and work out what the monthly demand charge would be if BMPI was able to maintain only its average demand (520 kW), and what they are being charged because the company has an occasional peak of 715 kW.

"It's more than double!" you exclaim, and you make one more claculation. "Why, the 715 kW demand peaks are costing you an additional $7566 a year!"

"And that's exactly why I need your help," Mr. Grabowski says. "I'd like to reduce the peaks and save as much as possible of that money. I can find a much better use for it!"

You explain to Mr. Grabowski that reducing the power demand by trimming consumption in various areas of his plant is not going to be enough in itself. "You have already cut consumption almost to the bone using routine methods," you say. "Yet there is a possibility you could use your standby generator to take up part of the peak load. I'd like to look into it for you. But first I should warn you that it may be more expensive than you expect. At this stage I just don't know."

"Why don't you write me a proposal," Mr. Grabowski suggests. "Outline what you plan to do and what it will cost, and then I can take it from there."

Part 2: Writing a Trip Report. On returning to your office you immediately go down to the library area, sit at the videotext terminal, and type in a request for a search of current information on "power shaving." The screen lists several articles, from which you identify the June 1984 issue of Plant Management & Engineering as a magazine carrying relevant information. The screen also identifies the magazine as volume 43, Issue 5, and types the title of the article as: PEAK POWER SHAVING WITH ON SITE GENERATORS, written by Dennis Pybus. You request a print of the article, which is delivered to you two minutes later.

From the article you learn that your assumption is correct: it will indeed be feasible for BMPI to use the company's standby generator during periods of peak demand. You also discover that:

1. It's not simply a matter of turning the standby generator on and phasing it in. The phasing-in has to be done very carefully and with special switching equipment.
2. There are two phase-in methods that can be used, each with its advantages and disadvantages.
3. The standby generator preferably should handle a constant load, such as lighting and office equipment, rather than a manufacturing load which is likely to fluctuate.

Next you telephone the Crestview Power Company engineering department and speak to Marvin Renato, who is the industrial service engineer.

"We encourage peak load shaving," he replies in answer to your inquiry. "It permits us to reduce the total power capacity we have to generate. But we have to inspect and approve each installation, to ensure that improper phasing does not damage either the customer's generator or our transformers."

Marvin gives you the names of two companies which are already using standby generators to cover peak demand, each using a different method. To see the peak load transferring method in use, he suggests that you visit Midwest Manufacturers on Cormorant Road. But to see utility paralleling in use you will have to go to the Walston Paper Mill in Norville, in a neighboring state, because no one has tried utility paralleling locally.

You feel you now have enough information to be able to propose an investigation project to Broderick Metal Products Inc. First, however, you want to get Vern Rogers' approval to go ahead.

YOUR ASSIGNMENT: Write a memorandum-form trip report to Vern Rogers. Briefly describe your visit to BMPI and your findings. Tell him you have a solution, that you need to do some research, and that you will be writing a proposal to Mr. Grabowski. Ask for Vern's assistance in working out the proposed costs.

Part 3: Writing a Proposal. Vern Rogers has your trip report in his hand when you call into his office a few days later. "It seems to me you have a two-part proposal here," he announces. "The first part will be a study to find out how to phase in the standby power generator at BMPI. The study will end with a technical report describing what should be done. The second part will be a proposal for us to do the installation for them. We can work out a firm price for part 1 but not for part 2; that will go into the report you write at the end of part 1."

You and Vern discuss what the proposal should contain. "I guess I need to tell Mr. Grabowski that power can be saved by using his standby generator," you say, "but that we need to carry out a study to determine how it can be done and what the cost will be."

"And you should give him an idea of what the power saving will be," Vern adds. "In both kilowatts and dollars per month."

"I have already calculated the power saving," you reply. "By using the standby generator and making certain economies they could shave the peak by about 150 kW; that is, from 715 kW to 565 kW. I haven't calculated the saving in dollars yet, but I will do so and then quote it in the proposal."

For the proposal, you work out the study costs based on a total 70 hours of your time at $25 per hour:

For work done so far:	6 hours
For investigation research:	
General research:	16 hours
Visit to Midwest Manufacturers:	5 hours
Visit to Walston Paper Mill:	18 hours
Report preparation:	20 hours
Further consulting with client:	5 hours

And then you add $200 anticipated expenses for travelling to Norville.

"All right," Vern says, when you show him your calculations. "Now, in your proposal you should also give him a ballpark figure for the conversion costs."

You tell him that Marvin Renato at Crestview Power Company said BMPI is probably looking at $5000 to $7000 in conversion costs, depending on whether they use the *peak load transferring* or the *utility paralleling* method.

"You should also tell him roughly how long it will take BMPI to recover their investment," Vern adds. "That is, how many months or years it will take for the money saved each month to equal how much it will cost for us to do the study and for them to install the conversion equipment. Of course, you can give Mr. Grabowski only an estimate in your proposal, but when you write the report at the end of your study you will be able to give him an exact cost recovery time."

You work out a schedule for the project:

Activity	Days required
Initial research	3
Visit to Midwest Manufacturers	1
Visit to Walston Paper Mill	2
Visit to CPC	1
Evaluation of results	2
Writing report	3
Typing and proofreading	4
Total number of working days	16

(The number of days does not equal the number of hours you calculated earlier because you will not be working full time on the project.)

YOUR ASSIGNMENT: Write a letter-form proposal to BMPI in which you: (1) tell Mr. Grabowski what you propose to do; (2) describe the gains to be achieved, expected problems, and the cost; (3) include a time schedule; (4) outline briefly what your report will contain; and (5) request approval for H. L. Winman and Associates to carry out the study. Assume that Vern Rogers's name will appear in the signature block at the foot of the letter, but that you will sign the proposal for him.

Part 4. Investigating the Alternatives. Mr. Grabowski approves your proposal and mails purchase order No. 11583 to H. L. Winman and Associates, authorizing you to carry out the study " . . . at a cost of $1950. The cost of the proposed equipment purchase, installation, and testing must fall within the range of $5000 to $7000."

When Vern Rogers assigns the study to you, he suggests you do it in three stages:

1. An initial discussion with Crestview Power Company to determine the utility's requirements.
2. Discussions with the plant engineers at Midwest Manufacturers and Walston Paper Mill, to identify what using each type of phase-in system involves, both in equipment and procedures.
3. An evaluation of the alternative systems to determine which would be most suitable for BMPI.

Your discussion with industrial service engineer Marvin Renato at CPC tells you only a little more than you learned earlier. "You will have to decide which method is most suitable for your client: peak load transferring or utility paralleling" he says. "Utility paralleling offers a few more advantages but demands more careful phasing-in and is more costly to install. When you have set up your plan, come and talk to me again because I'll need to approve it. And of course I'll have to inspect the installation and procedures when all the equipment is in place."

* * *

The plant engineer at Midwest Manufacturers is Chuck Chong, and the plant is at 414 Cormorant Road. Chuck suggests you spend an afternoon with him, going over their peak load transferring system.

"We've got a 100 kW generator," he announces when you arrive, "and it used to stand idle all the time, except when there was the occasional power failure. Now we use it about forty percent of the time, and the savings in power costs have been averaging $680 a month. The whole setup is automatic, and highly reliable."

Chuck explains that the key elements of the phase-in equipment are a power transducer and an automatic transfer switch. The transducer continuously measures the power "demand" (that is, the power in kilowatts the plant is drawing at any particular moment), and if the demand exceeds a certain level it sends a message to a control circuit which automatically actuates the power transfer switch. "In our case, we have set the cut-in level at 820 kW. As soon as the plant demands that amount of power, the transfer switch operates and the standby generator is immediately phased in. It stays phased in until the power demand drops below 820 kW."

"How does the system select which manufacturing equipment is to be fed by the standby generator?" you ask.

"It doesn't," Chuck replies. "With peak load transferring you cannot select a variable load such as one gets on a manufacturing line. It's much better to pick a stable load and let the standby generator consistently supply power to it."

"You mean, you can't just add the power from the standby generator to the power from CPC?" you ask.

"No way!" Chuck says, and he explains that with the load transferring method you *transfer* a particular load to the generator. "We use the plant lighting load, which is consistently between 90 and 95 kW. In that way we supply power to a continuously stable load and CPC supplies power to a continuously variable load."

Fuel costs to run Midwest Manufacturer's standby generator represent about 13 to 15% of the total savings in power costs. "That is," Chuck says, "each month we save about $800 in power costs, but it costs us about $120 for the fuel oil to run the generator."

* * *

The following afternoon you drive to Norville and check in at the Nor'Wester Inn, ready to spend all next day at Walston Paper Mill. The electrical engineering technologist at Walston is Carol Halaby.

"We use the utility paralleling method to replace some of the utility's power during peak demand periods," Carol explains. "We have a 24-hour production cycle, 365 days a year. In paper production we cannot afford a break in the production flow, so we have two standby generators, each capable of producing 90 kW."

In utility paralleling, she explains, the standby generator is run *in parallel* with the supply from the power utility. "We don't transfer a specific load to the generator," Carol says. "Instead we let the standby generators handle any load *in excess* of a predetermined level. In our case we occasionally have peaks as high as 1120 kW, but most of the time our demand is only about 950 kW. Consequently we have set 960 kW as the demand level at which the standby generator cuts in. It then provides the power for whatever the plant's demand is *above* 960 kW."

"Oh, I see!" you exclaim. "The power you draw from the utility remains stable once your demand reaches 960 kW, but the power you draw from your standby generators fluctuates depending on your total demand."

"Exactly," Carol replies. "It's the reverse of the setup you saw two days ago at Midwest Manufacturers. In peak load transferring the standby generator supplies power for a stable load and the power utility supplies power for a fluctuating load. In utility paralleling the utility supplies power for a stable load and the standby generator supplies power for a fluctuating load."

"As far as I can see," you say, "the results are different but the technology remains the same."

"Not entirely," Carol says. "For both methods you need a power transducer to measure the total power demand and an automatic transfer switch. But for utility paralleling you also need to control the governor and voltage regulator on your standby generator, because you are continually having to adjust the generator's output to suit the varying load required by your plant. Consequently the paralleling equipment is more costly than for peak load transferring. And synchronization with the utility's power supply is extremely tricky. It has to be set up very, very carefully, after which synchronization is automatic."

"Which does the power company prefer?" you ask.

"Utility paralleling," Carol replies. "Without any doubt, even though it means much closer coordination between the utility and the power user. They like utility paralleling because then they *know* our maximum demand will be 960 kW and they can plan on it."

Carol points out that fuel costs are higher than for a standby generator used for the peak load transferring method. At Walston Paper Mill the fuel oil costs represent about 22% of the power savings realized through peak shaving. "That is," she explains, "for the $1150 we save in power costs each month, we spend about $250 on fuel oil to run the standby generators."

* * *

When you talk to Marvin Renato (CPC's industrial service engineer) two days later, he confirms that CPC prefers utility paralleling even though it is more expensive to install, more difficult to coordinate the paralleling between the utility's supply and the standby generator, and more dangerous to use from a safety standpoint. "There's the possibility of personal danger to the operator and damage to our and the customer's equipment," he says. "The danger is low, however, if the equipment is properly installed and maintained. But so far no one in your area has tried it."

Marvin adds that CPC will be glad to work with H. L. Winman and Associates if Broderick Metal Products Inc. decides to go ahead with peak power shaving. "We encourage our commercial customers to use their standby generators to help even out the power demands we have to meet," he says, "regardless of whether they choose peak load transferring or utility paralleling."

"Have you got a copy of electrical specification AESC 342/85?" Marvin asks just before you leave. You say that you have not. "You'll need it," he continues. "It's titled Requirements for High Speed Power Transfer Switches. But I'll give you a bit of advice: the specification calls for a minimum of 2000 operations at 600 amperes. For your use you would be wise to increase that by 250%: to 5000 operations at 1500 amperes."

* * *

Having established that peak load transferring would be technically feasible for BMPI, you get initial cost estimates to determine whether the installation would be financially practicable. To your surprise and pleasure you discover that the cost will be some $300 to $400 less than the costs you quoted in your proposal to Mr. Grabowski.

At this point you are 10 days into the BMPI project. This morning you receive the following memo from Vern Rogers:

> Can you give me a brief progress report on the BMPI "peak power shaving" project? I'll be seeing Mr. Grabowski on another matter tomorrow afternoon, and it would be useful if I could tell him how the project is progressing.

YOUR ASSIGNMENT: Write a short memo-form progress report to Vern Rogers, in which you tell him your main findings (both technical and financial) and predict that your report will be ready to submit to BMPI 18 days from today.

Part 5: Preparing a Bibliography. You realize that in your report to BMPI you will need to refer to information you have gathered from other sources. So you jot down notes concerning each information source, from which you plan eventually to construct a bibliography. There are three persons from whom you gained spoken information:

1. Marvin Renato at Crestview Power Company (see parts 2 and 4).
2. Chuck Chong at Midwest Manufacturers (see part 4).
3. Carol Halaby at Walston Paper Mill (see part 4).

For each, you will have to work out the date you visited or spoke to that person. There are also five printed documents you refer to during the project:

1. Mr. Grabowski's letter (see start of project).
2. An article in Plant Management and Engineering (see part 2).
3. The specification for automatic transfer switches (see part 4). The specification was issued by the Associated Electrical Standards Committee (AESC) of Houston, Texas in May 1985.
4. An article in the weekly newspaper called The Financial Post of October 20, 1985. It refers to peak power shaving as an effective way to reduce electric power costs in states which are heavily dependent on generating power from fossil fuels (particularly coal, which has a dirty carbon effluent). The article is titled Using Natural Gas to Reduce Electric Power Costs and is authored by Cheryl Warkentin, a Financial Post independent correspondent. She draws attention to the Paragon Manufacturers and Distributors' plant in Ediston, Montana, which has just converted a 120 kW oil-fueled standby generating plant to natural gas, which will save $3800 annually. She comments that the generator could as easily be fueled by propane or methane gas.
5. A textbook by Douglas M. Horviss, printed in 1981, titled: "Electric Power Generators, Their Use and Purpose." Published by Aztec Press Ltd of Chicago, Illinois, it contains one chapter (Chapter 7) devoted entirely to synchronizing parallel generators, under the general heading of "Bringing Parallel Generators on Line." The relevant pages are 117 to 141. This is the second edition of the textbook.

YOUR ASSIGNMENT: Write a bibliography covering these eight information sources.

Part 6: Writing a Formal Project Report. You feel you have almost all the information you need to start preparing a project report for Mr. Grabowski. You decide to prepare a formal report rather than a letter report, because you realize Mr. Grabowski will most likely have to get the approval of the firm's principals (the owners) to spend the money needed to install equipment you are proposing. A formal report will be a much more convincing document for him to give them.

CALCULATING THE INSTALLATION COSTS

But you still have to get firm prices for purchasing, installing, and testing the phase-in (or generator transfer) equipment. So you prepare the attached specification sheet listing what needs to be done and mail it to three electrical contractors who indicate they would like to bid on the job.

The three bids you receive show that Mercury Electric Contractors Inc. of 1350 Pendlebury Road has the lowest price:

	Peak Load Transfer	Utility Paralleling
Mercury Electric Contractors Inc.	$3072	$4188
Mid-West Installers	$3381	$4865
Charles Hubert Electric	$3610	$5220

The prices Mercury Electric Contractors Inc. quote in their letter dated one week ago are listed below for each part of the work identified in the specification sheet.

Peak Load Transfer System		Utility Paralleling System	
1.1 (a)	440	2.1 (a)	1788
(b)	610	(b)	536
(c)	738	(c)	264
(d)	184	2.2	1600
1.2	1100		
	$3072		$4188

Next, you telephone Marvin Renato at Crestview Power Company and ask him if there will be a fee for inspecting and testing the finished work.

"There certainly will," Marvin replies. "Wait one minute while I check my fee schedule." He tells you that CPC's charge for inspecting and testing a peak load transferring system will be $300, and for the utility paralleling system it will be $700. "I'll send you a letter confirming these figures," he adds. When the letter arrives it is dated three days ago.

Finally, you calculate what H. L. Winman and Associates' fee will be for circuit design and project management (the time you will put in on the job). For the peak load transferring system the fee will be $900, and for utility paralleling the fee will be $1200.

You check your costs with Vern Rogers. "Your figures sound fine to me," he says. "But I think there are two other factors you have to consider: operator training, and annual service and maintenance of the system."

"BMPI's electrician will need to be trained to monitor the system and make minor adjustments," he continues. "For peak load transferring, training will cost only $400. But for utility paralleling the training will be more expensive, say $600. Training costs must be included as part of the total installation and testing prices you will be quoting in your report."

Vern explains that annual service and maintenance is not part of the total installation and testing cost, since it will not apply during the first twelve months (while the equipment is under warranty). "But in subsequent years there will be a cost, and this should be included in your report," he says. For the peak load transferring system he suggests an estimated annual service and maintenance cost of $300 and for the utility paralleling system an annual cost of $600.

(There is one other expense, which you cannot calculate until later. This is the cost of the diesel fuel to run the standby power generator. You plan to make these calculations when the total power savings have been calculated.)

CALCULATING THE SAVINGS

At this point you pay attention to the savings that BMPI will be able to achieve by peak load sharing. First, you calculate three ways for reducing the monthly power bill:

1. Phasing in the standby power generator at appropriate times will reduce the monthly peak demand by:
   * 105 kW if the peak load transferring method is used.
   * 120 kW if the utility paralleling method is used.
2. Making the economies you identified during your initial visit to BMPI will lower the peak load by a further 24 kW (see Part 1).
3. Rescheduling some of the manufacturing operations during peak demand periods will reduce the peak demand by a further 30 kW (see Note below).

This means the total peak load will drop from 715 kW to 556 kW (for peak load transferring) or 541 kW (for utility paralleling). Using the CRESTVIEW POWER CO. ELECTRIC POWER DEMAND CHARGES sheet (see Part 1), you work out the monthly power demand charge based on the basic rates of:

$1.40/kW for a peak demand between 501 and 550 kW
$1.50/kW for a peak demand between 551 and 600 kW

The results of these two calculations you deduct from the current monthly peak demand charge of $1358.50 (calculated by multiplying 715 kW by $1.90/kW).

Now you have two sets of figures representing the saving BMPI can achieve each month, one by peak load transferring and one by utility paralleling.

However, you discover there is an additional saving you had not previously considered. By using peak load shaving BMPI will also reduce the company's monthly power consumption bill (the kilowatt hours of electricity the company consumes each month). These kilowatt-hour-savings will be:

For peak load transferring:	$74/month
For utility paralleling:	$106/month

In each case you add these additional savings to the total savings you had calculated previously.

And, finally, from these total savings you deduct the cost of fuel oil which will be used to drive BMPI's standby generator, at 15% of total savings for peak load transferring and 22% of total savings for utility paralleling. Now you have your final actual saving that can be realized for each method.

NOTE: You work out a way for shaving an additional 30 kW off the peak load by ingeniously grouping machines that require similar power into three groups of six. For example, you group six machines that draw between 8 and 10 kW. Your plan is to bring each machine "on line" in sequence, so that each machine will operate for 50 out of every 60 minutes (thus, one of the six machines will stand idle for 10 minutes during each hour). Among the 18 machines, you form one group drawing 6 to 8 kW, another drawing 8 to 10 kW, and a third drawing 13 to 15 kW, for a total of about 30 kW. (You figure that in any hour an operator has to pause to study blueprints or arrange his or her work, and that these pauses could be limited to the machine's "off line" time once every hour.) Of course, this "sequencing" would occur only during the occasional peak demand period (about one day in every 10 to 12 days).

WORKING OUT THE COST RECOVERY TIME

Your last calculation is to work out how long it will take BMPI to recover its investment, using either the peak load transferring or utility paralleling method.

Vern Rogers says that you have to include the fee that H. L. Winman and Associates is charging BMPI (to conduct the study and write the report) as part of the project's total cost. Thus you have to add $1950 to the total installation expenses you have worked out ($1950 is the fee you quoted to BMPI in Part 3).

Your calculations for each peak load shaving method should look like this:

$$\left(\begin{array}{l} \text{Total cost} \\ \text{to install and} \\ \text{test equipment} \end{array} \right) + \left(\begin{array}{l} \text{Fee for} \\ \text{conducting} \\ \text{study} \end{array} \right) \div \left(\begin{array}{l} \text{Monthly savings} \\ \text{that will} \\ \text{be achieved} \end{array} \right)$$

The result will be the number of months that will elapse before BMPI saves enough electricity to cover the total cost of installing the equipment. (Sometimes this figure is referred to as the break-even point.)

WRITING YOUR REPORT TO BMPI

In your report to BMPI, Mr. Grabowski will expect you not only to present your findings but also to recommend which peak power shaving method you believe should be used. This means you will have to decide whether to recommend the more efficient (but also more costly and more technically difficult) method preferred by CPC, or the simpler, less efficient, and less costly method. You will also be expected to describe both methods in your report and to demonstrate why you are recommending one of them.

Your formal report should contain:

1. An introduction which describes the background, identifies the purpose, and outlines the scope of the project.
2. A discussion which contains descriptions of (not necessarily in the sequence listed here):
   2.1 Your approach to the project.
   2.2 What peak load shaving is.
   2.3 The two peak load shaving methods.
   2.4 What power cost saving can be achieved.

2.5 What the total costs will be.

2.6 A comparative analysis of the two peak load shaving methods.

3. Conclusions which summarize the main points you make in the discussion.

4. A recommendation which states what you think BMPI should do.

5. A list of *references* (endnotes), each keyed to a statement made in the discussion.

6. Appendixes containing specific technical details (tables, analyses, forms, etc).

YOUR ASSIGNMENT: Write the formal report and address it to Mr. Grabowski at BMPI. You are encouraged to research and use additional information, beyond that provided here, which will give greater depth to your project.

ATTACHMENT

SPECIFICATIONS FOR
SUPPLYING, INSTALLING, AND TESTING
POWER GENERATOR PHASE-IN EQUIPMENT AT
BRODERICK METAL PRODUCTS INC.

You are required to provide two bids, for supplying, installing, and testing: (1) a peak load transferring system; and (2) a utility paralleling system. Each bid is to cover the following:

(1) *Peak Load Transfer System*

1.1 Supplying the following equipment:

    (a) Double-pole, double-throw (DPDT) high speed switch with 6-cycle transfer time; 1500 Ampere; 8000 operations (minimum), to meet Spec C22.2, No. 178.

    (b) Power transducer, Camlock model T2820.

    (c) Computerized peak load demand control, Camlock model C1350.

    (d) Cabling and hardware.

1.2 Installing and testing the system.

(2) *Utility Paralleling*

2.1 Supplying the following equipment.

    (a) Items 1.1 (a), (b), and (c), above.

    (b) Controller and circuit breakers for controlling the governor and voltage regulator on the standby generator.

    (c) Cabling and hardware.

2.2 Installing and testing the system.

The prices you quote for each item must include federal and state taxes. There must also be a one-year warranty on all equipment and installation work.

7

Other Technical
Documents

In our technological era, with its increasingly complex range of equipment and processes, there is a growing need for manufacturers to write clear technical manuals to accompany their products. Most equipment is issued with a book of operating instructions that contains: (1) a brief description of the equipment; (2) instructions on how to use it; and (3) a list of likely replacement parts. Also available, but usually only to qualified repair specialists, is a set of maintenance instructions with detailed maintenance and repair procedures plus a comprehensive parts list describing every item in the equipment. Both publications perform the same task for a different type of reader. Operating instructions assume that the reader has only slight technical knowledge, whereas maintenance instructions assume that the reader is a technical expert.

This chapter describes how to write a technical description and instruction. It assumes that you will know your topic thoroughly and consequently may find it difficult to write in simple terms for someone who does not. It also offers suggestions for writing a scientific paper and for converting your knowledge of a process, equipment, or new technique into an interesting magazine article or technical paper.

TECHNICAL DESCRIPTION

The suggestions that follow apply to any technical description. Whether you are describing heavy construction equipment, a delicate instrument, a site plan for a building, or a schematic of an electronic circuit, you must organize your description so that it follows a distinguishable, coherent pattern.

WRITING PLAN

Most technical descriptions can be divided into three compartments:

1. A **summary statement** that briefly describes the equipment, view, process, or procedure, and states its main purpose:

> An overhead projector can project 8 in. × 10 in. film transparencies containing drawings and illustrations onto a screen, which normally is above and behind a speaker. The projector stands in front of or beside the speaker, so that the speaker can use the projector and view or point to parts of the illustration without facing away from the audience.

2. A list of **main parts,** presented in the sequence in which they will be described:

> The projector consists of three main parts: a base containing the light source, a ground glass plate for holding the transparency, and a lens assembly for projecting the light onto a screen.

If there are more than three parts, they may be presented in subparagraph or "point" form. There must be a logical pattern to the list, as described under **Spatial Arrangement** below.

3. A step-by-step **description** of the item or process:

> The light source is housed in a 12-inch square metal projection compartment and consists of a high-intensity bulb supported above a concave, highly polished mirror. The assembly includes a spare bulb, a cooling fan, and an interlock which disconnects the power when the cover plate that forms the upper surface of the compartment is opened.

> The cover plate is a metal frame attached by hinges to the back of the projection compartment, with an 11-inch square ground glass plate set into it. Suspended immediately above the glass plate is a lens-and-mirror assembly, which takes the projected light, magnifies it, and projects it onto a screen. The assembly is supported by an arm extending from a vertical post attached to the projection compartment. This arm is moved up and down the post to control focus, and the lens assembly is tilted to adjust the height of the projected image on the screen.

Notice particularly that a description *only describes;* it never tells the reader to do anything or to take specific action.

SPATIAL ARRANGEMENT

Suppose Martin Dawes asks you to write a description of a new business complex known as "Tower Twenty-One" which he can give to visitors to the construction site. The bottom five floors are ready for occupancy, while the remainder are in varying stages of completion, ranging from nearly ready at the sixth floor to the final concrete-pouring stages on the twenty-first floor. A logical approach would be to describe the floors in sequence from the basement up, or, alternatively, from the penthouse down. Readers would immediately understand your approach because it would be natural for them to think of a vertical building as a series of floors arranged in sequence from 1 to 21. But if you started by describing the bottom five floors, jumped to the top, then back to floors 10, 11, and 12, then up to 17 and 18, and so on, you would confuse them. They would find your description difficult to follow because it would be *incoherent*.

Coherence depends on arranging information in such an order that the logic is evident to the readers. There can be many patterns. A spatial arrangement (which means "arranged in space") depends on the shape of the subject:

Vertical Subjects	As indicated by the "Tower Twenty-One" example, tall, narrow subjects lend themselves to a vertical description, each item being described in order from top to bottom, or bottom to top. Vertically arranged control panels and electronic equipment racks fall into this category.
Horizontal Subjects	Long, relatively flat subjects demand a horizontal description, with each part being described in order from left to right, or right to left.
Circular Subjects	Round subjects, such as a clock, a pilot's altitude indicator, or a circular slide rule, suit a circular arrangement, with the description of markings starting at a specific point (for example, "12" on a clock) and continuing clockwise until the entire circle has been described. In effect, this is the same as the horizontal description, if you consider that the circle can be broken at one point and unwound into a straight line. Alternatively, if the subject has a series of items that are arranged more or less concentrically (such as the scales of a circular slide rule), they can be described in sequence starting at the center and working outwards, or from the outermost item inwards.

Regardless of the shape of the subject and the arrangement of its parts, there is almost always some pattern that can describe it. If it is not clearly left-to-right, top-to-bottom, or center-to-circumference, you may have to search for a logical arrangement. The irregular shape of the overhead projector described earlier poses such a problem. Here, the logical flow of information follows the flow of light, starting at the light source and following it through the glass plate and transparency to the lens assembly. Similarly, the description of the sound-level meter in Figure 7-1 uses an overall vertical pattern (top to bottom) with an occasional horizontal pattern (left to right) within it.

SEQUENTIAL ARRANGEMENT

Sometimes you may prefer to describe a subject sequentially, introducing items in a natural order that does not depend on physical layout. This method is particularly suitable for describing processes, techniques, or equipment that has to be operated. The two most common methods are operating order and cause to effect.

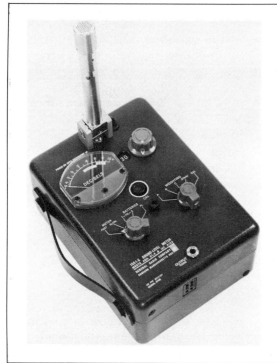

The GenRad 1551-C Sound-Level Meter measures sound levels from 25 dB to 150 dB. The microphone at the top doubles as an on-off switch: when the microphone is raised as shown, the unit is "on"; when the microphone is lowered, so that it rests downward along the front of the meter, the unit is "off." The DECIBELS scale at the upper left registers how much the recorded sound level deviates (from −6 to +10 dB) below or above a preset level visible in the small circular window beside the scale. This preset level is adjusted in 10-degree increments, from 30 dB to 140 dB, by the control knob at the upper right.

At the center is a pushbutton for calibrating the meter scale. Below it, in the lower half of the unit, are two control knobs for selecting various operating modes. The left knob selects either fast or slow meter needle response, or tests the unit's batteries. The right knob selects a weighting network suitable for the range of sound levels being measured, or the calibration function. At the bottom is a 7000 ohm output jack for connecting the sound-level meter to other instruments, or to a recorder.

Figure 7-1. Technical description: GenRad 1551-C Sound-Level Meter. (Photo and description courtesy GenRad Inc., Concord, Mass.)

Operating Order. The items are introduced in the order in which they are employed when the equipment is in use. This makes a very natural description which prepares the reader for the operating instructions that follow. Its major disadvantage is that the writer has to find a way to mention static items that the operator does not use. In the case of the overhead projector, a sequential description would differ significantly from the spatial description mentioned earlier:

Sequential Description (Operating Order) The transparency to be viewed is placed on a ground glass plate set into the hinged cover on top of the overhead projector's 12-inch-square metal base. From inside the base a light source, which sits above a highly polished concave reflector, is controlled by a switch on the front of the projection compartment. Focus and height of the image on the screen are controlled by adjusting the lens assembly, which is suspended 12 to 15 inches immediately above the glass plate. This assembly is held in position by an arm extending horizontally from a vertical post attached to the back of the projector's base. Focus is adjusted by rotating a knurled knob on the upright post, which raises or lowers the lens assembly. Image height is adjusted by tilting the lens assembly.

Cause to Effect. This method is used mainly to describe processes or operations which lead to a direct result. Each step in the process is described in sequence. It can be a simple description of the action that takes place when I depress a key on my typewriter to cause an imprint on the paper, or it can be a complex description of the effect of soil disturbance on permafrost in the discontinuous zones of Alaska and northern Canada.

Sequential Description (Cause to Effect)	When exploration crews in search of oil and mineral deposits first cleared long narrow strips of undergrowth for their seismic lines, they little knew the damage they were causing. Beneath its thin protective covering of vegetation, the soil had remained frozen for eons at a temperature only a degree or two below freezing. Then, with the stubby spruce gone and the insulating layer of moss torn up and thrown aside, the ice in the soil began to melt. Small pools of water formed which spread into shallow ponds that in succeeding summers became a narrow stretch of muskeg. The short summer months of the subarctic offer little time for vegetation to grow, so that now, 30 years later, pools of muskeg are still visible between the thin cover of poplar and coarse grasses that have slowly replaced the spruce and moss. The seismic lines have become a permanent scar upon the landscape.

Which method should you use? In some cases the topic itself, or its physical shape or arrangement of parts, will dictate the most suitable method. In other cases you may have a choice. When you do, try to find the method that will be of most help to readers unfamiliar with the equipment or process. And, for complex subjects, insert a photograph or drawing beside your description.

TECHNICAL INSTRUCTION

When H. L. Winman special projects engineer Andy Rittman wants a job done, he issues instructions in clear, concise terms: "Take your crew over to the east end of the bridge and lay down control points 3, 4, and 7," he may say to the survey crew chief. If he fails to make himself clear, the crew chief has only to walk back across the bridge to ask questions. But Fred Stokes, chief engineer at Macro Engineering Inc seldom gives spoken instructions to his electrical crews. Most of the time they work at remote sites and follow printed instructions, with no opportunity to walk across a project site to clarify an ambiguous order.

A technical instruction tells somebody to do something. It may be a simple one-sentence statement that defines what has to be done but leaves the time and the method to the reader. Or it may be a step-by-step procedure that describes exactly what has to be done and tells when and how. It is the latter type of technical instruction that will be described here.

Before attempting to write an instruction you must first define your readers, or at least establish their level of technical knowledge and familiarity with your subject. Only then can you decide the depth of detail you must provide. If they are familiar with a piece of equipment, you may assume that the simple statement "Open the cover plate" will not pose a problem. But if the equipment is new to them, you may have to broaden the statement to help them first identify and open the cover plate:

> Find the hinged cover plate at the bottom rear of the cabinet. Open it by inserting a Robertson No. 2 screwdriver into the narrow slot just above the hinge and then rotating the screwdriver half a turn counterclockwise.

START WITH A PLAN

A clearly written instruction contains four compartments of information: *what* has to be done; *why* it has to be done; *what equipment* is needed; and *how* the work is to be done.

1. A **summary statement** outlines briefly what has to be done:

> The 28 Vancourt Model AL-8 overhead projectors in rooms A4 to A32 are to be bolted to their projection tables . . .

2. The **purpose** explains why the work is necessary:

> . . . to reduce the current high damage rate caused by projectors being accidentally knocked onto the floor.

(A technician who understands *why* a job is necessary will much more readily follow an instruction.) Frequently, the summary statement and the purpose can be combined into a single paragraph or sentence, as has been done here.

3. A short paragraph or list describes **materials required,** which technicians performing the task can use as a checklist to ensure they have accumulated everything they will need before they start work:

> To carry out the modification you will require:
> - Modification kit OHP4, comprising
>   1 template, OHP4-1
>   4 bolts, flat head, 2 in. long, 1/8 in. dia
>   4 washers, 1 in. dia, with 5/32 in. dia central hole.
> - A 1/4 in. drill with a 3/16 in. drill bit
> - A Phillips No. 2 screwdriver
> - A slot-head No. 3 screwdriver
> - A sharp pencil

4. The steps that readers must follow take them through the whole process.

> Proceed as follows:

> 1. Disconnect the projector's power cord from the wall socket, then take the projector to another table and turn it on its side.

2. Using a slot-head No. 3 screwdriver, unscrew the four bolts that hold the feet onto the base of the projector. Remove but retain the bolts and feet for future use.

3. Place template OHP4-1 onto the projection table and position it where the projector is to stand. Using a sharp pencil, mark the table through each of the four holes in the template.

4. Drill four $\frac{3}{16}$ in. dia holes through the table top, at the places marked in step 3.

5. Place the overhead projector, right side up and with the lens assembly facing the screen, so that the four screwholes identified in step 2 coincide with the four holes in the table top.

6. From beneath the table, place a washer under each hole and insert a 2 in. flat head bolt up through the washer and hole until it engages the corresponding hole in the projector base. Tighten the four bolts in place, using a Phillips No. 2 screwdriver.

The writing plan embodying these four compartments, when combined with the suggestions below, will consistently ensure that any instructions you write will be clear, direct, and convincing.

GIVE YOUR READER CONFIDENCE

A well-written technical instruction automatically instills confidence in its readers. They feel they have the ability to do the work even though it may be highly complex and quite new to them. Consider these examples:

Vague   Before the trap is set, it is a good idea to place a small piece of cheese on the bait pan. If it is too small it may fall off and if it is too big it might not fit under the serrated edge, so make sure you get the right size.

Clear and   Cut a 17-inch length of 10-gauge wire and strip 1 inch of insulation
Concise   from each end. Solder one end of the wire to terminal 7 and the other end to pin 49.

The first excerpt is much too ambiguous. It only suggests what should be done, it hints where it should instruct (almost inviting readers to nip their fingers), and in 31 explanatory words it fails to define the size of a "small" piece of cheese. The second excerpt is assertive and keeps strictly to the point. Such clear and authoritative writing immediately convinces its readers of the accuracy and validity of the steps they have to perform.

The best way to be authoritative is to write in the imperative mood. This means you should begin each step with a strong verb, so that your instructions are commands:

Ignite the mixture. . . .   *Connect* the green wire. . . .
Mount the transit on its tripod. . . .   *Excavate* 3 feet down. . . .
Apply the voltage to. . . .   *Measure* the current at. . . .
Cut a 2 inch strip of. . . .   *Count* the number of blips. . . .

The imperative mood in the clear, concise excerpt quoted above keeps the instruction taut and definite. Notice how the verbs *cut, strip,* and *solder* make readers feel they have no alternative but to follow the instructions. The vague excerpt would have been equally effective (and much shorter) if it had been written in the imperative mood and if the vague verb "place" had been replaced by an image-conveying verb-adverb combination such as "wedge firmly:"

> Before setting the trap, wedge a ⅜ inch cube of cheese firmly under the serrated edge of the bait pan.

The following two statements clearly show the difference between an instruction written in the imperative mood and one that is not:

> Disengage the gear, then start the engine.
>   (*definite; uses strong verbs*)
> The gear should be disengaged before starting the engine.
>   (*Indefinite; uses weaker verbs*)

The first statement is strong because it tells readers to do something. The second is weak because it neither instructs them to do anything nor insists that anything need be done ("should" implies it is only *preferable* that the gear be disengaged before the engine is started).

In the imperative mood the first word in a sentence most often will be a strong verb; sometimes, however, the verb may be preceded by an introductory or conditioning clause:

> Before connecting the meter to the power source, *set* all the switches to "zero."

The imperative mood is maintained here because the main verb starts the statement's primary clause (in this case the clause that describes the action to be taken).

If you want to check whether a sentence you have written is in the imperative mood, ask yourself whether it *tells* the reader to *do* something. If it does, then you have written an *instruction*.

AVOID AMBIGUITY

There is no room for ambiguity in technical instructions. You have to assume that the person following your instructions cannot ask questions, and so you must never write anything that could be interpreted more than one way. This statement is wide open to misinterpretation:

> Align the trace so that it is inclined approximately 30° to the horizontal.

Each technician will align the trace with a different degree of accuracy, depending on his or her interpretation of "approximately." How accurate does "approximately" require the technician to be? Within 5°? Within 2°? Within ½°? Maybe even 10° either side of 30° is acceptable, but the reader does not know this and feels doubtful. Worse still, the reader's confidence in the technical validity of the whole instruction is undermined. Vague references like this must be replaced by clearly stated tolerances:

Align the trace so that it is inclined 30° (±5°) to the horizontal.

More subtle, but equally open to misinterpretation, is this statement:

Adjust the capstan handle until the rotating head is close to the base.

Here the offending word is "close," which needs to be replaced by a specific distance:

Adjust the capstan handle until the distance between the rotating head and the base is 25 mm.

Similarly, such vague references as "*relatively* high," "*near* the top," and "*an adequate* supply" must be replaced by clearly stated measurements, tolerances, and quantities.

Specifying Tolerances.   Many instructions call for readers to take readings and record the results in special places provided either on the instruction sheet or on a separate data sheet. This is particularly common in field testing and troubleshooting of electronic equipment, where a typical entry might read like this:

Connect the voltmeter to test points
8 and 17, then note the reading   5.5 V ——— V

The figure 5.5 V is the measurement that the technician should obtain, and the short line is provided for entering the voltage actually recorded. Since obtaining an exact voltage is difficult, a slight variation from the specified voltage is permissible. This tolerance can be stated:

As a percentage: 5.5 V (±6%)
As a specific voltage: 5.5 V (±0.33V)
As a voltage range: 5.17 to 5.83 V

The voltage range is the best choice because it does not force the reader to make a calculation (and, hence, possibly make an error).

Indefinite Words.   Weak words such as "should," "could," "would," "might," and "may" so weaken the authority of an instruction that they reduce the reader's confidence in the writer. Though their meaning is clear, they sound indefinite:

> Set the meter to the +300 V range. The needle should indicate 120 volts (±2 volts).

In this example, "should" *implies* that it would be nice if the needle indicated within 2 volts of 120 V, but it is not essential! No doubt the writer meant it *must* be within the specified voltage range, but has failed to say so. Nor has the reader been told to note the reading. The writer has forgotten the cardinal rule of instruction writing: *Tell* the reader to *do* something. To be authoritative, the instruction needs very few changes:

> Set the meter to the +300 V range, then check that its needle indicates 120 volts (±2 volts).

Notice how the steps in the sample instruction in Figure 7-2 are clear, concise, and definite. You need not be a specialist in the subject to recognize that they would be easy to follow.

WRITE BITE-SIZE STEPS

Technicians working on complex equipment in cramped conditions need easy-to-follow instructions. You can help them by writing short paragraphs, each containing only one main step. If a step is complicated and its paragraph grows unwieldy, divide it into a major step and a series of substeps. Instruction writing lends itself to subparagraphing like this:

> 3. List the documentary evidence in block J of Form 658. Check that blocks A to G have been completed correctly, then sign the form and distribute copies as follows:
>   3.1 Attach the documentary evidence to Copies 1 and 2 and mail them to the Chief Recording Clerk, Room 217, Civic Center, Montrose, Ohio.
>   3.2 Mail Copy 3 to the Computer Data Center, using one of the special pre-addressed envelopes.
>   3.3 File Copy 4 in the "Hold—Pending Receipt" file.
>   3.4 When Copy 2 is returned by the Chief Recording Clerk, attach it to Copy 4 and file them both in the "Action Complete" file.

To avoid ambiguity, use a simple paragraph numbering system. Unless your instruction is going to be very long, there is nothing better than a straightforward system that starts at 1. Subparagraphs can be assigned decimal subparagraph numbers as shown above. This system also builds in a means for easy cross-referencing between steps; for example:

> 18. Reassemble the unit, reversing the procedure of steps 6 to 9.

Heathkit®

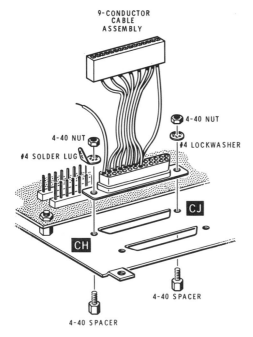

9-CONDUCTOR CABLE ASSEMBLY

4-40 NUT
#4 SOLDER LUG
4-40 NUT
4-40 NUT
#4 LOCKWASHER
CJ
CH
4-40 SPACER
4-40 SPACER

Detail 4-7B

( ) Refer to Detail 4-7B and mount the 9-conductor cable assembly at CH and CJ. Use a 4-40 spacer, a #4 solder lug, and a 4-40 nut at CH; use a 4-40 spacer, a #6 lockwasher, and a 4-40 nut at CJ.

( ) Position socket S401 at the other end of the assembly with the slotted side up, as shown and install it on plug P401.

( ) Connect the black wire coming from the assembly to solder lug CH (S-1).

( ) Similarly mount the 7-conductor cable assembly at CK and CL with a #4 solder lug at CL. Then install socket S402 on plug P402, again, with the slotted side up.

( ) Connect the black wire coming from the assembly to solder lug CL (S-1).

( ) Refer to Detail 4-7C and position the 6-conductor cable assembly and the connector bracket as shown. Slide the connector bracket onto the connector and mount it at CN with a 6-32 × 1/4″ pan head screw, #6 lockwasher and a 6-32 nut.

( ) Position socket S403 at the other end of the assembly with the slotted side as shown and install it on plug P403.

Refer to Pictorial 4-8 (Illustration Booklet, Page 13) for the following steps.

( ) Position the back panel as shown and insert the two end and center tabs into the slots in the back panel of the chassis.

( ) While holding the panel in place, install socket S404 on plug P404. Be sure to match up the lip on the socket and plug when you install it.

( ) Similarly install socket S405 on plug P405.

( ) With the panel still engaged in the chassis, rotate it to a vertical position and fasten it at CP with a #6 × 1/4″ sheet metal screw.

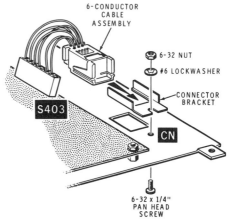

6-CONDUCTOR CABLE ASSEMBLY
6-32 NUT
#6 LOCKWASHER
CONNECTOR BRACKET
S403
CN
6-32 x 1/4″ PAN HEAD SCREW

Detail 4-7C

Figure 7-2. Excerpt from an instruction manual. (Courtesy of the Heath Company, Benton Harbor, Michigan.)

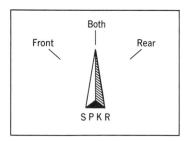

Figure 7-3. Part of a control panel.

IDENTIFY SWITCHES AND CONTROLS CLEARLY

Opinions differ as to the best method for identifying switches, controls, and operating positions marked on equipment. Some authorities recommend using capital letters throughout, while others prefer to enclose some of the words within quotation marks. I suggest you follow these two rules:

1. Identify switches, controls, and selector positions exactly as they appear on the equipment. The loudspeaker selector switch shown in Figure 7-3 would thus be described as the SPKR switch. This helps the reader to locate the correct control on a panel containing many dials and switches. As most equipment labeling is in capital letters, much of the time you will have to use capital letters.
2. Differentiate between switches, controls, handles, and valves, and the positions to which they can be set, by enclosing the position settings within quotation marks. The loudspeaker selector switch in Figure 7-3 would be described as being set to the "BOTH" position. If you want to tell readers to select only the front loudspeaker, instruct them to do so like this:

Set the SPKR switch to the "FRONT" position

or

Set the SPKR switch to "FRONT."

INSERT FAIL-SAFE PRECAUTIONS

Precautionary comments are inserted in instructions to warn readers of dangerous conditions or damage that can occur if they do not exercise care. There are three types:

Warnings: To alert readers of an element of personal danger (such as the presence of unprotected high voltage terminals).

Cautions: To tell them when care is needed to prevent equipment damage.

Notes: To make general comments (for example, "on some older models the valve is at the rear of the unit").

Draw attention to a precautionary comment by placing it in the middle of the text, indenting it on both sides, and preceding it with the single word WARN-

ING, CAUTION, or NOTE. Draw a box around the words WARNING and
CAUTION to give them extra prominence:

> WARNING

Disconnect the power source before removing the cover plate.

A precautionary comment must always *precede* the step to which it refers.
This will prevent an absorbed reader who concentrates on only one step at a time
from acting before reading the warning. Never assume that mechanical devices,
such as indentation and the box drawn around the precautionary word, are
enough to catch a reader's attention.

Warnings and cautions must be used sparingly. A single warning will
catch a reader's attention. Too many of them will cause a reader to treat them all
as comments rather than as important protective devices.

INSIST ON AN OPERATIONAL CHECK

The final test for any technical instruction is the reader's ease in following it. Since you cannot always peer over a reader's shoulder to correct mistakes,
you should find out whether users are likely to run into difficulty *before* you send
an instruction out. To obtain an objective check, give the draft instruction to a
technician of roughly equal competence to those who eventually will be using it,
and observe how well the task is performed.

As you watch (and you must watch with a zipped lip!), note every time
the user hesitates or has difficulty. When the task is finished, ask if any parts
need clarification. Then rewrite ambiguities, and recheck your instruction with
another technician. Repeat these steps until you are confident readers will be
able to follow your instruction easily.

THE SCIENTIFIC PAPER

Earlier chapters described how technical reports should be planned, organized,
written, and presented by engineers and engineering technicians working for
industry, business, and government. Within this context one additional report
remains to be described: the research report prepared by scientists and technologists working in industrial and university laboratories. Research reports are
most often prepared and published as *Scientific Papers* which, although their
parts are similar to those of an investigation report, differ in style, organization,
and emphasis. Hence, I am treating them separately here.

A scientific paper either identifies and attempts to resolve a scientific
problem, or it tests (validates) a scientific theory. It does so by describing the four
main stages of the research:

1. Identifying the problem or theory.
2. Setting up and performing the tests.
3. Tabling the test results (the findings).
4. Analyzing and interpreting the findings.

These four stages represent the major divisions of a scientific paper, with each stage preceded by a descriptive heading: **Introduction, Materials and Methods, Results,** and **Discussion.** These stages are similar to those used for the laboratory report in Chapter 4 and the investigation report in Chapter 5. There are, however, differences in a scientific paper's appearance and writing style.

APPEARANCE

A scientific paper straddles the borderline between semiformal and formal presentation. Normally the title is centered about two to three inches from the top of the first page. (These dimensions apply to a typewritten paper, as in Figure 7-4; for a typeset paper less open space would be used.) The author's name and the name of the company or organization the author works for can also be centered at the top of the page, about one inch below the title. Alternatively, the author's name and affiliation can be placed at the bottom left of the first page, or at the end of the paper.

The abstract (summary) appears next, and starts about 1¼ inches below the title or the author's name. For a typewritten paper the abstract should be indented about one inch from both side margins. For a typeset paper, the abstract normally is set in bolder type and may also be indented slightly on both sides.

The body of a scientific paper starts about ¾ inch beneath the abstract and looks much like the body of a semiformal report. Normally, a typewritten scientific paper is double-spaced throughout, including the abstract (that is, there is a blank line between each line of type), while a typeset report is single-spaced (there are no blank lines, except between paragraphs and before and after headings). The first line of every paragraph is indented five typewriter spaces.

WRITING STYLE

The rules for brevity, clarity, and directness suggested in Chapters 1, 3, and 11 for technical letters and reports apply equally to scientific papers. But there is one exception: Where I have recommended that you consistently write in the active voice, in some sections of a scientific paper—particularly the Materials and Methods section—convention requires that you use the passive voice. For example, in a technical report I have advised you to write:

I placed the sample in the chamber . . .

or

The technician placed the sample in the chamber . . .

But in a scientific paper you will more commonly write:

The sample was placed in the chamber . . .

In other words, the identity of the "doer" is consistently concealed in a scientific paper, and use of the first person ("I" or "we") is carefully avoided. This penchant for the passive voice is traditional and remains unaffected by the change to the active voice for technical reports and papers.

ORGANIZATION

A scientific paper has six main parts, which are described in detail below.

1. **Abstract** (Summarizes the paper, emphasizing the results)
2. **Introduction** (Provides background details and outlines problem or theory tested)
3. **Materials and Methods** (Describes how the tests were performed)
4. **Results** (States the findings)
5. **Discussion** (Analyzes and interprets the findings)
6. **References** (Lists the documents consulted)

Sometimes you may receive significant assistance or guidance from other people during your research and may want to acknowledge their help. Such acknowledgments normally are placed after the discussion but before the references.

If you are accustomed to writing laboratory reports or investigation reports, you have probably noticed that the "Conclusion" heading has been omitted from this list. Where the conclusions in a laboratory or investigation report serve as a *separate* terminal summary (a summing up of the results and their analysis), in a scientific paper they are embedded in both the introduction and discussion, and only rarely are preceded by separate headings.

Abstract. The rules for writing an abstract are almost identical to those for writing the summary of an investigation report. In an abstract you should: (1) outline the problem and the purpose of your investigation; (2) mention very briefly how you conducted the investigation or tests; (3) describe your main findings; and (4) summarize the conclusions that you have drawn. All this must be done in as few words as possible; ideally, your abstract will be about 125 words long and never more than 250 words. A typical abstract is shown in Figure 7-4.

From the abstract, readers must be able to decide whether the information you provide in the remainder of the paper is of particular interest to them and whether they should read further. Because a scientific paper is addressed to readers who generally are familiar with your technical or scientific discipline,

ACID RAIN TESTING IN THE CITY OF MONTROSE, OHIO, 19XX

Corrine L. Danzig and Mark M. Weaver
University of Montrose Research Laboratory

ABSTRACT

Acid rain is a growing concern in the United States, with its
effects becoming increasingly noticeable south of Lakes Erie and
Ontario. To determine what increases have occurred in the City
of Montrose, Ohio over the past nine years, acidity levels were
measured and compared to measurements recorded in 19XX, when the
average pH level was YYY. The current tests showed that the average
pH now is ZZZ, but that the acidity levels are not uniform. The
greatest toxicity was found to be at the University of Montrose,
in the southeastern area of the city, with the lowest toxicity in
the northwestern area of the city. The change was attributed
primarily to the increase in fossil fuel-burning heavy industries
in the Bluff Heights industrial park, which is three miles north-
west (generally upwind) of the principally affected areas.

INTRODUCTION

Over the past 20 years, observers in the states immediately south of the
Great Lakes have reported an increased incidence of crop spoilage and tree
defoliation, which has been attributed to acid rain created primarily by
fossil fuel-burning industries and automobile emissions. Tests were taken
initially in 19XX to assess the pH levels of the precipitation falling in and
around the city of Montrose, Ohio (1). These were compared with numerous

Figure 7-4. The first page of a scientific paper.

you may use technical terminology in the abstract. (This is the major difference between a report summary and a scientific abstract.) The abstract should be written last, when the remainder of the paper has been written, so that you can *abstract* the brief details you need from what you have already written.

Introduction. In the introduction you prepare readers so that they will readily understand the technical details in the remainder of the paper. The introduction contains four main pieces of information:

1. A definition of the problem and the specific purpose of the investigation or tests you conducted. (This should be a much more detailed definition than appeared in the abstract.)
2. Presentation of background information that will enable the reader to fully understand and evaluate the results. Often it will include—or sometimes may consist entirely of—a review of the previous scientific papers, journal articles, books, and reports on the same subject. (This is known as a literature survey.) Rather than simply list the pertinent documents, you are expected to summarize the main findings of each and their relevance to your investigation. These documents should be cross-referenced to your list of references at the back of the report. Here is an example:

> Previous measurements of acid rain in Montrose were recorded in 19xx by Gershwain (3), who reported an average acidity level of xxxx, and. . . .

3. A short description of your approach to the investigation and why you chose that particular method.
4. A concluding statement that outlines your main findings.

You will have noticed that the introduction of a scientific paper *includes a brief summary of the results,* which is much less common in a semiformal investigation report and rare in a laboratory report. Yet here there is a parallel with the alternative method of presenting a formal report, the "pyramid," in which the writer presents the results three times (see Figure 6-7). In a scientific paper the results also are presented three times: very briefly in the abstract and introduction, and fully in the results section.

Materials and Methods. This section has to be thoroughly prepared and presented because, from what you write here, readers must be able to replicate (perform) an identical investigation or series of tests. The materials list must include *all* equipment used and specimens or samples tested, which may range from a whole-body nuclear radiation counter to a tiny microorganism. For ease of reference they should be listed in representative groups, such as:

Equipment	Plants
Instruments	Animals
Chemicals	Birds or fishes
Specimens	Humans

This section should be preceded by the subheading "Materials." If the list is long, you should consider using a subordinate heading before each group of materials, instruments, or subjects.

The methods should be introduced by the subheading "Method" and then described chronologically in paragraph and subparagraph form, using a main paragraph to introduce a test or part of a test and short, numbered subparagraphs to describe the specific steps that were taken. Be careful to avoid writing in the imperative mood; that is, take care that you do not inadvertently start writing an instruction. For example:

Write: 6.   The test unit was connected to the X-Y terminals of the recorder.

Not: 6.   Connect the test unit to the X-Y terminals of the recorder.

Results.   The results often may be the shortest section of your paper. If your methods section has described clearly how the investigation or tests were conducted, the results section has only to state what the result was:

Acid rain is above average in the southeastern part of the city, below average in the northwestern part, and average in the southwestern and northeastern parts. Tables 1 through 8 show the measurements recorded at the eight metering stations.

Tables and charts depicting your findings often will be a major part of the results section. However, you should never comment on the results, since analysis and interpretation belong *only* in the discussion.

Discussion.   Your readers now will expect you to analyze and interpret the findings you announced in the results. (The emphasis here is on "analyze and interpret;" you should not simply repeat the results in the discussion.) You will be expected to discuss:

• The results you obtained, compared to the results obtained by previous researchers.
• Any significant correlation, or lack of correlation, between parts of your own findings.
• Factors that may have caused the differences.
• Any trends that seem to be evident or to be developing (ideally, by referring to graphs and charts you have presented in the results).
• The conclusions you draw from your analysis and interpretation.

The conclusions should show how you have responded to the problem stated at the start of your introduction, and identify clearly what your investigation, tests, and analyses show. As such, they should form a fitting close to the narrative portion of your research paper.

References.   The final section of your paper is a list of the documents you have referred to earlier or from which you have extracted information.

Chapter 6 provides general instructions on preparing a list of references either as endnotes, which is the preferred method for most technical reports, or as a bibliography. You can use Chapter 6 as a guideline, but you should check first with the editor of the journal in which your scientific paper is likely to be published to determine: (1) whether an endnote or bibliographical listing is preferred, and (2) the exact format the journal uses for listing authors' names, book and journal titles, publisher details, and so on.

TECHNICAL PAPERS AND ARTICLES

WHY WRITE FOR PUBLICATION?

The likelihood that one day you may be asked to write a technical paper for publication, or even want to do so, may seem so remote to you now that you might be justified in skipping this section. Yet this is something you should think about, for getting one's name into print is one of the fastest ways to obtain recognition. Suddenly you become an expert in your field and are of more value to your employer, who is happy because the company's name appears in print beneath yours. You become of more value to prospective employers, who rate authors of technical papers more highly than equally qualified persons who have not published. And you have positive proof of your competence, and sometimes a few extra dollars from the publisher.

Mickey Wendell has an interesting topic to write about: As a senior technician in H. L. Winman and Associates' Materials Testing Laboratory, he has been testing concretes with various additives to find a grout that can be installed in frozen soil during the Alaskan winter. One mixture that he analyzed but discarded contained a new product known as Aluminum KL. As a by-product of his tests he has discovered that mixing Aluminum KL with cement in the right proportions results in a concrete with very high salt resistance. He reasons that such concrete could prove invaluable to builders of concrete pavements in snow-affected areas of the United States and Canada, where salt mixtures are applied in winter to melt the snow.

Mickey has been thinking about publishing this particular aspect of his findings and has jotted down a few headings as a preliminary outline. Here are the four steps he must take before his ideas appear in print.

STEP 1: SOLICIT COMPANY APPROVAL

Most companies encourage their employees to write for publication, and some even offer incentives such as cash awards to those who do get into print. However, they expect prospective authors to ask for permission before they submit their manuscripts.

To obtain permission, Mickey must write a brief memorandum outlining his ideas to John Wood, his department head. He should ask for approval to

H L WINMAN AND ASSOCIATES

INTER-OFFICE MEMORANDUM

To: John Wood .... From: .... Mickey Wendell ....

.... Subject: .. Approval for Proposed

Date: June 24, 19xx .... Technical Article

May I have company approval to write an article on concrete additives for publication in a technical journal? Specifically, I want to describe our experiments with Aluminum KL, and the salt-corrosion resistance it imparted to the concrete samples we tested for the Alaska transmission tower project. I believe that our findings will be of interest to many municipal engineers in the northern United States and Canada, who for years have been trying to combat pavement erosion caused by the application of salt during snow removal.

I was thinking of submitting the article to the editor of "Municipal Engineering," but I'm open to suggestions if you can think of a more suitable magazine.

MW:kr

mw

Approval granted. Let me see an outline and the first draft before you submit them. Suggest you also consider preparing a technical paper for presentation at the Combined Conference on Concrete to be held in Chicago next March.

John Wood
July 6, 19xx

Figure 7-5. Soliciting company approval to publish an article or paper.

240

submit a paper, explain what he wants to write about and why he thinks the information should be published, and outline where he intends to send it. His proposal is shown in Figure 7-5. John Wood will discuss the matter at management level, then signify the company's approval or denial in writing. Mickey knows he must have *written* consent to publish his findings.

STEP 2: CONSIDER THE MARKET

Mickey must decide very early where he will try to place his paper. If he prefers to present his findings as a technical paper before a society meeting, as suggested by John Wood (see his comment in Figure 7-5), he will be writing for a limited audience with specialized interests. If he decides to publish in the journal of a technical society, he will be writing for a larger audience, but still within a limited field. If he plans to publish in a technical magazine, he will be appealing to a wide readership with a broad range of technical knowledge. His approach must therefore differ, depending on the type of publication and level of reader.

A guiding factor may be Mickey's writing capability. A paper to be published by a technical society requires high quality writing. The editor of a society journal normally does not do much prepublication editing, other than making minor changes to suit the format and style of the society's publications. A technical magazine article, however, will be edited—sometimes quite fiercely—by a professional editor who knows the exact style that readers expect. Such an editor prefers authors to approximate that style and expects them to organize their work well and to write coherently; but he or she is always ready to prune or graft, and sometimes even completely rewrite portions of a manuscript. Hence, the pressure on the authors of magazine articles is not so great.

A secondary consideration may be the state of Mickey's wallet: if it is thin and he needs a new set of tires for his Honda, he may choose to write a magazine article. Normally, there is no pay for writers of technical papers, other than recognition by one's peers.

Perhaps the most important factor is for Mickey to be able to identify a potential audience for his information. Readers of society journals and technical magazines may be the same people, but they expect different information coverage in a technical paper than they do in a magazine article.

Technical Paper. Readers of society journals are looking for facts. They neither expect nor want explanations of basic theories, and they can accept a strongly technical vocabulary. A technical paper can be very specific. It can describe a minute aspect of a large project without seeming incomplete, or it can outline in bold terms the findings of a major experiment. No topic is too large or too small, too specialized or too complete, to be published as a technical paper.

Technical Article. Most readers of magazine articles are looking for information that will keep them up to date on new developments. Some will have definite interest in a specific topic and would welcome a lot of technical details. Others will be looking mainly for general information, with no more than just the highlights of a new idea. Magazine authors must therefore appeal to a maximum number of readers. Their articles should be of general interest; their style can be brief and informal; their vocabulary must be understandable; and they should sketch in background details for readers whose technical knowledge is only marginal.

STEP 3: WRITE AN ABSTRACT AND OUTLINE

Many editors prefer to read either a summary of a proposed paper or an abstract and outline before the author submits the complete manuscript. They may want to suggest a change in emphasis to suit editorial policy, or even decline to print an interesting paper because someone else is working on a similar topic.

This type of summary is much longer than the summary at the head of a technical report; the abstract, however, usually is quite short. The summary contains a condensed version of the full paper in about 500 to 1000 words. An abstract contains only very brief highlights and the main conclusion (rather like the summary of a report), since it is supported by a comprehensive topic outline.

Some authors write the complete first draft of the paper before attempting to write a summary or abstract, then leave the revising and final polishing until after the paper has been accepted by an editor. Others prepare a fairly comprehensive outline, often using the freewheeling approach suggested in Chapter 1, and leave the writing until after acceptance. Both methods leave room for the author to incorporate changes before the final manuscript is written.

Since the summary or abstract and outline have to "sell" an editor on the newsworthiness of his topic, Mickey Wendell must make sure that the material he submits is complete and informative. In addition he must indicate clearly:

1. Why the topic will be of interest to readers.
2. How deeply the topic will be covered.
3. How the article or paper will be organized.
4. How long it will be (in words).
5. His capability to write it.

Mickey can cover the first four items in a single paragraph. The fifth he will have to prove in two ways. He can prove his technical capability by mentioning his involvement in the topic and experience in similar projects. He can demonstrate his ability to write well by submitting a clear, well-written summary or abstract.

If Mickey later decides to prepare his paper for presentation before the Combined Conference on Concrete, as John Wood has suggested, he will have to prepare a summary in response to a "call for papers" letter sent out by the society, and submit it to the chairman of the papers selection committee. His summary has to convince the committee that the subject is original, topical, and interesting, and that he has the technical capability to prepare it, if his paper is to be accepted.

STEP 4: WRITE THE ARTICLE OR PAPER

A good technical paper is written in an interesting narrative style that combines storytelling with factual reporting. Articles published in general interest magazines tend to be written like feature newspaper stories, whereas technical papers more nearly resemble formal reports. If the article deals with a factual or established topic, the writing is likely to be crisp, definite, and authoritative. If it deals with development of a new idea or concept, the narrative will generally be more persuasive, since the writer is trying to convince the reader of the logic of his or her argument.

Mickey will find that the parts of an article or paper are very similar to those of a report:

Summary	A synopsis that tells very briefly what the article is about. It should summarize the three major sections that follow. Like the summary of a report, it should catch and hold the reader's interest.
Introduction	Circumstances that led up to the event, discovery, or concept that is to be described. It should contain all the facts readers will need if they are to understand the discussion that follows.
Discussion	How the author went about the project, what he found out, and what inferences he drew from his findings. The topic can be described chronologically (for a series of events that led up to a result), by subject (for descriptive analyses of experiments, processes, equipments, or methods), or by concept (for the development of an idea from concept to fruition). The methods are very similar to those used for writing the discussion of the formal report (see Chapter 6).
Conclusion	A summing-up in which the writer draws conclusions from and discusses the implications of his major findings. Although he will not normally make recommendations, he may suggest what he feels needs to be done in the future, or outline work that he or others have already started if there is a subsequent stage to the project.

Illustrations are a useful way to convey ideas quickly, to draw attention to an article, and to break up heavy blocks of type. They should be instantly clear and usefully *supplement* the narrative. They should never be inserted simply to save writing time; neither should they convey exactly the same message as the written words. For examples of effective illustrating, turn to any major publication in your technical field and study how its authors have used charts, graphs, sketches, and photographs as part of the story. For further suggestions on how to prepare illustrative material, see Chapter 8.

Since an article or paper is to be read by many persons, considerably more revision time is required than for in-plant reports. Mickey Wendell should work closely with technical editor Anna King and should give himself time between major revisions to put the article aside so that he can return to it with a fresh mind. He will also be wise to call on at least one independent reviewer to read the paper when it is nearly ready for submission. (His final paper also may be read by independent reviewers selected by the magazine or journal editor.) He should respect his reviewer's comments, because they are likely to be similar to those of his eventual readers.

Finally, a comment on the editor's role: Mickey should not be surprised or disturbed if the editor who handles his manuscript makes some changes. These changes will affect only the arrangement of the information and will seldom alter the technical content (if they do, Mickey will have good reason for raising his voice). The editor knows the particular journal's readers well. If the editor feels the material is too lengthy, too detailed, or wrongly emphasized, he or she will make changes to bring it up to the expected standard. Mickey may feel that the alterations have ruined his carefully chosen phrases, but readers will not even be aware that changes have been made. They will simply recognize a well-written paper, for which Mickey rather than the editor will reap the compliments.

Assignments

DESCRIPTION WRITING

PROJECT 1: FAMILIARIZATION FOR A NEW STUDENT

An exchange student from Australia shortly will be joining your class.

Part 1. You are asked by a college administrator to prepare a 200- to 300-word description of your college, with particular emphasis on the faculty buildings the student will attend most classes in. The description will be sent to the newcomer in advance, to give him or her an idea of the institution's appearance, location, and layout. If possible, enclose a photograph.

Part 2. Write a cover letter to enclose with the description. Welcome the student to your college and course. His/her name and address are: Chris Larkin, 316 Toowoomba Crescent, Brisbane, Queensland, Australia.

PROJECT 2: RESISTORS

Mr. Wayne D. Robertson, general manager of Macro Engineering Inc, receives the following letter from his friend, Dave Kostyn, who owns a wholesale grocery company in Phoenix, Arizona:

Dear Wayne,

Can you help me? I have a dozen high school students in the Junior Achievement group I'm working with and they have planned a project to sell packages of "resistors" to radio amateurs and hobbyists around the city. I cannot help them very much because at the moment I have no idea what a resistor is or what it looks like! All I know is that a bunch of resistors will be coming to us loose and the youngsters will have to divide them into groups of 25, package them, and arrange to sell them.

What I need right now is a paragraph or two from you describing what a resistor looks like and what it's made of (plus anything else you think I should know).

Can you do that for me? I certainly would appreciate it!

Regards,

Dave

Mr. Robertson asks you to write the description. You may assume that Dave Kostyn will post your description on the wall, where it can be read by all the Junior Achievers.

Part 1. Write the description.

Part 2. Write a cover letter to attach to the description you will mail to Dave Kostyn. His address is 2125 Nantucket Boulevard, Phoenix, AZ 85023.

PROJECT 3: BLUEPRINTING PROCESS

Write a description of the blueprinting process, based on the information provided below or on a blueprinting machine available to you. Assume that your readers will be drafting students who have not yet used a blueprinter.

This is how a technician told me he made some blueprints:

The blueprinter was relatively simple to operate. Since the power was already on and the machine was operating, the steps involved in making my 20 blueprints were easy to follow. In the first place, for each print I took a sheet of blueprint paper and placed it on the feed table, face side up. (The face side is the yellow side.) The original tracing, right way up, was then placed onto the blueprint paper. Before pushing the two sheets of paper into the machine, I had to set the speed selector to the correct speed, which is 15 for medium speed blueprint paper. (Mine was medium speed.) The two sheets were then pushed toward the machine until they were grasped between the belts and the exposure roller. When the two sheets emerged I had to separate them and fold back the blueprint paper and feed it up and over a developing roller. The original tracing came right out and I removed it in readiness for making another blueprint. This printing procedure was repeated until I had made all the copies I needed. At the top of the machine there was a guide bar that directed the prints to one of two delivery trays. The front delivery tray is at the front of the machine immediately in front of the operator; the back delivery tray is used when you want to stack a succession of prints without removing them. Because I had 20 prints to make, I

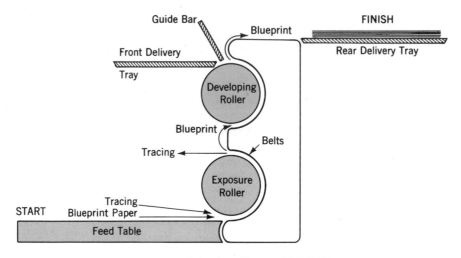

Figure 7-6. A typical blueprinter (side view).

selected "rear delivery." When I had finished printing I had to remember to turn the speed selector back to 5.

Figure 7-6 shows a blueprinter viewed from the side. Other details are:

1. The exposure roller is made of glass and has a high intensity light inside it.
2. The light burns off the yellow coating on the blueprint paper (except where the image of the tracing prevents light from reaching it).
3. The developing roller forces the paper through ammonia fumes (produced by heating liquid ammonia).
4. The blueprinter has a chimney to exhaust the fumes (because they are toxic).
5. Fumes "develop" the blueprint (the yellow areas turn blue).
6. The number set on the speed selector is the number of feet of paper the blueprinter produces in a minute.
7. The type of paper described here is more correctly known as "whiteprint" or "blueline" paper. In true blueprint paper the image appears white against a dark blue background.

PROJECT 4: TOOTHPICK TOWER

Describe the "toothpick tower" illustrated in Project 3 of Chapter 6. Read the assignment so you will understand why the tower was built and how it was used. Assume that your description will be given to students at other colleges who have not heard of the ASEI award.

Note: If you have built your own toothpick tower, then you may describe it instead of the tower illustrated in Chapter 6.

PROJECT 5: WATER CONSUMPTION OF MONTROSE

In Project 2 of Chapter 8 you are asked to prepare illustrations depicting quantities of water sold by the City of Montrose, Ohio, to various groups of consumers during the previous year. When this has been done, prepare an analysis of water consumption for the City Engineer, who will mail it with one of the illustrations to all water consumers. In your analysis identify when each segment of the community draws most and least water, and suggest reasons for these highs and lows.

PROJECT 6: RESEARCHING A NEW MANUFACTURING MATERIAL OR PROCESS

You are to research information on a topic allied to your technology and then prepare it for both written and oral presentation. The topic may be a new manufacturing material, method, or process. The written and spoken presentations must:

1. Introduce the topic.
2. State why it is worth evaluating.
3. Describe the material, method, or process.
4. Discuss its uniqueness and usefulness.
5. Show how it can be applied in your particular field.

To obtain data for your topic, you will have to research current literature and probably talk to industrial users, manufacturers, and suppliers. Typical examples of topics are: a new oil that can be used at very low temperatures; a method for supporting the deck of a bridge during concrete-pouring by building up a base on compacted fill, a new paint for use on concrete surfaces; and a new materials-handling system.

You may assume that both your readers and your audience are technicians to whom the topic will be entirely new.

INSTRUCTION WRITING

PROJECT 7: USING A PHOTOCOPIER OR BLUEPRINTER

Write an instruction for users of a photocopier or blueprinter that is accessible to you. Assume that the unit will be switched off when the user approaches it. Also assume that the instructions will be pinned to the wall beside the machine.

If neither a photocopier nor a blueprinter is available to you, base your instruction on the description and illustration included with Project 3 of this chapter.

Include a warning of the dangers of inhaling ammonia fumes, if you are writing an instruction for a blueprint machine.

PROJECT 8: TESTING CABLE CONNECTORS

Write an instruction to all installation supervisors at sites 1 through 17 (see Project 5 of Chapter 4) telling them to inspect all cable connectors on site. Additional information you will need is listed below.

1. Background to this project is contained in Project 5 of Chapter 4.
2. The instruction is to be written as an interoffice memorandum.
3. Don Gibbon, electrical engineering coordinator at H. L. Winman and Associates, will sign the memorandum.
4. Tell the site installation supervisors that they are to report the number of GLA connectors they find to Don Gibbon.
5. Tell them to replace all connectors marked GLA with connectors marked MVK.
6. All connectors on site are to be checked. (It would be best to check those in stock first, then use checked connectors to replace those in use that are found to be faulty.)
7. Inform them that faulty GLA connectors are to be sent to the contractor with a note that they are to be held for analysis under project HW44.
8. The site supervisors are to complete their tests within seven days.

PROJECT 9: INSTALLING A MINI-MINDER

You are employed in the construction department of Midstate Telephone System, and you have been asked to write an installation instruction sheet to be shipped with the "Mini-Minder" (an electronic intrusion detection device developed by MTS's engineering department). The Mini-Minder is a small detection unit which is mounted above windows, doors, or any other entries, where it automatically detects movement and transmits an alarm signal to a central unit. The central unit is concealed within the building and, when activated, both sounds an audible alarm and sends a message to police headquarters.

The Mini-Minder is shown in Figure 7-7. It is fixed to the wall above the door or window by removing the backplate and screwing the plate to the wall. The unit is then snapped onto the backplate (removing the unit from the backplate also sounds the alarm). The unit is battery operated.

All materials and hardware are supplied with the Mini-Minder but the installer will need an electric drill with a $\frac{3}{16}$ inch masonry drill bit, a Robertson No. 2 screwdriver, and a sharp pencil to do the job. The sequence in which the installation should occur is shown by the circled numbers in Figure 7-7.

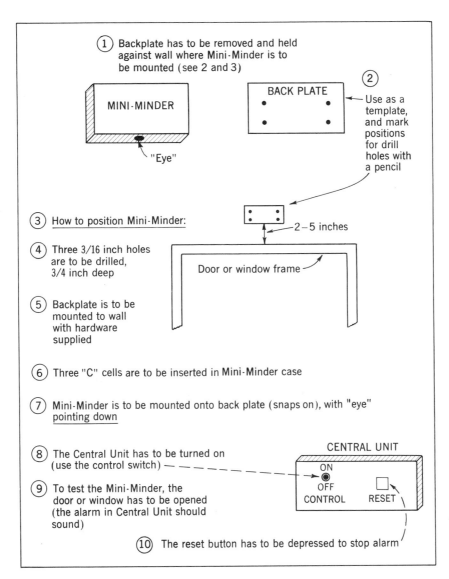

Figure 7-7. Installing and testing a Mini-Minder.

PROJECT 10: INSTALLING ENCODER EC7

You work for Midstate Telephone System and you have been asked to write instructions for installing an EC7 encoder at all MTS microwave transmission sites. The instructions are to accompany the encoder, which is the box illustrated in Figure 7-8. The encoder removes unwanted signals and improves transmission performance by 8% to 10%.

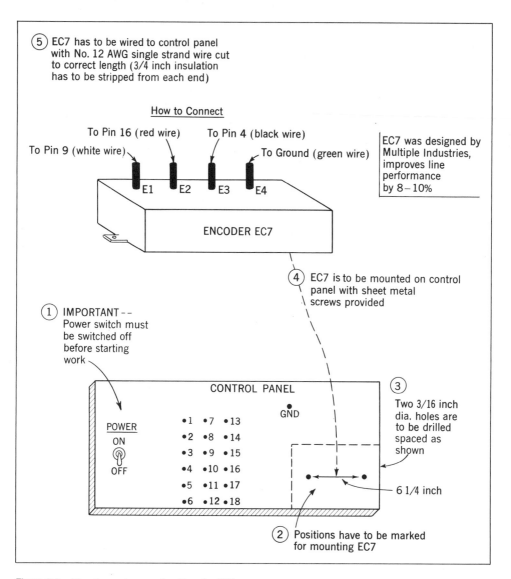

Figure 7-8. Mounting and connecting Encoder EC7.

In your instructions, tell the site technicians that they must: find a suitable location for the encoder at the bottom right of the control panel (see Figure 7-8); drill two holes for mounting the encoder; and connect the encoder with four wires (the connections are shown on the diagram). All materials (such as mounting hardware and wire) are supplied with each encoder but the technician will need a soldering iron, some Ersin 60/40 resin core solder, a drill with a ³⁄₁₆-inch bit, and a Robertson No. 2 screwdriver to carry out the work. The sequence in which the installation is to be carried out is shown by the circled numbers on the diagram.

When the installation is complete you should remind the technician to turn on the power to the control panel.

TECHNICAL ARTICLE

PROJECT 11:

This project assumes that you have built and tested the toothpick tower described in Project 3 of Chapter 6. Now you believe the topic would make an interesting article for a journal, magazine, or newspaper to publish.

Part 1.   Select a publication which you feel would be interested in printing your article. Describe why you have selected this particular magazine.

Part 2.   Prepare an outline of your proposed article. Describe your planned audience.

Part 3.   Before submitting an article for publication, you should always obtain management permission (a "release") to write and print it. Write a memorandum to your department head at the college you are attending, requesting approval to submit an article. Outline briefly what the article will contain and where you plan to submit it.

Part 4.   Write a 500-word summary of the article, plus a letter introducing it to the editor of the magazine.

8

Illustrating Technical Documents

If you open any well-known technical magazine you will immediately notice that illustrations are an integral part of most articles. Some are photographs that display a new product, a process, or the result of some action; others are line drawings that illustrate a new concept; some demonstrate how a test or an experiment was tackled; while another group may consist of charts and graphs that illustrate progress or show technical data in an easy-to-visualize form.

Good illustrations are not limited solely to magazine articles. They serve an equally useful purpose in technical reports, where their primary role is to help readers understand the topic. Interesting illustrations attract a reader's eye and encourage him or her to read a report. They can also break up dull-looking pages of narrative that lack eye appeal.

This chapter discusses the types of illustrations seen most often in technical reports, indicates the overlaps that exist between types, and suggests occasions when they can be used most beneficially.

PRIMARY GUIDELINES

The criterion for any illustration is that it should help explain the narrative; the narrative should *never* have to explain an illustration. Hence, an illustration must be simple enough for readers to understand quickly and easily. When I turn over the pages of *Scientific American* and stop at a particular page, usually the first things I look at are the drawings or photographs. If they immediately tell a story and catch my interest (and in *Scientific American* they regularly do!), I am encouraged to read the article.

To help *your* readers readily understand the illustrations you insert into your reports, you should:

1. Consider the audience the report is being written for, and what you
   readers to learn from each illustration.
2. Keep every illustration simple and uncluttered.
3. Let each illustration depict only one main point.
4. Position each illustration as near as possible to the narrative it supports
   Positioning the Illustrations, at the end of this chapter).
5. Label each illustration clearly with a figure or table number and a title (with th
   figure number and title centered *beneath* a figure or chart, and the table number
   and title centered *above* a table).
6. Add a caption (that is, comments or remarks) beneath a figure title, to draw
   attention to significant aspects of the illustration.
7. Refer to every illustration at least once in the report narrative.

GRAPHS *single & multiple*

Graphs are a simple means for showing a change in one function in relation to a
change in another. A function used frequently in such comparisons is time. The
other function may be temperature, erosion, wear, speed, strength, or any of
many factors that vary as time passes.

Suppose, for instance, that engineering technician John Greene wants to
determine how long it takes a newly painted manometer case to cool down after
it comes out of the drying oven. He goes to the paint shop armed with a stopwatch and a special thermometer. When the next manometer case comes out of
the oven he starts taking readings at half-minute intervals, and records the
results as in Table 8-1.

These readings are part of a study he is undertaking into the cooling
rate of different components manufactured by Macro Engineering Inc. The
information will also be used by the production department to establish how
long manometer cases must cool before assemblers can start working on them

Table 8-1 Cooling Rate, Manometer Case MM-7

TIME ELAPSED (min:sec)	TEMPERATURE (deg C)	TIME ELAPSED (min:sec)	TEMPERATURE (deg C)
:30	152.9	5:30	43.9
1:00	123.4	6:00	41.1
1:30	106.7	6:30	38.9
2:00	91.2	7:00	37.2
2:30	77.8	7:30	35.6
3:00	69.5	8:00	34.5
3:30	61.7	8:30	33.4
4:00	55.6	9:00	32.8
4:30	51.5	9:30	31.7
5:00	47.3	10:00	31.1

Ambient temperature 22.8°C   Oven temperature 180°C

with bare hands (the maximum bare-hand temperature has been established by management/union negotiation to be 38°C).

A quick inspection of this table shows that the temperature of the case drops continuously, is within 8.3 degrees of the ambient temperature after ten minutes (*ambient* means "surrounding environment"), and is down to the bare-hand temperature after seven minutes. A much closer examination determines that the temperature drops rapidly at first, then progressively more slowly as time passes.

Single Curve. John Greene can make the data he has recorded in Table 8-1 much more readily understood if he converts it into the graph in Figure 8-1. Now it is immediately evident that the temperature drops very rapidly at first, then slows down until the rate of change is almost negligible. There is no point in measuring or showing further drops in temperature unless John wants to demonstrate how long it takes the component to cool right down to the ambient temperature (probably 30 minutes or more).

Multiple Curves. In Table 8-2 John compares the temperature readings he has recorded for the cooling manometer case with measurements he has

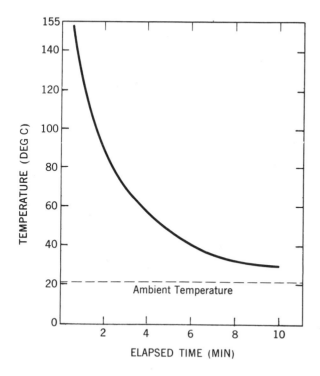

COOLING RATE – MANOMETER CASE MODEL NO. MM-7   **Figure 8-1.** Graph with a single curve.

Table 8-2 Cooling Rates for Three Components

TIME ELAPSED (MINUTES)	TEMPERATURE (DEG C)		
	COVER PLATE	PANEL BOARD	MANOMETER CASE
0:30	154.5	145.1	152.9
1	136.7	97.3	123.4
2	112.3	67.8	91.2
3	95.1	51.7	69.5
4	82.3	42.3	55.6
5	71.7	36.1	47.3
6	63.9	32.2	41.1
7	55.6	29.5	37.2
8	49.5	27.2	34.5
9	43.4	26.1	32.8
10	38.9	25.0	31.1

taken under similar conditions for a cover plate and a panel board. This time, however, he simplifies the table slightly by showing the temperatures at one-minute intervals.

What can we assess from this table? The most obvious conclusion is that in ten minutes the cover plate has cooled down less than the manometer case, and even less than the panel board. We can also see that the initial rate at which the components cooled varied considerably (the panel board, very quickly; the manometer case, fairly quickly; the cover plate, seemingly quite slowly), but it is difficult to assess whether there were any *changes in rate of cooling* as time progressed.

This can be shown more effectively in a graph, which John Greene has plotted in Figure 8-2. The rapid initial drop in temperature is evident from the initial steepness of the three curves, with each curve flattening out to a slower rate of cooling after two to four minutes. The difference in cooling rates for the three components is much more obvious than in the table.

Graphs will be essential when John Greene presents this data in a future report. If his intent is to present a general description of temperature trends, his narrative can be accompanied only by graphs like these, or by charts. But if he also wants to discuss exact temperature at specific times for each material, then the narrative and graphs will have to be supported by figures similar to those in Table 8-2. Factors that both you and John should consider when preparing graphs, charts, and tables are outlined below.

Constructing a graph usually offers no problems to technical people because they recognize a graph as a logical means for conveying statistical information. But constructing a graph that also tells a story is an aspect they may easily overlook. In order to illustrate your reports with graphs designed to emphasize the right information, you must first know the tools you have to work with.

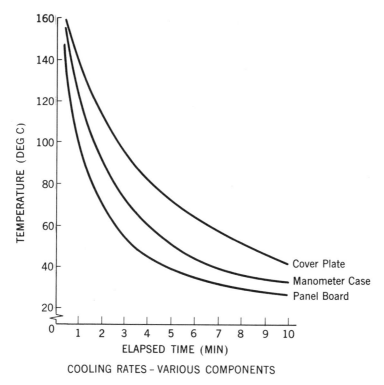

Figure 8-2. Graph with multiple curves.

Scales. The two functions to be compared in the graph are entered on two scales: a horizontal scale along the bottom and a vertical scale along the left side. (On large graphs the vertical scale is sometimes repeated on the right side to simplify interpretation.) The scales meet at the bottom left corner, which normally—but not always—is designated as the zero point for both. The curved lines in Figures 8-1 and 8-2 are the actual graphs; they are known as curves even though they may be straight lines or a series of short straight lines joining points plotted on the graph.

The functions represented by the two scales are commonly known as the dependent and independent variables, so named because a change in the dependent variable *depends* on a change in the independent variable. This can be demonstrated best by an example. If I wanted to show how the fuel consumption of my automobile increases with speed, I would enter speed as the independent variable along the bottom scale, and fuel consumption as the dependent variable along the left side (see Figure 8-3). Fuel consumption *depends* on speed (or, if you prefer, speed *influences* fuel consumption); the speed does not depend on the fuel consumption. The same applies to John Greene's temperature measurement graphs: Temperature is the dependent variable because it depends on the

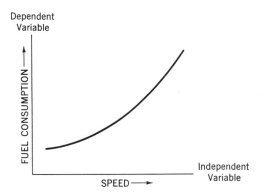

Figure 8-3. The dependent variable depends on the independent variable.

time that has elapsed since the components came out of the oven (the independent variable).

When you construct a graph, the first step is to identify which function should form the horizontal scale and which the vertical scale. Table 8-3 lists some typical situations which show that the same function can be an independent variable in one situation and dependent in another. Selection of the independent

Table 8-3 Identifying Dependent and Independent Variables		
Graph Illustrates	*Dependent Variable (vertical scale)*	*Independent Variable (horizontal scale)*
1. The effect that fequency has on the gain of a transducer	Gain	Frequency
2. How attendance at a ball game varies with temperature	Attendance	Temperature
3. How much a motor's speed affects the noise it produces	Noise	Speed
4. The changes in temperature brought about by changes in pressure	Temperature	Pressure
5. How much an increase in payload re- duces an aircraft's range by limiting the amount of fuel it can carry	Aircraft range (or fuel load)	Payload
6. How much increasing the fuel load of an aircraft to achieve greater range reduces its effective payload	Payload	Fuel load (or aircraft range)
Note: A function can be either dependent or independent, depending on its role in the comparison (see temperature in examples 2 and 4, and both functions in examples 5 and 6).		

variable depends on which function can be more readily identified as influencing the other function in the comparison.

The second factor to consider is scale interval. Poorly selected scale intervals, particularly scale intervals that are not balanced between the two variables, can defeat the purpose of a graph by distorting the story it conveys. Suppose John Greene had made the verticle scale interval of his time vs. temperature graph in Figure 8-1 much more compact, but had retained the same spacing for the horizontal scale. The result is shown in Figure 8-4(a). Now the rapid initial decrease in temperature is no longer evident; indeed, the impression conveyed by the curve is that temperature dropped only moderately at first, remained almost constant for the last three minutes, and will never drop to the ambient temperature. The reverse occurs in Figure 8-4(b), which shows the effect of compressing the horizontal scale: Now the curve seems to say that temperature plummets downward and it will be only a minute or two until the ambient temperature is reached. Neither curve creates the correct impression, although technically the graphs are accurate.

I said earlier that both scales normally start at zero, which would be the case when the resulting curve is comfortably balanced in the graph area. If it is crowded against the top or right side, then a zero starting point is unrealistic. In

Figure 8-4(a). Effect of compressed vertical scale.

Figure 8-4(b). Effect of compressed horizontal scale.

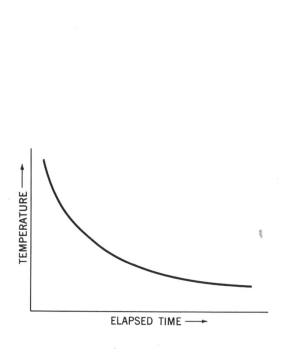

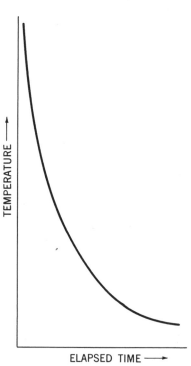

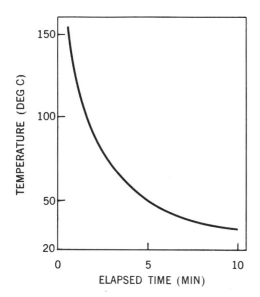

Figure 8-5. A correctly centered curve.

Figure 8-1 the curve occupies the top 75% of the graph area. Since no points will ever be plotted below the ambient temperature (which will hover around 23°C), the bottom portion of the vertical scale is unnecessary. This can be corrected by starting that scale at a higher value (say 15°, as in Figure 8-5), or by breaking the scale to indicate that some scale values have been omitted (Figures 8-2 and 8-6).

A graph should be easy to read. Multiple-curve graphs should not have too many curves or they will be hard to interpret. For example, a graph containing more than three curves is probably getting too crowded, particularly if the curves cross each other. In such instances you can help a reader identify the most important curve by making it heavier than the others (Figure 8-6), and can differentiate among curves that cross by using different symbols for each (Figure 8-7):

Most important curve	—bold line
Next most important curve	—light line
Third curve	—dashes
Least important curve	—dots

Colored lines should *not* be used, because the average copier will reproduce all lines in only one color (normally black).

Simplicity. Simplicity is important in graph construction. If a graph illustrates only trends or comparisons, and the reader is not expected to extract specific data from it, then a grid is not necessary. But if the reader will want to extrapolate quantities, a grid must be included. In the graphs we have examined, Figure 8-5 has no grid, Figure 8-1 has an implied grid that only suggests the grid

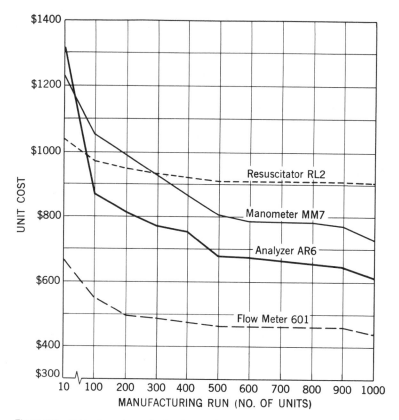

Figure 8-6. Bolder lines draw attention to most important curves (those showing maximum benefit from quantity manufacturing). Graphs normally should carry no more than three curves. (Courtesy Macro Engineering Inc, Phoenix, Arizona.)

pattern, and Figures 8-6 and 8-7 have full grids so that quantities can be extracted from them. Graphs without grids do not need top and right borders (Figure 8-3).

Plot points should be omitted and all captions should be horizontal. Captions for the curves should appear at the end of the curve whenever possible (Figure 8-2), or, alternatively, above or below the curve (Figures 8-6 and 8-7). They should never be written along the slope of the curve. The only caption that may be entered vertically is the identification for the vertical scale function.

CHARTS

Most charts show trends or compare only general quantities. Although it can be said that many graphs fall within this definition, I choose to separate them from charts because graphs have the ability to provide accurate interpretation, where-

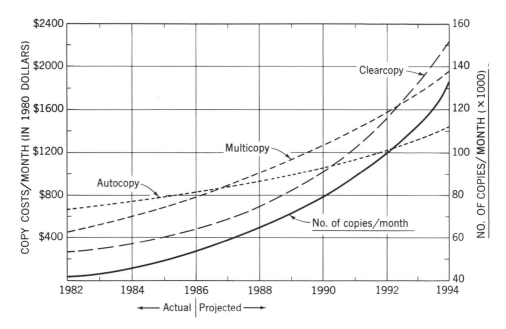

Figure 8-7. Different symbols distinguish between curves showing current and projected copying costs for three copiers. Note the two vertical scales, which permit three functions to be shown on one graph. (Courtesy H. L. Winman and Associates, Cleveland, Ohio.)

as charts generally do not. Hence, charts are more often seen in reports, technical articles, and papers intended for a general readership.

BAR CHARTS

You can use bar charts to compare functions that do not necessarily vary continuously. In the graph in Figure 8-1, John Greene plotted a curve to show how temperature decreased continuously with time. He could do this because both functions were varying continuously (time was passing and temperature was decreasing). For the production department, however, he has to prepare a report on how long it takes various components coming from the oven to cool to a safe temperature for bare-hand work. He prepares a bar chart to depict this because he knows the report will be read by both management and union representatives, and some of the readers may need easy-to-interpret data. He also has only one continuous variable to plot: elapsed time. The other variable is noncontinuous because it comprises the various components he has tested. In this case elapsed time is the dependent variable, and the components the independent. The bar chart John constructs is shown in Figure 8-8.

Scales for a bar chart can be made up of such diverse functions as time, age groups, heat resistance, employment categories, percentages of population,

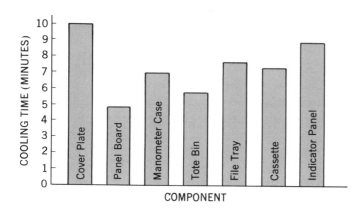

Figure 8-8. Vertical bar chart with one continuous variable (cooling time).

Time Required Before Oven-Dried Components
Reach Bare-Hand Handling Temperature (38°C)

types of soil, and quantities (of products manufactured, components sold, oscilloscopes in use, and so on). Charts can be arranged with either vertical or horizontal bars depending on the type of information they portray. The bars normally are separated by spaces the same width as each of the bars.

In a complex bar chart, the bars may be shaded to indicate comparisons within each factor being considered. The horizontal bar chart in Figure 8-9 uses two shades to describe two factors on the one chart, and has a legend to help the reader identify what each shade represents. Sometimes individual bars can be shaded to show proportional content, like those in Figure 8-10.

Horizontal bar charts can be used in an unconventional way when information can be arranged naturally on either side of a zero point, as when comparing negative and positive quantities, satisfactory and defective products, or passed and failed students. The chart in Figure 8-11 divides products returned for repair into two groups: those that are covered by warranty, and those that are not. Each bar represents 100% of the total number of items repaired in a particular product age group and is positioned about the zero line depending on the percentage of warranty and nonwarranty repairs.

HISTOGRAMS

A histogram looks like a bar chart, but functionally it is similar to a graph because it deals with two continuous variables (functions that can be shown on a scale to be increasing or decreasing). It is usually plotted like a bar chart because it does not have enough data on which to plot a continuous curve (see Figure 8-12). The chief visible difference between a histogram and a bar chart is that there are no spaces between the bars of a histogram.

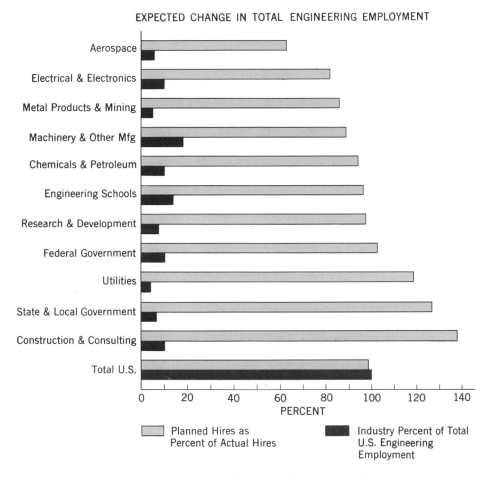

EXPECTED CHANGE IN TOTAL ENGINEERING EMPLOYMENT

Figure 8-9. This horizontal bar chart does three things: it shows the percentage of engineers employed in major industries, it predicts hirings, and it demonstrates the reduction in employment for engineers in the aerospace industry. (Courtesy Machine Design.)

SURFACE CHARTS

A surface chart (Figure 8-13) may look like a graph, but it is not. Its construction may seem so awkward that a technical person might wonder when it would be necessary to use one. Yet as a means for conveying information pictorially to nontechnical readers, it can serve a very useful purpose.

Like a graph, a surface chart has two continuous variables that form the scales against which the curves are plotted. But unlike a graph, individual curves

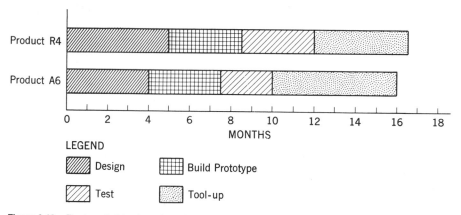

Product R4

Product A6

0 2 4 6 8 10 12 14 16 18
MONTHS

LEGEND

▨ Design ▦ Build Prototype

▨ Test ▦ Tool-up

Figure 8-10. The bars in this chart show development times for proposed new products. The legend is included with the chart.

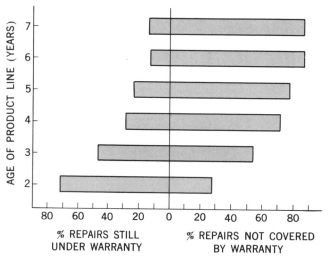

AGE OF PRODUCT LINE (YEARS)

80 60 40 20 0 20 40 60 80

% REPAIRS STILL UNDER WARRANTY   % REPAIRS NOT COVERED BY WARRANTY

Figure 8-11. A bar chart constructed on both sides of a zero point.

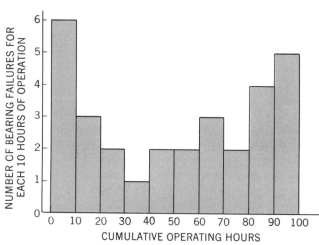

NUMBER OF BEARING FAILURES FOR EACH 10 HOURS OF OPERATION

0 10 20 30 40 50 60 70 80 90 100

CUMULATIVE OPERATING HOURS

Figure 8-12. This histogram shows the number of bearing failures for every 10 hours of operation. Considerably more data would have been required to construct a curve.

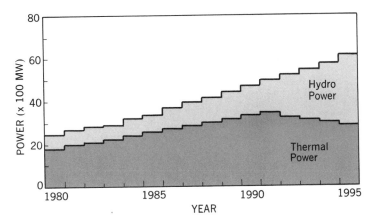

Figure 8-13. Surface chart adds hydro data to thermal data to show actual and projected energy resources of a power utility. (Courtesy Crestview Power Company.)

cannot be read directly from the scales. The uppermost curve on a surface chart shows the *total* of the data being presented. This curve is achieved as follows:

1. The curve containing the most important or largest quantity of data is drawn in first, in the normal way. This is the Thermal curve in Figure 8-13.
2. The next curve is drawn in above the first curve, using the first curve as a base (or "zero") and adding the second set of data to it. For example, the energy resources shown as being variable in 1990 are:

    Thermal Power: 33,000 MW
    Hydro Power: 14,500 MW

In Figure 8-13, the lower curve for 1990 is plotted at 33,000 MW. The 1990 data for the next curve is 14,500 MW, which is added to the first set of data so that the second curve indicates a *total* of 47,500 MW. (If there is a third set of data, it is added on in the same way.)

The area between curves is shade to indicate that the curves represent the boundaries of a cumulative set of data. Normally, the lowest set of data has the darkest shade, and each set above it is progressively lighter.

PIE CHARTS

A pie chart is aptly named, because it looks like a whole pie viewed from above with cuts in it ready for people of varying appetites. It is a pictorial device for showing approximate divisions of a whole unit. The pie chart in Figure 8-14 depicts the percentage of electronic equipment manufactured in four major product categories.

If a pie chart has a lot of tiny wedges which would be difficult to draw and hard to read, some of them are combined into a larger single wedge and given a general heading, such as "miscellaneous expenses," "other services," or "minor effects." All the wedges must add up to a whole unit, such as 100%, $1.00, or 1 (unity).

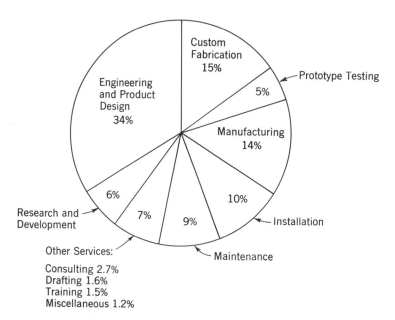

Figure 8-14. Pie chart shows Macro Engineering Inc's products and services. "Slices" add up to 100%.

DIAGRAMS

Under this general heading is included any illustration that helps the reader understand the narrative yet does not fall within the category of graph, chart, or table. It can range from a schematic drawing of a complex circuit to a simple plan of an intersection. I must add one restriction: If included in the narrative part of a report, it must be clear enough to read easily. This means that complex drawings should be placed in an appendix and treated as supporting data.

Diagrams should be simple, easy to follow, and contribute to the story. They can comprise organization charts (see Chapter 2), flow diagrams (Figures 2-4, 5-3, 5-5, 8-15, and 9-1), site plans, and sketches.

PHOTOGRAPHS

A photograph can do much to help a reader visualize shape, appearance, complexity, or size. For example, the photographs of Harvey Winman, Tina Mactiere, and Wayne Robertson add depth to the narrative description of the two engineering companies in Chapter 2. The criterion when selecting any photograph is that it be clear and contain no extraneous information that might distract the reader's attention.

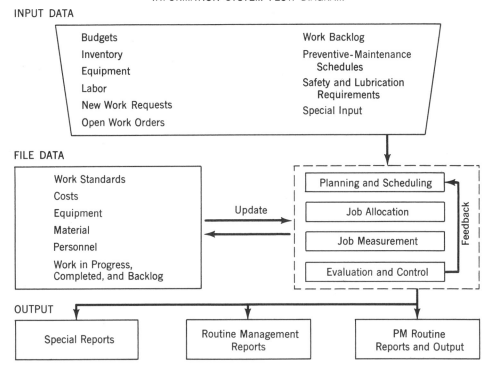

Figure 8-15. Flow diagram of a Maintenance Information System. (Courtesy Mechanical Engineering.)

Photographs are more difficult to copy than drawings, since most photocopiers do not print them clearly. For really good reproduction they need professional services often beyond the capability of an in-plant printing department. Consequently, before deciding to use photographs you should find out if your printer can reproduce them clearly and economically. If the printer cannot handle them easily, you may have to glue individual copies of the photographs into your report (an awkward process), place them in a sleeve or envelope at the back of the report (a cumbersome arrangement), have copies printed professionally (an expensive proposition), or replace the photographs with good sketches.

TABLES

The criterion for inserting a table into a report is whether the reader will need to refer to it. If it is going to be used as the report is read, then it should be included in the discussion. If the reader will be able to understand the report without referring to it, but may want to consult it later, it should be included as an appendix. If the information in the table can be expressed more simply by words, a graph, or a chart, then the table should be omitted.

A table inserted as an illustration should be as short as possible so that it can be read easily. If the information compiled during a series of tests is lengthy, the essentials should be summarized and built into a short table. The table should also refer to the main body of information, which is placed in the appendix. Similarly, a table should have as few columns as possible, and each column should contain only data that the reader will need.

From the design viewpoint, I prefer a table that is "open," that is, without lines between the vertical and horizontal columns. (For comparison, Tables 8-1 and 8-2 are open, while Table 8-3 is closed.) The captions at the head of each column should be clear and specific and should include units of measurement (such as decibels, volts, and seconds) so that the units need not be repeated throughout the table. This has been done in Tables 8-1 and 8-2.

It is not enough simply to insert a table and assume that the reader will know what to infer from it. The narrative should refer to the table and comment on its reason for being there. This entry draws the reader's attention to a specific area of the table:

> The voltage fluctuations were recorded at ten-minute intervals and entered in column 3 of Table 7, which shows that fluctuations were most marked between 8:15 and 11:20 a.m.

Alternatively, a similar comment can be inserted as a note beneath the table.

So far we have assumed that tables contain only technical information, such as the results of tests. As Table 8-3 shows, this is not necessarily true. Frequently, the best way to show comparative data or summaries of analyses is to insert informative abstracts or comments in a table, as has been done at the end of Fred Stokes's evaluation report in Figure 5-6 of Chapter 5.

POSITIONING THE ILLUSTRATIONS

Not only must the narrative refer to every illustration in a report, but each illustration should be on the same page as or facing the narrative it supports. A reader who has to keep flipping pages back and forth between narrative and illustrations will soon tire, and your reasons for including the illustrations will be defeated.

When reports are printed on only one side of the paper, full-page illustrations can become an embarrassment. The only feasible way to place them conveniently near the narrative is to print them on the back of the preceding page, facing the words they support. But this in turn may pose a printing problem. A more logical solution is to limit the size of illustrations so that they

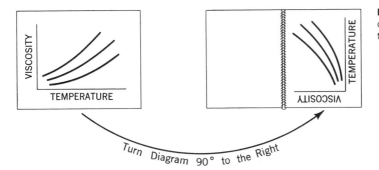

can be placed beside, above, or below the words, and then to make sure that the typist keys them in at the right points.

Horizontal full-page illustrations may be inserted sideways on a page but must always be positioned so that they are read from the right (see Figure 8-16). This holds true whether they are placed on a left or right page.

When an illustration is too large to fit on a normal page, or is going to be referred to frequently, you should consider printing it on a foldout sheet and inserting it at the back of the report. If the illustration is printed only on the extension panels of the foldout, the page can be left opened out for continual reference while the report is being read. This technique is particularly suitable for circuit diagrams and flow charts.

If the equipment used to print your reports cannot reproduce large foldout sheets, you can prepare the illustrations on tracing paper and make blueline prints of them. An ideal size is 11 × 22 inches, which folds conveniently into a standard 8½ × 11 inch report, as shown in Figure 8-17. Such a foldout can be cut from a standard size C (17 × 22 inches) sheet of blueprint paper. Even larger foldouts can be made in this way, but they tend to be unwieldy.

ILLUSTRATING A TALK

An illustration for a talk must be utterly simple. Its message must be so clear that the audience can grasp it in seconds. An illustration that forces an audience to read and puzzle out a curve detracts from a talk rather than complements it.

If you want to convert any illustration to a large visual aid or slide, you will have to observe the following limitations:

1. Make each illustration *tell only one story*. Avoid the temptation to save preparation time by inserting too much data on one chart. Be prepared to make two or three simple charts in place of a single complex one.
2. Use bold letters large enough to be read easily by the back row of your audience.
3. Use very few words, and separate them with plenty of white space.
4. Give the illustration a short title.

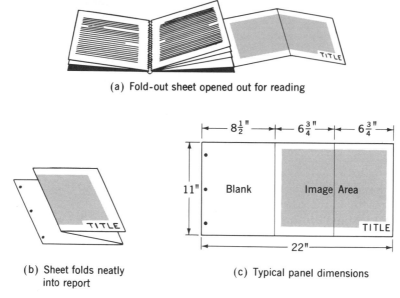

(a) Fold-out sheet opened out for reading

(b) Sheet folds neatly into report

(c) Typical panel dimensions

Figure 8-17. Large illustrations can be placed on a fold-out sheet at rear of report.

5. Insert only the essential points on a graph. Let the curves tell the story, rather than be buried in construction detail.
6. Accentuate key figures and curves with a bold or colored pen (but remember that from a distance some colors look very similar).
7. Avoid clutter—a simple illustration will draw attention to important facts, whereas a busy one will hide them.

Diagrams that are to illustrate an in-plant briefing may be hand-lettered with a felt pen on large sheets. But those to be used for a technical paper presented before a large audience should be prepared more professionally. In the first instance you are expected to do a reasonably good job at no great expense. In the second, you are conveying an image of yourself and the company you represent; in effect, your diagrams are demonstrating the technical quality of your company's products.

WORKING WITH AN ILLUSTRATOR

Although you might draw your own illustrations, in industry you are much more likely to work with a company draftsperson who will prepare your drawings, graphs, and charts according to your requirements. Good communication between you and the illustrator is essential if the drawings you want are to appear in the form you visualized when you wrote your report.

An illustrator needs to have much more than a bare, roughed-out sketch to work from. You will have to describe your project and its outcome in detail, so that the illustrator will know:

- The background to and purpose of the report.
- Who the readers will be, what their technical knowledge is, and how they will use the information contained in the report.
- What each illustration is to portray, and what particular aspects are to be emphasized.
- What size each illustration is to be (vertical and horizontal dimensions).
- What printing method is to be used.
- Whether the report and its illustrations will be reduced in size during printing, and by how much. (A drawing that is to be reduced must have lines that are not too fine.)
- When you need the illustrations.

You can help an illustrator even more by providing a sketch of the proposed illustration and a copy of the words the illustration is to support. Better still, talk to the illustrator *before* you write your report, describe what illustrations you plan to use, and ask for suggestions for their preparation.

Similarly, always discuss printing methods with the person who will be making copies of your report before you start making reproduction copy. Certain reproduction equipment cannot handle some sizes, materials, and colors, and few can reproduce photographs clearly.

Assignments

PROJECT 1: ASSESSING RELAY LIFE

Eight years ago the Interstate Telephone Company bought and installed 90,000 relays, to be used in a long-range testing program that would assess failure rates. The relays purchased were:

Type	Quantity
Nestor 221	40,000
Vancourt 1200	20,000
Macro R40	20,000
Camrose Series 8	10,000

As relays failed they were replaced and the failures were recorded and totalled for each year:

Number of Failed Relays

YEAR	NESTOR 221	VANCOURT 1200	MACRO R40	CAMROSE 8
1	1,123	901	1,180	105
2	1,080	805	690	124
3	1,007	762	321	131
4	1,076	813	279	306
5	1,140	878	322	402
6	1,656	910	415	545
7	2,210	956	478	609
8	3,303	1,012	584	891
Totals	12,595	7,037	4,269	3,113

Prepare a graph or chart depicting these failure rates and predicting failures for the next five years.

PROJECT 2: WATER CONSUMPTION IN MONTROSE

The table below is a record of the average daily water consumption for the City of Montrose, Ohio, over the past calendar year. The City Engineer has to prepare two reports in which he will identify how much water has been consumed by different segments of the community, and analyze when and why variations in consumption occurred. He asks you to prepare a graph, chart, or diagram to accompany each of these reports. They are to comprise:

1. An illustration for a technical report that will be read by engineers and technicians involved in water supply and distribution.
2. An illustration for a water consumption analysis that will be sent to all consumers. (This part of the project should be done in conjunction with the writing assignment outlined in Project 5 of Chapter 7.)

City of Montrose, Ohio Average Daily Water Consumption (gallons)

MONTH	BUSINESS & INDUSTRY	PRIVATE HOMES	SCHOOLS & COLLEGES
January	4,256,000	3,608,000	1,910,000
February	4,310,000	3,673,000	2,296,000
March	4,318,000	4,127,000	2,501,000
April	4,325,000	4,980,000	2,507,000
May	4,331,000	5,641,000	2,610,000
June	4,417,000	6,775,000	2,192,000
July	4,484,000	8,926,000	807,000
August	4,491,000	9,681,000	762,000
September	4,369,000	6,810,000	1,418,000
October	4,323,000	5,604,000	2,298,000
November	4,298,000	4,254,000	2,304,000
December	4,243,000	3,672,000	1,641,000

PROJECT 3: MEASURING RADAR RANGE

During the past six months you have been conducting tests to assess the range of a new radar system known as the Search R20. Your tests were taken with a radar set on the ground at various elevations, and an aircraft approaching at various altitudes. Two sets of tests were taken, each with the radar set positioned at four different elevations but equipped with two different magnetrons:

MAXIMUM RANGE IN MILES MEASURED WITH AIRCRAFT AT:

	Altitude of Radar Set (ft)	5,000 ft (mi)	10,000 ft (mi)	20,000 ft (mi)	40,000 ft (mi)	60,000 ft (mi)
MAGNETRON QA	Sea Level	48	86	112	134	161
	1200	73	105	136	151	179
	2400	87	138	166	182	189
	4800	101	159	193	211	220
MAGNETRON KB	Sea Level	54	101	127	156	188
	1200	79	118	160	175	201
	2400	96	151	183	198	213
	4800	111	170	214	229	244

Prepare graphs or charts to demonstrate:

1. The effect that transmitter height has on the range measured.
2. The differences between magnetrons QA and KB.

PROJECT 4: TRAFFIC COUNTS AT TARNAPIN

Prepare graphs or charts depicting traffic flow at the primary locations along highway 271 as it passes through the town of Tarnapin (see Project 1 of Chapter 6). Assume that your illustrations will become part of a report being prepared for Wendy Kibarre, president of the Tarnapin Business Owners' Association.

PROJECT 5: NOISE LEVELS AT MANSASK INSURANCE CORPORATION

Prepare graphs or charts to show how noise levels differ under the four sets of conditions at Mansask Insurance Corporation:

1. When only Mansask is working.
2. When both Mansask and Artistic Novelties (the noisy neighbor) are working.
3. When only Artistic Novelties is working.
4. When neither office is working.

Include a "normal" office working noise level. Remember that your illustrations should *compare* the noise levels, not simply show the levels recorded for the four sets of conditions. (For details, refer to Project 2 of Chapter 6.)

PROJECT 6: SHAVING PEAK POWER CONSUMPTION

Prepare an illustration to depict the power savings that can be achieved by phasing in Broderick Metal Products Incorporated's standby power generator during peak electricity demand periods (see Project 4 of Chapter 6).

Part 1.   Prepare a graph or chart to depict present consumption against savings to be achieved by using peak power transferring or utility paralleling.

Part 2.   Prepare a graph or chart to depict the points at which a break-even point will be reached with either method.

Part 3.   Prepare a table depicting factors to be compared between the two methods.

9

Technically—Speak!

eval. form   *podium*   *Martin's letter*

This chapter covers two facets of public speaking, both concerned with the oral presentation of technical information. The first is the oral report, sometimes called the technical briefing, delivered to a client or to colleagues. The second is the technical paper presented before a meeting of scientific or engineering-oriented persons. Both depend on public speaking techniques for their effectiveness, although neither requires vast experience or knowledge in this field. Also included in this chapter are some suggestions on how to contribute properly to meetings you attend.

THE TECHNICAL BRIEFING

Your department head approaches your desk, a letter in hand, and says:

> Mr. Winman has had a letter from the RAFAC Corporation. They're sending in some representatives next Tuesday. I'd like you to give them a rundown on the project you're working on.

Every day visitors are being shown around industrial organizations, and every day engineers and technicians are being called upon to stand up and say a few words about their work. On paper, this sounds straightforward enough, but to those who have to make the oral presentation it can be a traumatic experience. Much of their nervousness can be reduced (it can seldom be entirely eliminated, as any experienced speaker will tell you) if they are given some hints on public speaking. The best training, of course, is practical experience, which can be gained only by standing up and doing the job.

ESTABLISH THE CIRCUMSTANCES

As soon as you have been informed that you are going to deliver an oral report, you should establish some of the circumstances surrounding the visit—factors that will have a bearing on your approach. Jot down what you need to know, then ask your department head questions like these:

Who are the visitors?
You will probably be introduced to them, but at the critical moment before you speak you don't want to be concentrating on names. If you have heard them before they will be easy to remember.

How much will they know already?
You don't want to bore your visitors by repeating unnecessary details. Find out if they will come to you with no knowledge of your project, or whether management will have given them some preliminary information before you speak.

How long do you want me to talk?
Find out if management wants you to describe the project in detail, or simply touch on the highlights. It could be that the stop in your area is only a two-minute pause on a plant tour, or it may be a specific visit to study your project. The intent will directly influence your subject coverage.

Where is the briefing to take place?
Are you to address the visitors in the board room? Or will they be coming down to the project area? Availability of equipment may dictate how you tackle your subject and whether you need to make drawings to illustrate your talk.

Only when these factors have been firmly established can you start making notes. Jot down the topics you intend to discuss, and arrange them in an interesting, logical order. Think of an unenlightened person sitting in front of the equipment and try to look at it in the same way that he or she will. Don't let your familiarity with the project blind you to characteristics that are unimportant to you but would be interesting to the observer.

FIND A PATTERN

The best technical briefings follow an identifiable pattern, just as written formal reports do. You can establish a pattern for your briefing by answering three questions that you would be likely to ask if you were a visitor to another plant.

1. What Are You Trying To Do?   Use the answer to this question to build an *Introduction,* as you would for a formal report. Offer your listeners some background information, which may comprise:

> How your company became involved in the project (with, perhaps, a comment on your own involvement, to add a personal touch).

Exactly what you are attempting to do (in more formal terms, your objectives).
The extent or depth of the project.

2. What Have You Done So Far? This would be the *Discussion* section
of the formal report. The answers to this question should cover:

How you set about tackling the project.
What you have accomplished to date (work done, objectives achieved, results
obtained, and so on).
Preliminary conclusions you have reached as a result of the work done (if the
work is complete, these will be the final conclusions).

3. What Remains to Be Done? (or **What Do You Plan to Do
Next?**) This question is relevant only if the project is still in progress, in which
case it is equivalent to the *Future Plans* section of a written progress report.
Answers to this question should cover:

The scope of future work.
Results you hope to achieve.
A time schedule for reaching specific targets and final completion.

If the project is complete, this question is not relevant and is replaced by an
alternative question: **What Are the Results of Your Project?** The answers would
then be combined with the final answer to question 2 and would be equivalent to
the *Conclusions* section of a written report.

Now we have a pattern for the main part of your briefing. But you still
have to give it a beginning and an ending. The beginning should be a quick
synopsis of the project in easy-to-understand terms—the equivalent of a report
Summary. The ending can be a quick summing up (a Terminal Summary) that
leads into an opportunity for your listeners to ask questions. The complete
pattern is shown as a flow diagram in Figure 9-1.

PREPARE TO SPEAK

Make brief speaking notes on prompt cards no smaller than 6 × 4
inches. If it is not convenient to hold the cards, place them on a makeshift
speaker's stand, or even mount them on the back of a piece of equipment where
you can read them easily without straining. Write in large, bold letters that you
can see at a glance, using a series of brief headings to develop the information in
sufficient detail. A specimen card is shown in Figure 9-2. Prompt cards like this
are scaled-down versions of the speaking notes I recommend for technical paper
presentation. For comparison, a typical page of notes is illustrated in Figure 9-3.

Don't overlook the practical aspects of the briefing. If you have equipment to demonstrate, consider its layout in relation to a logically organized

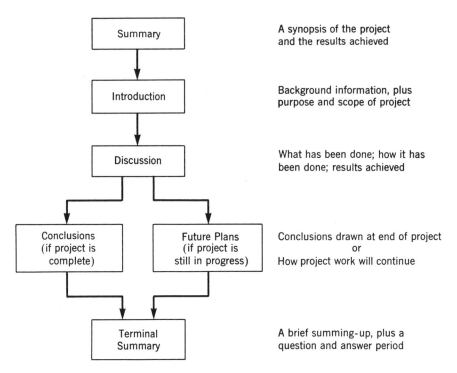

Summary	A synopsis of the project and the results achieved
Introduction	Background information, plus purpose and scope of project
Discussion	What has been done; how it has been done; results achieved
Conclusions (if project is complete) **Future Plans** (if project is still in progress)	Conclusions drawn at end of project or How project work will continue
Terminal Summary	A brief summing-up, plus a question and answer period

Figure 9-1.   Flow diagram for a technical briefing.

Figure 9-2.   Prompt card for an oral report.

```
                                              3
INSTRUCTOR'S CONSOLE -
      TAPE RECORDER - 4 TRACK
                         MODIFIED
      SLIDE PROJ - 35 MM
      MONITOR HEADPHONES
      MASTER COUNTER

STUDENTS' POSITIONS - (AT EACH)
   HEADPHONES
   VOL CONTROL
   SWITCHES - PUSHBUTTONS A-B-C
   COUNTER - TOTAL RESET TYPE
```

description. Try to arrange the briefing so that you will move progressively from one side of the display area to the other, instead of jumping back and forth in a disorganized way. If the display is large and easy to see, let it remain unobtrusively at the back of the area. If it is small, consider moving it forward and talking from beside or behind it.

If visual aids will help you give a clearer, more readily understood briefing, then prepare as many as you will need. They may range from a series of steps listed as headings on a flip chart to a working model that demonstrates a complex process. Strive for simplicity; let the visual aid support your commentary rather than make the commentary explain an overly complex aid. Some suggestions for preparing graphs, charts, and diagrams are contained in Chapter 8.

Take a leaf from the experienced technical speakers' notebook and practice your briefing. Run through it several times, working entirely from your prompt cards, until you can do so without undue hesitation or stumbling over awkward words. If the cards are too hard to follow, or contain too much detail, amend them. Then ask a colleague to sit through your demonstration and give critical comments.

The time and effort you invest in preparing for a briefing will depend on your confidence as a speaker and your familiarity with the topic. The more confident you are, the less time you will need. No one will expect you to give a professional briefing, but everyone—visitors and management alike—will appreciate a carefully prepared talk presented in an interesting manner.

THE TECHNICAL PAPER

Chapter 7 discussed the steps Mickey Wendell would have to take to publish a magazine article or a technical paper. (He is a senior technician in H. L. Winman and Associates' Materials Testing Laboratory, and he has discovered that an additive called Aluminum KL mixed with cement in the right proportions produces a concrete with high salt resistance.) This chapter assumes that the papers committee of the Combined Conference on Concrete liked Mickey's abstract and summary, and the chairman of the committee has notified him that his paper has been selected for presentation at the forthcoming Chicago conference. Mickey has four months to prepare for it.

Mickey must recognize immediately that presenting a paper before a society meeting is much more demanding than delivering the same information at a technical briefing to company visitors. The occasion is more formal, the audience is much larger, and the speaker is working in unfamiliar surroundings. Many experienced engineers and scientists duck their responsibility to the audience when faced with such a situation and simply read their papers verbatim. This can result in a dull, monotonous delivery that would turn even a superior technical paper into a dreary, uninteresting recital. If Mickey is to avoid this trap, he must start preparing early.

PREPARATION

The key to an effective oral presentation is to have good speaker's notes and to practice with them. This takes time. Mickey must write the publication version of his paper soon after he hears that it has been accepted (not leave it to the very last week, as so often happens), because he will need it to prepare his speaking notes.

The spoken version of a technical paper does not have to cover every point encompassed by the written version. In the 15 to 20 minutes allotted to speakers at many society meetings, there is time to present only the highlights—to trigger interest in the listeners so that they will want to read the published version. Mickey has to consider how he is to stimulate and hold this interest.

Selecting Topic Headings. His speaking notes will consist mainly of topic headings extracted from the written version of his paper. He should jot these headings onto a sheet of paper and then study them with four questions in mind:

1. Which points will prove of most interest to the audience?
2. Which are the most important points?
3. How many can I discuss in the limited time available?
4. In what order should I present them?

When Mickey was writing his paper he was preparing information for a reader. Now he is preparing the same information for a listener and the rules that guided him before may not apply. The logical and orderly arrangement of material prepared for publication is not necessarily that which an audience will find either interesting or easy to digest.

Mickey may assume that the audience at a society meeting is technically knowledgeable, has some background information in the subject area, and is interested in the topic. Most of his audience will likely be civil engineers and technologists, with a sprinkling of sales, construction, and management people. He must keep this in mind as he examines his list of headings, identifies which points he intends to talk about, and arranges them in the order he feels will most suit his listeners.

Preparing Speaking Notes. There are many ways in which Mickey can prepare his speaking notes. He can type them onto prompt cards similar to those illustrated in Figure 9-2, print them in bold letters on 8½ × 11 inch sheets, or enter them in a notebook. But he should never take the shortcut of simply entering the headings in the margin of the typed copy of his paper. The temptation to read the paper may become too great if he is very nervous, and once he starts reading he will find it difficult to return to extemporaneous speech.

Anna King suggested that Mickey use a notebook because its pages are bound together. It would be reassuring to know that if he inadvertently drops his notes, he only has to turn to the correct page to continue speaking. It could be a catastrophe to find his carefully prepared prompt cards scattered around his feet!

The notebook she recommended should be about 9 × 7 inches when closed (slightly wider than this textbook), should lie flat when opened, and should have wide-spaced horizontal lines. Its left page should carry speaking notes, and its right page demonstration notes (see Figure 9-3). The left page is divided into four columns, the first for "time elapsed" and the remaining three each containing progressively more information. The right page is a storage area for notes indicating when demonstrations are to be carried out and slides or diagrams are to be presented. It also carries comments and excerpts to be read to the audience.

In effect, Mickey Wendell would be wise to have two notebooks: one for his initial speaking notes and for speaking practice, the other, which he will prepare just before the presentation, for his final speaking notes. He should take his list of topics, abbreviate them as much as possible, and enter them in the "Topics" column of the initial notebook. He should then expand each topic into a series of brief general headings, and enter these in the "Headings" column. Finally, he should support these two columns of cryptic notes with details (information he may need to refresh his memory) which he should enter in the "Details" column. On the right page he should enter comments on the visual aids he plans to use, and key them into the "Details" column at the proper points.

The amount of information he provides in these columns will depend on the complexity of the subject, on Mickey's familiarity with it, and on his previous speaking experience. An experienced speaker familiar with the subject needs less information than an inexperienced speaker such as Mickey. As a general rule, the notes should not be so lengthy that Mickey cannot extract pertinent points at a glance, for then there will again be a tendency to read. Neither should they be so brief that he has to rely too much on his memory, which could cause him to stumble haltingly through his talk.

PRACTICE

Mickey's next task is to practice speaking from the notes at least once a day for several days. It will not be enough simply to scan or read the notes and assume that he is thus becoming familiar with them. He must speak from them aloud, as though presenting the paper to an audience.

Using Preliminary Notes. After each reading he should modify the notes, including more information or deleting unnecessary words. As he grows

DEMONSTRATIONS

Time	Topics	Headings	Details
6	Preliminary Notes	Format	Author's choice Easiest for *him* This recommended: flexible
		Layout	Notebook—lies flat Left p.—columns Right p.—Slides, diagrams —DEMONSTRATE →
	Topics		Abbreviate—enter col 2 Divide into headings
		Headings	Abbreviate—enter col 3
		Details	Support headings—col 4 Amount depends on: Complexity of subject Author familiarity

SLIDE 3:

5

TIME	TOPICS	HEADINGS	DETAILS
6	NOTES	FORMAT	SPEAKER'S CHOICE MUST BE EASY TO USE MUST BE FLEXIBLE
		LAYOUT	IN NOTEBOOK LEFT PAGE - COLUMNS RIGHT PAGE - SLIDES, DIAGRAMS ← DEMO
8	TOPICS	CHOOSE	ABBREVIATE • ENTER IN COL. 2
		HEADINGS	CHOOSE • WRITE IN COL. 3
		DETAILS	SUPPORT HEADINGS • WRITE IN COL. 4 AMOUNT DEPENDS ON- • COMPLEXITY OF SUBJECT • KNOWLEDGE OF SUBJECT • EXPERIENCE AS SPEAKER
10	SPEAKING PRACTICE	SPEAK	ALOUD - 4 TIMES TIME - USE STOPWATCH MARK WEAK SPOTS

4

Figure 9-3. Speaking notes in notebook form.

familiar with the notes his confidence will increase, and certain sentences and apt phrases will spring readily to mind at the sight of a single word or short heading. Thus he will soon find that he can easily maintain oral continuity between headings.

As he rehearses his paper, Mickey should time himself and enter time marks at suitable intervals in the left column. He should aim to speak for less time than the Papers Committee allows; the ideal is to allow 17 to 18 minutes speaking time for a paper scheduled for a maximum of 20 minutes. Then, if he wishes to make some previously unanticipated remarks when he presents his paper (possibly referring to statements in papers presented prior to his), he will have the time.

If he is using visual aids—and he should whenever possible, to insert variety into his presentation—he should practice using them, first on their own and then as part of the whole paper. This will give him a chance to check whether he has keyed them in at the right places, and whether the entries are sufficiently clear to permit him to adjust from speech to visual aid and back again without losing continuity.

Using Final Notes.   When he is satisfied that his preliminary notes are satisfactory and will not need any major changes, he can prepare his final speaking notes. These should be typed with a large typeface or hand-lettered (in ink) in clearly legible capital letters.

Mickey should plan to have these final notes ready at least three days before leaving for the conference. To a certain extent the headings in the first set of notes have helped to trigger familiar phrases and sentences. Now he has to familiarize himself with new pages. During these last practice sessions he should attempt a full dress rehearsal by presenting the paper before some of his colleagues, among whom there may be someone qualified to comment on his platform techniques. If this is not possible, he should at least try speaking the paper alone, standing at a rostrum or desk to simulate actual conditions. This "dry run" will also give him the opportunity to check his speaking time.

PRESENTATION

Overcoming Nervousness.   There are very few speakers who are not at least a little nervous when the time comes to present their paper. Some nervous tension is perfectly normal, and can even help a speaker give a better performance. Mickey will find that once he starts speaking most of this nervousness disappears. A lot will depend on his speaking notes: If he has done his job well, knows that they are reliable, and is confident using them, he will find the familiar phrases and sentences form easily. Then he will begin to relax and speak with even greater confidence.

Improving Platform Manner.   Mickey was lucky; he was able to learn some elementary platform techniques from Ron Brophy, a member of the Winman electrical engineering staff who had presented papers previously. Ron was present at Mickey's "dry run" and spent some time afterward telling him how he could improve his presentation. Here are some of Ron's suggestions:

> Speak at a moderate rate—120 to 140 words per minute is recommended.
>
> Speak up. If possible, try speaking without a microphone, since this gives you much greater freedom of movement and tonal flexibility. If a microphone has to be used, try to maintain a moderately constant speaking level.
>
> Pause occasionally to study your speaker's notes. Never be afraid to stop speaking for a few moments while consolidating your position and establishing that every major topic has been covered. This also gives you an opportunity to check elapsed time.
>
> Look at your audience. Try to speak to individuals in turn, rather than the group as a whole, picking them out in different parts of the room so that every listener will feel he or she is being addressed personally.
>
> Use humor, but only if it fits naturally into the paper and you are adept at speaking humorously. Make sure that your audience laughs *with* you and not *at* you.
>
> Avoid distracting habits that tend to divert audience attention. Examples are pacing back and forth or balancing precariously on the edge of the platform (the audience will be far more interested in seeing whether you fall off than in following your paper). Nervous afflictions, such as jingling keys or coins in your pocket (put them in a back pocket, out of reach), playing with objects on the speaker's table (remove them before you start speaking), or cracking your knuckles, should also be avoided.

Ron Brophy also pointed out to Mickey that although adequate pre-platform preparation and knowledge of platform techniques can give an author confidence, they are not sufficient in themselves to break down the initial barrier between speaker and audience. Successful speakers and instructors develop a well-rounded personality which they use continuously and unconsciously to establish a sound speaker-audience relationship. For the author of a technical paper who faces an audience once, and then only briefly, the most important personality attribute is enthusiasm.

Enthusiasm is contagious. If a speaker likes his subject, really enjoys describing it, his enthusiasm will be demonstrated in his presentation by the vigorous manner in which he tackles his material. If he is also businesslike and cheerful he will quickly reach his audience, who will respond to his approach by listening attentively.

In summary, a technical person who wants to make a good presentation before an audience should:

1. Prepare the material thoroughly.
2. Practice speaking.
3. Learn a few platform techniques.

If you are willing to devote the time and have the interest to try these methods, you will discover that you are speaking *to* your listeners; not *at* them, as you would have done had you simply read your original paper. Even more important, you will win the support and respect of your audience, who will applaud your attempts to speak extemporaneously.

TAKING PART IN MEETINGS

We all have occasion to attend meetings. In industry you may be asked to sit on a committee set up for a multitude of reasons, from resolving technical problems that are tying up production to organizing the company's annual picnic. The effectiveness of such meetings is controlled entirely by those taking part. Meetings attended by persons *aware of their role* as participants can move quickly and achieve good results; those attended by individuals who seize the opportunity to air personal complaints can be deadly dull and cripple action. Unfortunately, cumbersome, long-winded meetings are much more common than short, efficient ones.

Meetings can be either structured or unstructured, depending on their purpose. A structured meeting follows a predetermined pattern: Its chairperson prepares an agenda that defines the purpose and objectives of the meeting and the topics to be covered. The meeting then proceeds logically to each point. An unstructured meeting uses a conceptual approach to derive new ideas. Only its purpose is defined, since its participants are expected to introduce suggestions and comments which may generate new concepts (this approach is sometimes known as "brainstorming"). I will discuss the structured meeting here, because you are much more likely to encounter it in industry.

THE PARTICIPANT'S ROLE

You can contribute most to a meeting by arriving prepared, stating clearly your facts, ideas, and opinions when called on, and keeping quiet the remainder of the time. If you observe these three basic rules you will do much to speed up affairs. Let's examine them more closely.

Come Prepared. If a meeting is scheduled to start at 3:00 p.m. do not wait until 2:30 to gather the information you need. Arriving with a sheaf of papers in hand and shuffling through them for the first 15 minutes creates a disturbance and makes you miss much of what is being said. Start gathering information as soon as you know what is required of you, sort it out to identify the items you need, then jot down topic headings and specific data you will have to quote. Take into the meeting only those papers you will need.

Be Brief. In your opening remarks summarize what you have to say, then follow with facts and details. Present only those items your listeners need to know. If you have statistical data to offer, print copies and distribute them when you begin to speak. If you have a lot of information to distribute, ask the chairperson to distribute copies with the agenda so that everyone can look it over before coming to the meeting. Be ready to answer questions and analyze your facts in greater depth, but be sure to keep to the main topic. Finally, address your remarks to the chairperson.

Keep Quiet. There are many parts of a meeting when your role is to be only an interested observer. At these times you should keep quiet unless you have a relevant question, an additional piece of evidence, or an educated opinion. Avoid the annoying habit of always having something to add to the information others are presenting (recognize what others already know: you are not an authority on everything). At the same time, do not withhold information if it would be a genuine contribution. Be ready to present an opinion when the chairperson indicates that a topic should be discussed, but only if you have thought it out and are sure of its validity. Recognize, too, that a discussion should be a one-to-one conversation between you and the chairperson, or sometimes between you and the topic specialist. It should never become a free-for-all with each person arguing a point with his or her neighbor.

THE CHAIRPERSON'S ROLE

Good chairpersons are difficult to find. A good chairperson controls the direction of a meeting with a firm hand, yet leaves ample room for the participants to feel they are making the major contribution. The chairperson must be a good organizer, an effective administrator, and a diplomat (to smooth ruffled feathers if opinions differ too widely). Much of the success of a meeting will result from the chairperson's preparation before the meeting starts and ability to maintain control as it proceeds.

Prepare an Agenda. Some time before the meeting starts, the chairperson should prepare an agenda of topics to be discussed and circulate it to all committee members. If certain members have specific contributions to make, the agenda should indicate by name who will be presenting the information. A typical agenda might follow the pattern in Figure 9-4, which also reminds committee members of the time and date of the meeting.

Run the Meeting. The chairperson's first responsibility is to start the meeting on time: a person with a reputation for being slow in getting meetings started will encourage latecomers. The second responsibility is to keep the meeting as short as possible without seeming to "railroad" decisions. The third responsibility is to maintain adequate control.

The meeting should be run roughly according to the rules of parliamentary procedure. (Since most in-plant meetings are relatively informal, full parliamentary procedure would be too cumbersome.) The chairperson should introduce each topic on the agenda in turn, invite the person specializing in the topic

---

MACRO ENGINEERING INC

INTEROFFICE MEMORANDUM

TO: Members, Electronic Facsimile Research Committee

FROM: Daniel K. Tomashewski, Chair

DATE: September 15, 19xx

REF: Agenda for Next Meeting

The monthly meeting of the EFRC will be held in Conference Room B at 3 p.m. on Friday September 18. The agenda will be:

1. Progress: installation of DPS-2A ultra-narrowbeam system.
                                                (R. Taylor)

2. Test program report: high speed production units. (C. Bundt)

3. Proposal for multiple remote production units.   (W. Frayne)

4. Plans for annual Research Division banquet.
                                       (C. Tripp; J. Kosty)

5. Other business. (Please submit topics to me by 10 a.m. on Friday September 18.)

Secretary for this meeting will be J. Kosty.

D. Tomashewski

Dan Tomashewski

Figure 9-4. A typical agenda for a meeting.

to present a report, then open the topic for discussion. The discussion offers the greatest challenge, for the chairperson must permit a good debate to generate among the members, yet be able to steer a member who digresses back to the main topic. He must be able to sense when a discussion on a subject has gone on too long, and be ready to break in and ask for a decision. Similarly, he must know when strong opinions are likely to block resolution of a knotty problem, and assign a person or subcommittee to investigate further.

Sum Up.   Before proceeding from one topic to the next, the chairperson should summarize the outcome of the discussion on the first topic. The outcome can be a general conclusion, a consensus of members' opinions, a decision, or a statement of action defining who is to do what, and when. In this way all members will be aware of the outcome, and the secretary will know what to enter into the minutes.

The chairperson should also sum up at the end of the meeting, this time reviewing major issues which were discussed and the main results. The chairperson should also point the way forward by mentioning any important actions to be taken and the time, date, and place of the next meeting (presuming there is to be one).

The best way to learn to be a good chairperson is to watch others undertake the role. Study those who seem to get a lot of business done without appearing to intrude too much in the decision-making. Learn what you should not do from those whose meetings seem to wander from topic to topic before a decision is made, have many "contributors" all speaking at the same time, and last far too long.

THE SECRETARY'S ROLE

Sometimes a stenographer is brought in to act as secretary and record the minutes of a meeting, but more often the chairperson appoints one of the participants to take minutes. If you happen to be selected, you should know how to go about it.

Recording minutes does not mean writing down everything that is said. Minutes should be brief (otherwise they will not be read), so there is room only to mention the highlights of each topic discussed. Items that must be recorded are: (1) main conclusions reached; (2) decisions made (with, if necessary, the name[s] of the person[s] who made them, or the results of a vote); and (3) what is to be done next and who is to do it. The best way to get this information quickly is to write the agenda topics on a lined sheet of paper, spacing them about two inches vertically. In these spaces jot down the highlights in note form, leaving room to write in more information from memory immediately after the meeting.

The completed minutes should be distributed to everyone present, preferably within 24 hours. They should be a permanent record on which the chairperson can base the agenda for the next meeting (if there is to be one), and participants can depend for a reminder of what they are supposed to do. I like the format shown in Figure 9-5, which provides an "action" column to draw participants' attention to their particular responsibilities.

Assignments

Speaking situations you are likely to encounter in industry will develop from projects on which you are working. Hence assignments for this chapter are assumed to grow naturally out of the major writing assignments presented in other chapters.

MACRO ENGINEERING INC

ELECTRONIC FACSIMILE RESEARCH COMMITTEE

<u>Minutes of Meeting</u>

Friday September 18, 19xx, 3:00 p.m.

In attendance: C. Bundt   R. Miller
                V. Feldman   R. Taylor
                W. Frayne   D. Tomashewski (Chair)
                J. Kosty (Secretary) C. Tripp

<u>Minutes</u>	<u>Action</u>
1. The DPS-2A ultra-narrowbeam system is not yet operational because the S-76 interface requires further modification. R. Taylor expects modifications to be complete September 24, and to have a system operational date of September 28.	R. Taylor
2. C. Bundt reported that phase 1 of the MS-4 test program was completed August 27, but phase 2 was interrupted September 16 when the high speed paper feed ignited and burned. He requested:	
2.1 A budget increase of $6800 to cover purchase and modification of an MS-2 paper feed to replace the damaged prototype.	C. Bundt
2.2 Rescheduling of phase 2 to October 6 to 28, and phase 3 to November 1 to 16. Approved.	R. Miller C. Bundt
The vacant LMS-4 test time (Sept 21 to Oct 3) will be used to reprogram the test rig and to install a miniature TV camera behind the paper feed.	V. Feldman J. Kosty
7. Proposals for papers to be presented at the International Electronics Conference in New York, May 25 to 27, 19xx, are to be submitted to D. Tomashewski. Deadline: October 9.	All

John R. Kosty
Secretary

Figure 9-5. Minutes of a meeting. Note the "Action" column, which draws individuals' attention to their post-meeting responsibilities.

TECHNICAL BRIEFINGS

Many of the projects in Chapters 5 to 7 offer opportunities for you to brief a client, management, or other members of your department on the results of a technical investigation. Projects that particularly suit oral reporting are:

Chapter	Project	
5	1	Selecting New Trucks and Automobiles
	2	Developing a Production Line Training System
6	1	A Traffic Study for the Town of Tarnapin
	2	Coping With a Noisy Neighbor
	4	Shaving the Peak
7	6	Researching a New Manufacturing Material or Process

MEETINGS

Some of the projects in Chapters 5 and 6 offer excellent opportunities for group participation. If you are assigned one of these projects, try tackling it this way:

1. Install a class manager to coordinate the project (he or she should be voted into office).
2. Hold a meeting to discuss the project, and assign different individuals or teams to undertake specific aspects. Appoint a secretary to keep minutes, and set up a schedule for the project.
3. Let each team research its part of the project separately.
4. During the research or investigation phase, hold a meeting for the teams to report progress and compare ideas.
5. At the end of the research phase, hold a meeting for the teams to report their initial results.
6. Give the teams time to write their final reports and to prepare technical briefings.
7. Let the class manager and secretary write a management report that outlines the project as a whole and summarizes the teams' general findings.
8. Hold a briefing for the teams to present their findings orally (to individuals defined by your instructor). Let the class manager act as chairperson and introduce the speakers.

Three projects in Chapter 5 and two in Chapter 6 particularly suit group participation:

Chapter 5
Project 1: Selecting New Trucks and Automobiles
Project 3: New Space for the Drafting Department
Project 4: A Visit to Wakeling Processors Inc

Chapter 6
Project 1: A Traffic Study for the town of Tarnapin
Project 4: Shaving the Peak

For each assignment, after you or your group has made an initial decision, hold a meeting to discuss the various choices. Be sure to arrive at a consensus for the best choice.

Project 3 of Chapter 6 (Building a Toothpick Tower) provides you with an opportunity to hold two meetings:

1. After construction but before testing, let each participant describe his or her design, discuss why it was chosen, and predict the maximum load it will sustain.
2. After testing, let each participant describe how his or her tower failed and discuss what should have been done to prevent premature failure.

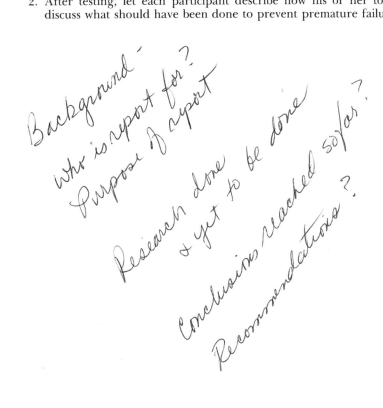

10

Communicating with Prospective Employers

As head of the Administration and Personnel Department of H. L. Winman and Associates, Tanys Young is responsible for hiring new staff. Recently she advertised for an engineer to coordinate a new project, and received 48 applications from across the country. Since it would have been impractical to interview all the applicants, she narrowed the field down to the nine applicants she felt had the best qualifications, basing her selection on the information contained in the 48 application letters and resumes she had received.

One of the applicants was Eugene Koenig of Indianapolis, Indiana, who believed—quite rightly—that he probably was the most qualified person for the job. But his name was not one of the nine on Tanys's short list, and so he was not even interviewed. Tanys has since met the nine selected applicants and has offered the job to the most promising person. She will never know that Eugene would have been a better person to hire. And Eugene will never know why he was not even considered.

What went wrong? The fault was entirely Eugene's, who has yet to learn that his bland, inadequately developed and presented job application severely inhibited his chances of obtaining employment.

Today, a job seeker such as Eugene has to tailor his resume and application letter to capture the interest of a *particular* employer rather than simply send copies of a general resume and letter to every employer. In a highly competitive job market, extremely careful orchestration of the whole employment-seeking process is essential, from resume preparation to personal presentation during an interview.

This chapter describes the six stages in the job-seeking process and the careful steps each applicant must take to ensure his or her proper consideration as a potential employee. The steps start with the preparation of a personal data record and then proceed through preparing a resume, writing an application letter, completing an application form, attending interviews, and accepting or declining a job offer.

THE EMPLOYMENT-SEEKING PROCESS

Figure 10-1 illustrates the five steps a *successful* applicant has to take before being employed. At each step the number of contenders for a particular job is reduced until only one person remains. Tanys Young received letters and resumes from 48 potential employees, but asked only 25 of them—those she felt most nearly met the company's requirements—to complete a company application form. Thus in one step she cut the field in half. Then, from the application forms and resumes, she selected nine persons to interview.

The significant factor here is that Tanys narrowed the field down to only 19% of the original applicants *based solely on their written presentations*. Like Eugene Koenig, you cannot afford to be one of the over 80% who were eliminated from the interview stages because they prepared inadequate written credentials.

The five steps of the job-seeking process are outlined briefly below. At each step you have to present a confident, positive image of yourself if you are to proceed to the next step.

1. Initial Contact. Your first step as a job seeker is to approach a prospective employer and ask to be considered for employment. You may be responding to an advertisement or approaching the employer "cold" (in the hope that the employer either has or shortly will have an opening). Historically, job seekers most often have responded to employers' advertisements. Today,

Figure 10-1. The five stages or steps of the job-seeking process.

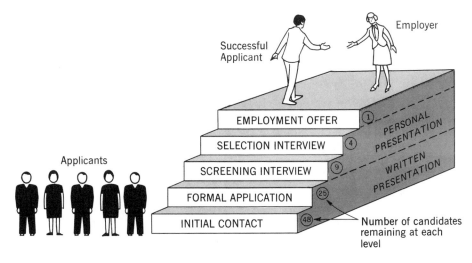

however, with a much tighter job market employers advertise far less frequently, preferring to take a cheaper and simpler approach. "If I advertise a position," Tanys Young explains, "I have to cope with anywhere from 50 to 100 applications. So I have come to rely on speculative applications from job seekers who take the initiative and approach us. And sometimes I accept referrals from people I know."

You may make an initial contact by presenting yourself at an employer's door, by writing a letter, or by telephoning. The personal visit and the letter are preferred because both provide you with an opportunity to place a resume containing your name, address, and qualifications in the employer's hands. Telephoning is not recommended for the initial contact because it provides no opportunity for you to leave information about yourself with the employer.

2. Formal Application. In step 2 of the job-seeking process the employer asks you to complete a company application form. This normally is a requirement because employers prefer to have all applicants documented in the same way.

3. Screening Interview. The first interview you attend helps the employer identify which applicants have the strongest potential. In a large firm such as H. L. Winman and Associates, the screening interview may be conducted only by the employment manager.

4. Selection Interview. Having identified the most promising candidates, the employment manager now arranges for second interviews. This time the manager of the area where the successful applicant will be employed also is present, sometimes with technical specialists from the department. They will ask questions particularly relevant to the technical aspects of the applicant's background and the position to be filled.

5. Job Offer. The employer makes a formal offer of employment to the successful applicant, often first by telephone and then by letter. The applicant responds, also by telephone and letter, to confirm his or her acceptance (or, sometimes, rejection) of the offer.

Not all employment-seeking processes follow exactly along these lines. Sometimes an astute applicant will obtain a company application form beforehand and submit it in step 1, with his or her letter and resume. At other times there may be only one job interview or, in special cases, there may be three or even more interviews.

DEVELOPING PERSONAL DATA SHEETS

Before you take any steps toward writing a resume or filling in an application form, you should first prepare a personal data file. A data file contains all the information about yourself that you may be called upon to supply when applying for a job. Unfortunately, making up a data file is a laborious, seemingly unrewarding task that seldom gets done. Not until a job applicant is suddenly faced with the need to insert partly forgotten information into a resume or application form does the value of a data file become apparent!

At this stage you probably are asking: "Why do I need to do all this? Surely I already know everything there is to know about myself!" True, at the moment you do. But in future years you will find your ability to recall the names of specific people, and dates and addresses, will become vague, not because of approaching senility but because time and events piled one on another make the details more difficult to remember.

A data file comprises five sets of information concerning your:

Education
Work experience
Extracurricular activities
Associates and supervisors (those who may act as referees)
Personal information

For each of these topics you will need a separate sheet of lined paper or, better, an 8 in. × 5 in. lined card. (For some of the topics, such as work experience, you will need more than one sheet or card.) On each of the cards write one of the above headings and then list everything you can remember about yourself and that topic, as outlined below. Ideally, develop each list chronologically, so that it will be easy to add more information later on.

EDUCATION

On the first card list everything you can remember about the schools, colleges, and universities you have attended or are attending. Start at junior high school and list the name of each school, the address and telephone number, and the dates you were there. For high school also list your graduation date and your area of specialization. For college and university, particularly note dates, courses taken, special options, and the full name of the degree, diploma, or certificate you were awarded. If you were intensively involved in a school's special activities or sports, make note of the details and your particular role, especially if you were captain of a team or president of a society.

WORK EXPERIENCE

For each job you have held in the past, list comprehensive details regardless of whether the work you were doing is directly related to the work you are now seeking. (Remember that this is your *personal* file, from which you will draw information for future resumes and application forms; it is not a file you will show to prospective employers.) For each previous employer—and for your present employer, if you are currently employed—write down:

- The full name, address and telephone number of the company or organization, and the full name and title of each supervisor you worked for.
- The dates you started and finished employment and, if you held several positions within the company, the name of the position and the date you were appointed to it.
- Your job title, or titles if you held several positions.
- Your specific responsibilities and duties for each position, paying particular attention to the supervisory aspects and responsibilities of any job that you carried out without supervision.
- Any special skills you learned on the job.
- Special commendations you received, or results you achieved.

EXTRACURRICULAR ACTIVITIES

Here you list your activities in organizations that were not necessarily part of the jobs you have held or your education, but which show your participation and leadership qualities. For each activity you should include the dates of your involvement and the name, address, and telephone number of a person who can vouch for your participation. The entries on your list should cover:

- Membership in a club, society, or group, particularly noting your responsibilities as an active participator or committee member. (For example, member of sports committee or secretary of administrative committee.)
- Participation in community activities such as the Big Brother or Sister Organizations, 4-H Club, Red Cross Society, Parent-Teacher Association (PTA), or local community club. Particularly describe any executive or administrative positions you have held, with special responsibilities and dates.
- Involvement in a technical society on a local or national level, with particular mention of any conferences you have attended or papers you have presented or published.
- Participation as a sports enthusiast, with special mention of your role as a team leader or coach.
- Involvement in hobby activities such as stock car racing or rebuilding, a computer club, or dog breeding.
- Awards you have received for any activities you have been involved in.

For each job applicant this list will differ, depending on the person's background and interests. The important thing is to jot down everything you might want to draw on in the future, even if you think it is inconsequential at the moment.

REFEREES

Referees are the people you feel are best fitted to speak on your behalf. They fall into two groups: people who can vouch for your *capabilities* (as an employee, student, or committee member), and people who can speak for your *character*. Those who can speak well of your capabilities you can name on a resume; those who can speak for your character you can name on application forms where an employer *specifically* asks for a character reference. Preferably, neither list should contain the names of direct family members, or even of aunts, uncles, or cousins, because they would have less credibility as unbiased referees in an employer's eyes. You will need to write down each person's:

- Full name, professional title (such as chief engineer), place of employment, and job position.
- Employer's address and telephone number.
- Home address and telephone number.

(If a referee has changed jobs since you worked for him or her, you should list details of both the previous and current employer.)

For a person you worked with in an extracurricular activity, you will also need to list:

- The name of the organization you both were involved with, and the referee's position within that organization.
- Whether the person prefers to be called or written to at home or at work.

PERSONAL INFORMATION

On this last sheet or card you record all kinds of information about yourself, information you *may* need when applying or being accepted for a job position. (Under human rights legislation, an employer cannot ask for personal information such as your sex, age, weight, and ethnic origin until after you are employed; you may, however, choose to offer such information yourself.) Facts you should list are:

- Date and place of birth
- Marital status and number of children
- Height and weight
- Social security number
- Addresses you have lived at, and the dates
- Details of medical problems you have encountered

All this information becomes part of your permanent private file. But, once you have completed your data file, you should never consider it as complete. Once a year, while memories of dates, names, and addresses are still fresh, you should bring the file up to date. In this way you will have a continuous file ready for use whenever you need it.

PREPARING A RESUME

A resume contains cogent information about yourself, carefully assembled and presented so that prospective employers will be impressed not only by your qualifications but also by your ability to display your wares effectively. (The correct spelling is "résumé," but common usage in the United States has made the accentless "resume" acceptable.) Although resumes are known by several other names, with biography of experience and curriculum vitae being the more common alternatives, they all serve the same purpose: to provide brief yet comprehensive details of a job applicant's background, education, and experience.

Although resumes can be organized in many different ways and presented in varying formats, two designs are preferred by engineers and technologists. For convenience, and to differentiate between them, I will refer to these as the traditional resume and the focused resume. Both are widely used, so you will have to decide which format best suits your background and experience.

THE TRADITIONAL RESUME

For decades the accepted approach to resume writing has been to divide a job applicant's information into five parts, each preceded by an appropriate heading and listed in this order:

Personal Information
Education
Experience
Extracurricular Activities
References

This traditional arrangement is well recognized and still widely used, with one small change: in recent years the personal information section has been deleted. In the past, under this heading job applicants commonly listed their address, telephone number, marital status, age, state of health, and social security number. Today, only the applicant's address and telephone number are retained, and these are combined with the person's name to form a centered "title block" at the top of the page, as Alison Witney has done in her resume in Figure 10-2.

Eliminating the personal information section has had a secondary advantage. Employers are more interested in an applicant's accomplishments than the rather dull data describing personal information. With the personal details removed, the applicant's education and work experience are lifted higher on the page where they are more prominent.

The traditional resume is particularly suitable for recent college graduates with limited work experience, or for college students who shortly expect to graduate. The resume in Figure 10-2 has been prepared by a biological sciences undergraduate with two previous jobs totalling three years full-time employment. Guidelines for preparing a resume in this format are presented below and keyed to the circled numbers beside Alison Witney's biography.

1. The words "Biographical Details" can be replaced by the single word "Resume," or you may even omit the title because your name and address appear immediately beneath it to form a suitable heading for the top of the resume.

BIOGRAPHICAL DETAILS

ALISON V. WITNEY

210, 1670 Fulham Boulevard
Amiento, Florida   32704
Tel: (305) 474 6318

EDUCATION AND TRAINING

Graduate of Morton Stanley High School, Corisand, FL, 1981.
Will graduate with Diploma in Biological Science from Amiento Technical
College, June 1986.

WORK EXPERIENCE

1983 to date   Animal Treatment Center, Amiento, FL. One year full time,
                  two years part time, as veterinary assistant. Responsible
                  for reception, grooming, and exercising of animals, assis-
                  ting veterinarian during operations, changing dressings,

                  administering injections and anaesthetics, and performing
                  administrative duties such as accounting and ordering of
                  supplies.

1981 to 1983   Remick Airlines, Orlando, FL.   Accounts clerk in air
                  freight department; coordinating billings, preparing
                  invoices, following-up lost shipments, assisting clients,

                  and writing monthly reports. For nine months, assisted
                  in payroll preparation.

1978 to date   Bar None Riding Stables, Corisand, FL. Part time employ-
                  ment teaching the care and handling of horses and basic
                  riding techniques to young riders. Assisted in grooming,
                  cleaning, feeding, and saddling-up.

ADDITIONAL INFORMATION

Winner of two educational awards: Morton Stanley Science Scholarship (1980),
and Amiento Technical College Biology Scholarship (1985).
Member of YWCA since 1971, where I now teach swimming and lifesaving.
Interests: teaching horseback riding and jumping, swimming, water skiing.

REFERENCES

The following persons have agreed to act as references on my behalf:

Dr. Alex Gavin   Mr. Charles Devereaux
Veterinary Surgeon   Owner-Manager
Animal Treatment Center   Bar None Riding Stables
2230 Wolverine Drive   2881 Westshore Drive
Amiento, FL 32704   Corisand, FL 32726
Tel: (305) 474 1260   Tel: (305) 632 2292

Figure 10-2. A traditional resume or biography of experience.

2. Each line of your name, address and telephone number should be centered about the page centerline. This gives the top of the resume a nicely balanced appearance.

3. There is no need to list all the schools you have attended. Simply state the name of your last school, the highest level you attained, and the year of graduation. Add to this the name of each college or university you have attended, the type of course you enrolled in, the diploma or degree you received, and the year that you graduated (or expect to graduate). This information may be presented in chronological or reverse sequence.

4. Experience is usually presented in reverse order, with an applicant's most recent work experience appearing first and earliest experience last. Generally, say most about recent experience, providing that the length of employment has been long enough to warrant it, and about earlier work that was similar to that of the position being sought.

5. For each employer, state the name of the company or organization first, underline it, and then identify the city and state in which it is located. Then describe the position you held (give the job title), and what the work involved. Particularly draw attention to the *responsibilities* of the job rather than merely list the duties you performed. Use words that create strong images of your self-reliance, such as:

    coordinated

    monitored

    presented

    planned

    organized

    implemented

    supervised

    directed

   For several short, part-time jobs, describe them together and draw attention to the most important, like this: "Several after-school jobs, primarily as a stock clerk in a groceteria."

6. The two-column arrangement of dates and work experience is important because it gives a much "lighter" appearance to the page. If the job descriptions were carried to the left—under the dates—the job details would appear as much heavier, less visually appealing blocks of information.

7. Employers are particularly interested in an applicant's activities and interests outside normal work. They want to know if the person is more than a routine employee who arrives at 8 a.m., works until 4:30 p.m., then drives home, eats supper, and watches television all evening. Information on your hobbies, interests, and participation in sports and community activities tells prospective employers that you recognize your role in society, are not too rigid or too narrow, and adapt well to your environment. Employers reason that such an applicant will make an interesting, active employee who will not only contribute much to the company, but also take part in social and sports functions.

8. Try to draw your list of references from a cross section of people you have worked for, been taught by, or served with on committees, and ensure that their relevance is apparent (their connection to one of your previous jobs or activities must be clear). Before including them in your list, check that all are willing to act as referees.

Both Alison Witney and Dennis Hammond (whose resume appears in Figure 10-3) are well aware of the important role a resume's appearance plays in a prospective employer's readiness to consider an applicant. A carefully arranged and printed resume, typed by a professional typist on a high-quality typewriter, implies that the applicant has high-quality credentials and should be interviewed.

THE FOCUSED RESUME

Job applicants who have more extensive experience to describe do better if they use a resume format that focuses an employer's attention on their particular strengths and specific aims. In this arrangement the person's employment objectives are summarized at the top of the page, immediately beneath the person's name and address:

OBJECTIVE

Following graduation as an engineering technician I spent seven years installing and testing transmission line towers in Minnesota, North Dakota, and Alaska. I now hold a Bachelor of Science degree in Civil Engineering and want to apply my experience and education to researching grouts for tower anchors in permafrost areas.

An assertive statement such as this at the start of a resume is similar to the summary statements recommended for the letters and reports described in Chapters 3 through 6. Its intent is to draw the employer's attention rapidly to the applicant's primary experience and education, and to the employment direction the applicant wants to pursue. If the resume "hits its mark" successfully, the employer automatically reads further into the resume to learn more about the applicant.

To be of most value the opening statement is focused to suit the needs of a particular employer, or sometimes a group of employers engaged in similar work. This factor alone highlights a significant difference between traditional and focused resumes. Whereas the traditional resume can be prepared for and distributed to many employers, the focused resume often may be prepared for only one employer. The implications for job applicants are far-reaching: now they have to invest much more time, care, and research into resume preparation to ensure their resumes are clearly directed toward a very specific audience.

The summary statement (usually titled either **Objective** or **Aim**) presents two main pieces of information the applicant believes an employer will most want to know: (1) what the job applicant is particularly qualified to do and already has extensive experience in doing; and (2) what work the job applicant wants to do. Ideally, there is a logical connection or development between these two pieces of information, and they are presented in a short paragraph of no more than two or three sentences.

At this point the focused resume diverges even further from the traditional resume. If you have started your resume with your **Objective,** you now have to ask yourself what a prospective employer is most likely to ask or want to know after reading your opening statement. Probably it will be: "What have you done that specifically qualifies you to achieve this objective?" To answer this question means focusing next on your work experience rather than your education, and particularly on those aspects of your work experience which are relevant to the position being sought. This in turn means moving details of your education so that they *follow* the experience section of the resume.

If you have mixed work experience—and many job applicants have—then you can go one step further and divide your experience into two parts: work related to the position you are seeking, and work in other areas. Thus the focused resume has six headings:

> **Objective** (or **Aim**)
> **Related Experience**
> **Other Experience**
> **Education**
> **Extracurricular Activities**
> **References**

A two-page resume prepared in this way is shown in Figure 10-3. The circled figures beside the resume refer to the comments below.

1. Dennis Hammond has sufficient information to warrant his preparing a two-page resume. Generally, job applicants with only limited work experience should try to keep their resumes down to one page. Applicants like Dennis can spread their information onto a second page, but preferably should not run onto a third page. (A third page can be used, however, if an applicant has published papers and articles or has obtained patents for new inventions, which can be listed on a separate sheet, which is identified as an attachment.)
2. Dennis's summary statement clearly shows his thrust toward fiber optics engineering and his desire to obtain employment in that field.
3. The positions described with each Experience section should be listed in reverse order, the most recent experience being described first and the earliest experience described last. The most recent and most relevant experience should be described in considerably greater depth than early or unrelated experience (compare the descriptions of Dennis's Southcentral Contractors' experience with his Bowlands Stores' experience). An applicant should take great care to discriminate between factors he or she feels are most interesting and those an employer would particulary like to read about, and should concentrate on the latter.
4. As in the traditional resume, each employer's name is listed first, underlined, and followed by the city and state. The person's position or job title is identified next, and then a description of what the job involved. If several positions have been held within the same firm, each is named and its duration stated so that the applicant's progress within the firm is clear.
5. Each position should draw particular attention to the personal responsibilities

Resume

DENNIS G. HAMMOND, P.E.

310 -- 408 Medwin Street
St. Cloud, Minnesota   56301
Tel: (612) 548 1612

OBJECTIVE

After four years supervising the installation and testing of wire and
fiber optic telephone communication systems, I returned to college to
obtain an M.S. in electronics engineering with a major in fiber optics.
I am now seeking employment where I can apply my knowledge and experience in fiber optics engineering.

RELATED WORK EXPERIENCE

June 1982 to September 1983 and May 1984 to date	Ebby, Little and Associates, Engineering Consultants, St. Cloud, Minnesota. Supervising engineer, respons- ible for installation, testing and analysis of tandem wire and fiber optic telephone communication links between Brainerd and Little Falls, Minnesota. Currently carrying out performance tests on installed links.
September 1973 to August 1979	Southcentral Installation Contractors Inc., Lincoln, Nebraska. For first four years, member of team in- stalling high voltage transmission lines and trans- former stations along power grid between Weekaskasing Falls, Nebraska and Bismarck, North Dakota. After 18 months appointed crew chief in charge of team installing interconnecting and distribution systems to townsites along the route; responsible for: hiring, training and supervising local labor; ordering and monitoring delivery of parts and materials; arranging and supervising subcontract work; and preparing progress and job completion reports. From June 1977 to August 1979, assigned as supervisor of team working under contract to Ohio Utilities Corporation, installing and testing fiber optic links between towns up to 28 miles apart.

OTHER WORK EXPERIENCE

January 1967 to February 1971	United States Air Force. Enlisted serviceman with Construction and Maintenance Directorate. For first two years, member of crew installing basic antenna systems and associated structures. For final two years, site technician responsible for maintenance of transmission lines and antennas at a midwestern USAF base. Attained rank of corporal.

2/...

Figure 10-3. A focused resume for a job applicant with a varied background.

Dennis G. Hammond - page 2

| June 1963 to December 1966 | Bowlands Stores Inc., Duluth, Minnesota. Stock clerk in grocery store No. 16. Full time for two summers and June to December 1966; part time while attending high school. |

EDUCATION

⑦ Master of Science in Electronics Engineering, with major in fiber optics, University of Minnesota, 1984.
Bachelor of Science in Electrical Engineering, University of Montrose, Montrose, Ohio, 1982.
Graduate Electrical Engineering Technician, Walter Halstadt Community College, Reece, Minnesota, 1973.
Graduate of Winona Collegiate, Duluth, Minnesota, 1967.

ADDITIONAL ACTIVITIES/INFORMATION

Member, Institute of Electrical and Electronics Engineers Inc. (IEEE), 1973 to date. Secretary, St. Cloud, Minnesota Section, 1982-83.
Awarded Orton R. Smith Scholarship for proficiency in applied mathematics, Walter Halstadt Community College, 1972, and Ohio Power and Light ⑧ Scholarship for achievement in communications engineering, University of Montrose, 1980.
Technical paper: "Accuracies of Computer Data Transmissions Attainable at High Baud Rates Over Fiber Optic Communication Links," in Communications Technology, 13:07, July 1984. (Paper based on thesis written as part of M.S. program, University of Minnesota.)
Military courses attended while in USAF:
  * Transmission Line Installation Techniques, 1967.
  * Supervisory Skills Development, 1969.
  * First Aid and Safety Methods, various courses, 1968-71.
Junior Leader, Duluth, Minnesota, YMCA, 1962-67, teaching swimming and aquatic activities to boys age 9-15. Awarded Red Cross Bronze Medallion, 1965.

REFERENCES

The following persons will provide information regarding my qualifications and work capabilities:

⑨

Martin F. Ebby, P.E.	Philip M. Karlowsky
Project Coordinator	Contracts Manager
Ebby, Little and Associates	Southcentral Installation
360 Rosser Avenue	Contractors Inc.
St. Cloud, Minnesota 56302	1335 Westfair Drive
Tel: (612) 544 1867	Lincoln, Nebraska 68528
	Tel: (402) 632 1450

and supervisory aspects of the job, rather than just list specific duties. Verbs should be chosen carefully so they make the position sound as comprehensive and self-directed as possible. If the paragraph grows too long (and Dennis's paragraph here is rather long), it can be broken into subparagraphs like these:

> . . . appointed crew chief responsible for
> - installing interconnecting and distribution systems
> - hiring, training, and supervising local labor
> - ordering and monitoring delivery of parts and materials
> - arranging and supervising subcontract work
> - preparing progress and job completion reports.

6. Single-spaced typing should be used as much as possible to keep the resume compact. At the same time there should be a reasonable amount of white space on each side and between major paragraphs to avoid a crowded effect.
7. Education can be listed either in chronological or reverse sequence. If a resume is to be sent out of state, or if the applicant was educated out of state, he or she should identify the city and state of each educational institution attended.
8. Employers are *interested* in a job applicant's accomplishments and extracurricular activities, particularly those describing community involvement and awards or commendations. This part of a resume can be preceded by a heading such as "Additional Information" rather than "Extracurricular Activities."
9. Both of the people Dennis has chosen as referees can be cross-referenced to his previous work experience. Telephone numbers are important, because most employers prefer to talk to rather than receive a letter from a referee.

Never be afraid to use a display technique for your resume that will enhance its professional quality and make it stand out among other resumes. (I do not mean you should make your resume "flashy," because an overdone appearance can spark a negative reaction from a reader.) An engineer with technical editing experience recently prepared a two-page resume which he had printed side-by-side on 11×17 inch paper, and then folded the sheet so that the resume was inside. On the outside front he printed only his name and the single word "Resume" (see Figure 10-4). When his resume was placed in a file among other resumes submitted for a particular job opening, its unusual, dignified appearance caught the employment manager's eye and resulted in the engineer being called in for an interview. Originality of presentation and approach can help capture a reader's attention, providing the information within the resume focuses on the particular employer's requirements.

WRITING A LETTER OF APPLICATION

Although some resumes may be delivered personally, the majority are mailed with a covering letter. Because potential employers will probably read the letter first, it must do much more than simply introduce the resume. It needs to state your purpose for writing (that you are applying for a job) and demonstrate that

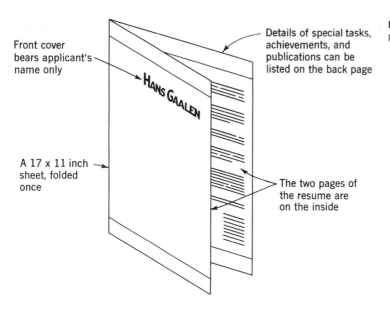

Front cover bears applicant's name only

HANS GAALEN

A 17 x 11 inch sheet, folded once

Details of special tasks, achievements, and publications can be listed on the back page

The two pages of the resume are on the inside

Figure 10-4. An imaginatively prepared resume.

you have some very useful qualifications that the reader should take the time to consider. As such it becomes a letter of application to which you have attached a resume containing more definitive information.

A strong, interesting, well-planned application letter can prompt an employer to place yours among those whose authors he or she would like to see in person. On the other hand, a dull, unemphatic letter may cause the same employer to drop it onto the pile of "also-rans" because its style and approach seem to imply you are a dull, unemphatic person. In today's highly competitive employment environment you cannot afford to let your letter be dropped onto the "also-ran" pile.

Every business letter—and a letter of application essentially is a business letter—should follow the "pyramid" method of writing. That is, it should open with a brief summary that defines the purpose of the letter, and then offer strong, positive details to support the opening statement. Finally, it should close with a brief remark that identifies what action is to be taken next.

This means that an application letter can be divided into three parts:

- An **initial contact,** which states that you are applying for a job and briefly identifies what special qualifications you have that make you a particularly suitable candidate. (The intent should be to capture the reader's interest in the first sentence or two.)
- An **evidence** section, which provides details and solid facts to support your contention that you are well qualified to hold the position. The facts you quote should be selected for their relevance to the position sought and their pertinence to the reader. (Remember that the most interesting experiences from your point of view may not be the most interesting to the reader.)
- A **closing statement** which, rather than just closing the letter with a polite remark, *opens the door* to the next step (the employment interview).

There are two types of application letter: Those written in response to an advertisement for a job that is known to be open, or at the employer's specific invitation, are called "solicited" letters. Those written without an advertisement or invitation, on the chance that the employer might be interested in your background and experience even though no job is known to be open, are referred to as "unsolicited" letters. The overall approach and shape of both letters are similar, but the unsolicited letter generally is more difficult to write.

THE SOLICITED APPLICATION LETTER

The main advantage in responding to an advertisement, or applying for a position that you know to be open, is that you can focus your letter on facts that specifically meet the employer's requirements. This has been done by Alison Witney in the letter illustrated in Figure 10-5, which she has written in response to an advertisement in a Florida local newspaper.

The following comments are keyed to the circled letters in Alison's application letter.

A. For a letter that will have a personal address at the top, you would be wiser to use the modified block style shown here rather than the full block style in which every line starts at the left-hand margin. (See Figure 3-4 of Chapter 3 for information on the modified block letter.) Because this style helps balance a personal letter on the page, it provides a more pleasant initial impression. Each line of the applicant's name, address, and telephone number, and the signature block at the end of the letter, should start at the page centerline.

B. Whenever possible, personalize an application letter by addressing it by name to the personnel manager or the person named in the advertisement. This gives you an edge over applicants who address theirs impersonally to the "Personnel Manager" or "Chief Engineer." If the job advertisement does not give the person's name, invest in a telephone call to the advertiser and ask the receptionist for the person's name and complete title. (You may have to decide whether to send your letter and resume to someone in the personnel department or to a technical manager who is more likely to be aware of the quality of your qualifications and how you could fit into his or her organization.)

C. This is the **initial contact,** in which Alison Witney summarizes the key points about herself that she believes will most interest her reader and states that she is applying for the advertised position. Note particularly that she creates a purposeful image by stating confidently "I am applying. . . ." This is much better than writing "I wish to apply . . . ," "I would like to apply . . . ," or "I am interested in applying . . . ," all of which create weak, wishy-washy images because they only imply interest rather than purposefully apply for a job. An equally confident opening is "Please accept my application for. . . ."

D. The **evidence** section starts here. It should offer facts drawn from the resume and expand on the statements made in the first paragraph. Broad generalizations such as "I have 13 years experience in a metrology laboratory" should be replaced with shorter-term descriptions that indicate the applicant's exact role and responsibilities and stress the supervisory aspects of each position. The name of a person for whom an applicant worked on a particular project can be

Alison V. Witney
210 - 1670 Fulham Boulevard
Amiento, FL 32704

Tel: (305) 474 6318

March 23, 1986

Dr. Eugene Cartwright, Director
Animal Science Experimental Institute
Mount Ashburn University
Three Hills, Alabama 35107

Dear Dr. Cartwright:

I am applying for the position of Research Technician (Animal Sciences) advertised in the March 18, 1986 Amiento County Herald. I have been involved with animals and their care and treatment for many years, and shortly will receive my Diploma in Biological Science.

My interest in animals dates back to 1973, when I first learned to care for, groom, and ride horses. I now teach horse riding in my spare time. For the past three years my employer has been Dr. Alex Gavin, veterinary surgeon at the Amiento, Florida, Animal Treatment Center, where I assist in the medical treatment of small animals. It was my interest in horses, plus Dr. Gavin's influence, that led to my enrollment in the two-year Biological Sciences course at Amiento Technical College, from which I will graduate in early June. The attached biographical details provide further information on my education, employment background, and work experience.

I will be visiting your research station from April 21 to 23, as part of my college term research project. May I call on you then, while I am at Three Hills?

Regards,

Alison Witney.

Alison V. Witney

enc

Figure 10-5. A solicited letter of application prepared by an undergraduate.

usefully inserted here because it adds credibility to the role and responsibilities the applicant describes.

E. The first line of each paragraph may be indented about five spaces, as shown here, or may start flush with the margin.

F. The **evidence** section should cover the key points an employer is likely to be interested in and draw the reader's attention to the attached resume. It may be divided into two paragraphs if a single paragraph seems to be too long (as Dennis Hammond has done in Figure 10-6.)

G. This paragraph is Alison Witney's **closing statement,** in which she effectively opens the door to an interview by drawing attention to her imminent visit to the advertiser's premises. She avoids using dull, routine remarks such as "I look forward to hearing from you at your earliest convenience" or "I would appreciate an interview in the near future," both of which tend to close rather than open the door to the next step.

H. Contemporary usage suggests that most business letters should end with a single-word complimentary close such as "Regards," "Sincerely," or "Cordially," rather than the more formal but less meaningful "Yours very truly."

THE UNSOLICITED APPLICATION LETTER

An unsolicited application letter has the same three main parts as a solicited letter and looks very much the same to the reader. To the writer, however, there is a subtle but important difference, in that it cannot be focused to fit the requirements of a particular position an employer needs to fill. This means the job applicant has to take particular care to make the letter sound both positive and directed. Here are some guidelines to help you shape a letter that you are submitting "blind."

• Make a particular point of addressing your letter by name and title to the person who would most likely be interested in you. This may mean selecting a particular department or project head, who will immediately recognize the quality of your qualifications and how you would fit into the organization, rather than applying to the personnel manager. Never address an unsolicited letter to a general title such as "Manager, Human Resources," because, if the company does not use such a title and you have not used a personal name, it will likely be the mail clerk who decides who should receive your precious letter.

• Try to find out enough information about a firm so that you can visualize the type of work it does and how you and your qualifications would fit the company's needs. This will enable you to focus your letter on factors likely to be of most interest to the employment manager or selected department head.

• Try to make your initial contact positive and interesting even though you are not applying for a particular position, as Dennis Hammond has done in his unsolicited letter in Figure 10-6.

Like Alison Witney, Dennis has used the modified block format for his letter. It is longer than Alison's because he has more information to present, and to do so he has created two **evidence** paragraphs.

```
                              Dennis G. Hammond, P.E.
                              310 - 408 Medwin Street
                              St. Cloud, MN 56301

                              Tel:  (612) 548 1612

                              August 1, 1986


    Mr. Cory D. Richardson, P.E.
    Chief Engineer
    Minnesota Data Transmission Systems Inc.
    440 Barker Tower
    1600 Winston Drive S
    Minneapolis, MN 55426

    Dear Mr. Richardson:

         Can you use a project engineer who has specialized in fiber optic
    transmission systems for the past five years?

         My experience evolves from two periods of employment and my area
    of specialization at the University of Minnesota.  For two years I was
    responsible for installing and testing fiber optic communication links
    for Ohio Utilities Corporation, and then for three years I supervised
    the installation, testing, and analysis of parallel wire and fiber
    optic telephone transmission lines for Ebby, Little and Associates of
    St. Cloud, Minnesota.  For my M.S. in Electronics Engineering I majored
    in fiber optic transmission of information, analysing signal losses at
    high baud rates over fiber optic lines up to 12 miles long.

         I also hold a B.S. in Electrical Engineering from the University
    of Montrose, Ohio, and have had six years experience installing high
    voltage transmission lines and antenna systems.  The enclosed resume
    describes my experience and responsibilities in greater detail.

         I would welcome the opportunity to meet you and learn more about
    your projects in fiber optic communications.  I travel frequently
    between St. Cloud and Minneapolis, and will call on you the next time
    I am in your city.

                              Sincerely,


                              Dennis G. Hammond

                              Dennis G. Hammond


    enc
```

Figure 10-6. An unsolicited letter of application prepared by an experienced engineer.

COMPLETING A COMPANY APPLICATION FORM

Filling in company application forms can become a boring and repetitive task, yet any carelessness on an applicant's part can draw a negative reaction from readers. Each company or organization usually uses its own specially designed form which, although it asks for generally the same basic information, may vary in detail. Consequently the suggestions below apply primarily to the *approach* you should take rather than suggest what you should write.

- When visiting prospective employers, always carry your personal data file with you so that you can readily search for details such as dates, telephone numbers, and names of supervisors.
- Treat every application form as though it is the *first* one you are completing—write carefully, neatly, and legibly. Never let an untidy application form subconsciously prepare an employer to meet an untidy worker.
- Complete *every* space on the form, entering N/A (not applicable), "Not Known," or a short horizontal line in spaces that do not apply to you or for which you genuinely do not have the information. This will prevent an employer from thinking you carelessly (or, worse, intentionally) omitted answering the question.
- Take care that your familiarity with your city and street names does not cause you to abbreviate or omit them. If you write "Mpls" for Minneapolis or "St. Pete" for St. Petersburg, or omit the "St.," "Ave.," or "Crescent" from behind your street name (because you *know* it is a street, avenue, or crescent), you may create the impression that your approach to work is to take shortcuts whenever possible.
- Use words that describe the responsibility and supervisory aspects of each job you have held rather than list only the duties you performed.
- Particularly describe extracurricular activities that show your involvement in the community or in which you held a teaching or coaching role.
- Pay particular attention if there is a section on the form that asks you to comment on how your education and past experience have especially prepared you for the position. Think this through very carefully before you write so that what you say shows a natural progression from past experience to the job you are applying for. If you can, and if they fit naturally, add a few words to demonstrate how the position fits your overall career plan. This can be a particularly difficult section to write so do not be afraid to obtain an opinion of its effectiveness from another person.

ATTENDING AN INTERVIEW

This is the third step in the job application process and the first occasion when you meet a prospective employer (or, more often, the employer's representative) face to face.

PREPARING FOR THE INTERVIEW

The key to a good interview is thorough preparation. If you have prepared yourself well, the interview will most likely run smoothly and you will present yourself confidently.

As soon as you are invited to attend an interview—or, better still, before you are called—start researching facts about the company (or organization, if it is a government establishment). Presumably, you will have done some research before submitting your letter of application. Now you need to identify additional information, such as the number of persons the company employs, specific fields in which it is involved, work for which it is particularly well known, its major products and services, important contracts it has received (news of which has been released to the media), locations of branch offices, and the company's involvement in community activities. Such knowledge can be extremely useful during the interview, because it permits you to ask intelligent questions at appropriate places—questions which indicate to the interviewer that you have done your homework.

You also need to prepare for difficult questions an interviewer may pose to test your readiness for the interview and the sincerity of your application. You may be asked:

> *Why do you want to join our organization?*
>
> *How do you think you can contribute to our company?*
>
> *Why do you want to leave your present employer?* (Asked only of persons who are already employed.)
>
> *Why did you leave such-and-such a company on such-and-such date?* (Asked of persons whose resumes show no explanation for a previous employment termination.)
>
> *What do you expect to be doing in five years? Ten years?*
>
> *What salary do you expect?*

If lack of preparation for such questions causes you to hesitate too long before answering, the interviewer may interpret your hesitation to mean you find the questions difficult to answer or that there are factors you would rather conceal. In either case, you may inadvertently be providing an entirely misleading impression of yourself.

An interviewer who asks what salary you expect is partly testing your preparation for the interview and partly assessing how accurately you value yourself. For an undergraduate at a university or college, the question is largely academic: undergraduates compare notes and quickly learn what starting salaries are being offered. But for a person who recently has been or currently is employed, the question is important and must be anticipated. Always know the salary you would like to receive and think you are worth. Avoid quoting a salary range, such as "between 19 and 22 thousand dollars," because it seems to indicate unsureness. Quote a definite figure, such as $21,000, and you will sound much more confident. If you fear that the salary you want to quote may be too high, you can always add the qualification "... depending, of course, on the opportunities for advancement and the fringe benefits your company offers."

You should be ready to ask questions during the interview. Just as the interviewer wants to acquire information about you, so should you want to learn things about the company and the opportunities it can offer. Consider what questions you would like answered, and jot them onto a small card to be stored in a convenient pocket or your purse. Then when the interviewer asks, "Now, do you have any questions?" you can pull out the card.

Make the entries on your card brief and clearly legible; ideally, use capital letters. Your list should be short, because you need to scan it quickly and the interviewer won't have time to answer a lot of questions. So limit your choice to the really important topics. Remember, too, that the quality of your questions will demonstrate how carefully you have given thought to the interview.

CREATING A GOOD INITIAL IMPRESSION

When you enter the interview room, the interviewer will have read the documents (letter of application, resume, application form) you submitted previously, and so will have some knowledge of you. You may be asked, however, to fill in some details orally, partly to set the interview in motion, and partly to refresh the interviewer's memory. This occurs particularly when you are being interviewed by several persons, who jointly are called an interview board. Usually, one member of the board has read your information thoroughly, but the others may have had time for only a quick glance at it just before you enter the room. For them, some repetition of your background can be useful.

Remember that you are being evaluated from the moment you step into the interview room. Consequently you should

- Walk in briskly and cheerfully.
- Shake hands firmly, because a limp handshake creates an image of a limp, indefinite applicant (an image you cannot afford to create).
- Repeat the person's name as you are introduced and look him or her directly in the eye.
- Sit when invited to so do, pushing yourself well back in the chair, making yourself comfortable, and avoiding folding your arms across your chest (which psychologically suggests that you resist questioning).

PARTICIPATING THROUGHOUT THE INTERVIEW

An interview normally falls into three fairly easy-to-distinguish parts. The initial part is an exchange of pleasantries between yourself and the interviewer, who wants you to be at ease. To help you adjust to the interview environment, he or she may ask questions on topics you can answer confidently, such as a major news item or something from the hobbies and interests section in your resume. This initial part of the interview normally is short.

In comparison, the middle part of the interview is quite long. During this part the interviewer tries to find out as much as possible about you. He or she will want to hear your opinions and have you demonstrate your knowledge on certain topics. The interviewer will want to control the direction the interview takes but will expect you to develop your answers and to comment on each topic in sufficient depth to establish that you have real knowledge and experience, backed up by well-thought-out opinions.

The closing portion of the interview also is short. The interviewer will ask if you have questions to ask and will discuss details about the company and employment with it. By this stage the interviewer should have a pretty good impression of you, and you should know whether you want to be employed by the company he or she represents.

To learn as much as possible about how you think and react, an interviewer expects you to give thoroughly developed answers to questions. An effective interviewer will pose questions and subsequent prompts in such a way that you are carried easily from one discussion point to the next and are automatically encouraged to provide comprehensive answers. But if you find yourself face to face with an inexperienced or inadequately prepared interviewer, the responsibility is solely yours to develop your answers in greater depth than the questions seem to call for.

For example, the interviewer may ask, "How long did you work in a mobile calibration lab?"

You might be tempted to reply "Three years," and then sit back and wait for the next question. You would do much better to reply: "For three years total. The first year and a half I was one of four technicians on the Minneapolis-to-Sioux City circuit. And then for the next year and a half I was the lab supervisor on the Fort Westin-to-Manomonee route."

An answer developed in this depth often provides the prompt (that is, piece of information) from which the interviewer can frame the next question.

Sometimes you will face a single interviewer, while at other times you may face an interview board of two to five people. In a single-interviewer situation you will naturally direct your replies to the interviewer and should make a point of establishing eye contact from time to time. (To maintain continuous eye contact would be uncomfortable for both you and the interviewer.) In a multiple-interviewer situation you should

1. Direct most of your questions, and your responses to general questions, to the person who apparently is the chairperson. (But if an answer is long you should occasionally take time to look briefly at and talk momentarily to other board members.)
2. If a particular board member asks you a specific question, address your response to that person.
3. If a board member has been identified as a specialist in a particular discipline, direct questions to that board member if they especially apply to that field.

In certain interviews—often when applicants are being interviewed for a high-stress position—you may be presented with a "stress" question. A stress question is designed to place you in a predicament to which there may be two or even more answers or courses of action that could be taken. You are expected to think *briefly* about the situation presented to you and then to select what you believe is the best answer or course of action. Often you will be challenged and expected to defend the position you have taken.

The secret is not to let yourself be rattled and to defend your answer rationally and reasonably even though the questioner's challenging may seem harsh or unreasonable. Remember that the interviewer is probably more interested in seeing how you cope in the stress situation than in hearing you identify the correct answer.

Here are seven additional factors to consider:

- Control your voice carefully, and make sure everyone can hear you. Speak at a moderate speed, carefully thinking out your answers. Don't be afraid to let your enthusiasm for a topic show.
- Don't be afraid to ask questions, but have a clear idea of what you want to ask before you pose them. The interviewer will recognize a good question and the clarity of thought behind it.
- If you do not know the answer to a question, say you don't know rather than try bluffing your way through. *Say you need time to study the issue.*
- If you do not understand the question, again don't bluff. Either say you do not quite understand or, if you think you know what the interviewer is driving at, rephrase the question and ask if you have interpreted it correctly. (Never imply that the interviewer posed the question poorly!)
- Use humor with great care. What to you may be extremely funny may not match the interviewer's sense of humor.
- Bring demonstration materials to the interview if you wish (such as a technical proposal or report you authored, or a drawing of a complex circuit you designed) but be aware that you may not have an opportunity to display them. If the topic they support comes up during the interview, introduce them naturally into your response to a question. But remember that the interviewer does not have time to read your work, so the point you are trying to make should be readily identifiable simply by viewing the demonstration item. Never force demonstration materials on an interviewer.
- Do not smoke unless the interviewer also smokes and invites you to do so.

Finally, try to be yourself. Remember that interviewers want to see the kind of person you really are. If you relax and answer questions comfortably and purposefully, they will gain a good impression of you. If you try too hard to be the kind of person you think the interviewers want you to be, or to give the kind of answers you think they want rather than the answers you really believe in, they may detect it and judge you accordingly.

ACCEPTING A JOB OFFER

The telephone rings and the personnel representative you met during your interviews tells you that the company is offering you employment at a salary of $xxxxx. You accept the offer! And then she asks when you can start work. (Employers recognize that if you are attending college there will be a waiting period until your course is finished and you have graduated; similarly, an employed engineer or engineering technician has to resign from his or her present position, normally giving either two weeks' or one month's notice.) You quote a starting date to the personnel representative, which she agrees to, and then she says she will confirm the offer in writing. She also asks you to write a letter confirming your acceptance of the position.

The two letters become, in effect, a mild contractual agreement: the employer offers you work under certain conditions, which you agree to. The letters can also prevent any misunderstandings from developing, which can occur if arrangements are made only by telephone. Consequently your acceptance letter should:

- Announce that you are accepting the offer of employment.
- Repeat any important details, such as the agreed salary and starting date.
- Thank the employer for considering you.

The following acceptance letter conforms to this pattern:

> Dear Ms. Tataryn:
>
> I am confirming my telephone acceptance of your May 19 offer of employment as an engineering technician in the Controls Department. I understand that I am to join the company on June 15 and that my salary will be $21,500 annually.
>
> Thank you for considering me for this position. I very much look forward to working for Magnum Electronics.
> Sincerely,

Sometimes an applicant may receive two offers of employment at the same time, and will have to decline one. The letter declining employment should follow roughly the same pattern:

- Decline the offer.
- Briefly explain why.
- Thank the employer.

Here is an example:

> Dear Mr. Genser:
>
> I regret I will be unable to accept your kind offer of employment. Since my interview with you I have been offered employment elsewhere and now must honor my commitment to the other company.
>
> Thank you for considering me for this position.
>
> Regards,

Declining a job offer pleasantly and formally in a carefully worded letter like this is insurance for the future: one day you may want to be employed by that employer!

Assignments

PROJECT 1: PREPARING A RESUME

You are to prepare a resume describing your background, education, work experience, extracurricular activities, and other interests. Do it in two parts:

Part 1. Prepare a personal data file, using five 8×5 in. cards.

Part 2. Prepare a resume which you could use when seeking employment at the end of your course. You may prepare either a traditional or a focused resume.

PROJECT 2: APPLYING FOR A LOCALLY ADVERTISED POSITION

From your campus student employment center or your local newspaper, identify a company currently advertising a position you could apply for at the end of your course.

Part 1. Write a letter applying for the position (assume that you will be graduating in six weeks). Also assume that you are attaching a resume to your letter. If you are replying to a newspaper advertiser, attach a copy of the advertisement to your letter.

Part 2. Assume that the company you wrote to in Part 1 sends you an application form. Obtain a standard application form from your campus student employment center and complete it as though it is the advertiser's form.

Part 3. Now assume that the company has telephoned and asked you to attend an interview next Tuesday. On a sheet of paper write down five questions you would ask during the interview. After each question explain why the question is important and what answer you hope it will elicit from the interviewer.

PROJECT 3: REPLYING TO OTHER ADVERTISERS

This project assumes that you are seeking permanent employment at the end of a technical or engineering-oriented training program. You are to reply to any one of the following advertisements, using your present situation and actual background. If it is still early in your training program, you may update the time and assume that you are now two months before graduation date.

ROPER CORPORATION (OHIO DIVISION)
requires a
CHEMICAL TECHNICIAN

to join a project group conducting research and development into the organic polymers associated with the coating industry. The successful applicant will also assist in the development of control techniques for producing automated color tinting. Apply in writing, stating salary expected to:

Mr. P. Rassmusen
Personnel Manager
P. O. Box 1728
Montrose, Ohio 45287

(Advertisement in Montrose Herald, April 17)

ARCHITECTURAL DRAFTSPERSON
required by
a well-established firm of architects and planners

Applicants should have completed a Design and Drafting course at a community college or similar training institution, in which strength of materials and structural design were a curriculum requirement. Experience in preparation of reports and proposals will be considered in selection of successful applicant.
Write to:

The Corydon Agency
Room 604—300 Main St.

(Advertisement of March 28 on College Notice Board)

H. L. WINMAN AND ASSOCIATES
475 Reston Avenue
Cleveland, Ohio 44104

CIVIL ENGINEERING TECHNICIANS

This established firm of consulting engineers needs three Engineering Technicians to assist in the construction supervision of several grade separation structures to be built over a two-year period. Graduation from a recognized Civil Engineering Technology course is a prerequisite. Successful applicants will be hired as term employees. Opportunities are excellent for transfer to permanent employment before the project ends.

Apply in writing to:

Mr. A. Rittman, Head
Special Projects

(Advertisement on College Notice Board)

MACRO ENGINEERING INC
invites applications from
ENGINEERS, TECHNOLOGISTS, AND TECHNICIANS
interested in Metrology

We operate a first-class Standards Laboratory that is a calibration center for precision electrical, electronic, and mechanical measuring instruments used in our military and commercial equipment maintenance programs.

Applicants for senior positions should hold a B.S. in Electrical Engineering. Laboratory Technicians should be graduates of a recognized electronics or mechanical technology course who have specialized in precision measurement.

Applications are invited from forthcoming graduates, who will work under the direction of our Standards Engineer. All applicants must be able to start work on July 1.

Apply in writing to:

Mr. F. Stokes
Chief Engineer
Macro Engineering Inc
600 Deepdale Drive
Phoenix, Arizona 85007

(Advertisement in Arizona Republic, March 18)

MONTROSE PAPER COMPANY

Offers excellent opportunities for recent graduates to join an expanding manufacturing organization in the pulp and paper industry.

ENGINEERS AND ENGINEERING ASSISTANTS

Positions are available for mechanical engineers and technicians to assist in the design, installation, and testing of prototype production equipment. Previous experience in a manufacturing plant would be helpful. Innovative ability will be a decided asset.

ELECTRICAL ENGINEERING TECHNICIAN

This person will assist the Plant Engineer in the maintenance of power distribution systems. Applicants should be graduates of a two-year course in Electrical Technology with good knowledge of automatic controls and machine application. Ability to read blueprints and working drawings is essential.

COMPUTER ENGINEERING TECHNICIAN

This position will suit either a graduate of a Computer Technology course or an Electronics Engineering Technician who has specialized in Computer Electronics.

Duties will consist of installation, maintenance, and troubleshooting of existing and future computer equipment.

ENVIRONMENT SPECIALISTS

Persons selected will test air pollutants and water effluents from our paper mill and production plant, and assess their environmental impact. Applicants should be graduates of a recognized course in the environmental or biological sciences.

Salaries for the above positions will be commensurate with experience and qualifications. Excellent fringe benefit program available. Write in confidence to:

Manager of Industrial Relations
MONTROSE PAPER COMPANY
Montrose, Ohio 45287

(Advertisement in Montrose Herald, March 10)

NORTH AMERICAN WILDLIFE INSTITUTE
requires
SCIENTISTS AND BIOLOGISTS
for its wildlife preservation projects in Florida,
New Mexico, Manitoba, Quebec, and Alaska.

Applicants should have a degree in the environmental sciences or an appropriate diploma from a junior college or technical institute. Persons anticipating graduation are particularly encouraged to apply.
Write to:
Dr. Frederick M. Hauser
Chairman, North American Wildlife Institute
Room 1620 — 385 East 47th Street
New York, N.Y. 10017

(Advertisement in last month's issue of American Wildlife)

ENGINEERING TECHNICIAN

required by manufacturer of
automated car wash equipment

Duties: Assist plant engineer design and test mechanical and electrical components; supervise installations in cities across U.S. and Canada; troubleshoot problems at existing installations.

Applicants must be free to travel extensively. Excellent salary and promotion possibilities.
Apply in writing to:
Personnel Manager
DIAL-A-WASH Incorporated
2020 Waskeka Drive
Montrose, Ohio 45287

(Advertisement in Montrose Herald, April 23)

We require a
CIVIL ENGINEERING TECHNICIAN

with an interest in building construction to prepare estimates, do quantity takeoffs, and assume responsibility for company sales and promotion activities.

Apply in writing to:

Personnel Department
PRECASCON CONCRETE COMPANY
227 Dryden Avenue

(Advertisement in your local newspaper, February 26)

PRODUCTION TECHNICIAN/DRAFTSPERSON

Duties:

Under the general direction of the Chief Engineer, to be responsible for detailed product design of prototype machines. Also will investigate and recommend improvements to existing product lines.

Qualifications:

Graduate of a recognized Mechanical Engineering degree or diploma course, or a Drafting course with emphasis on Mechanical Drawing. Experience with or knowledge of agricultural equipment would be an asset.

Please apply to:

Chief Engineer
Agricultural Manufacturing Industries
3720 Harvard Avenue

(Advertisement in your local newspaper, February 26)

PROJECT 4: APPLICATION FOR UNSOLICITED EMPLOYMENT

This project assumes that you are seeking permanent employment at the end of a technical training program but few job openings have been advertised in your field. Write to Macro Engineering Inc in Phoenix, or to H. L. Winman and Associates (for the attention of one of the department heads in Cleveland, or local branch manager Vern Rogers), applying for employment. Use your knowledge of the company and your real background. If it is still early in your training program, you may update the time and assume that it is now two months before graduation.

PROJECT 5: APPLICATION FOR SUMMER EMPLOYMENT

Assume that you are looking for summer employment. Reply to either one of the following advertisements:

CITY OF MONTROSE, OHIO
TRAFFIC DIVISION

The city of Montrose is carrying out an Origination-Destination (OD) Study throughout June and July, with computation and analysis to be carried out in August.

Approximately 40 students will be required for the OD Study, and 20 for the computation and analysis. For the latter work, some knowledge of statistical analysis will be an advantage.

Students interested in this work are to apply in writing to Mr. G. Braithwaite, Room 310, Civic Center, City of Montrose, Ohio 45287.

(Notice on College Notice Board)

REMICK AIRLINES
offers
SUMMER EMPLOYMENT
to a limited number of college students.

Duties: To work as part-time dispatchers, baggage handlers, aircraft cleaners, cafeteria assistants, etc, during the peak summer period (June 15–September 10) at Remick Airlines terminals at:

Fort Wiwchar, Michigan
Weekaskasing Falls, Wisconsin
Lake Lawlong, Nevada

Good pay; ample free time; free transportation to destination and return.
Write, describing previous work experience (if any), to:

Carl Nickerson   or: Mavis Chandler
Flight Operations Manager   305-5870 LaFrance Avenue
Room 301   Minneapolis, MN 55404
Montrose Municipal Airport
Montrose, OH 45286

(Notice on College Notice Board)

NORTHERN POWER COMPANY
BOX 1760, FAIRBANKS, ALASKA 99706

Applications are invited from students in two-, three-, or four-year technical programs who are seeking three months of well-paid summer employment. The work will include construction of living quarters and associated facilities for mining camps near Prudhoe Bay, Alaska. Terms of employment will be:

    13 full weeks—June 10 to September 10
    Excellent pay: $425 per week
    Transportation: Paid
    Accommodation: Paid
    Health: Applicants should be in topnotch physical condition.

Students interested in this work are to write to Mr. G. R. Cardinal at the above address. Although previous experience in construction work is not essential, knowledge of general construction methods will be considered an asset for some positions.

(*From* Construction—North, *February 20*)

PROJECT 6: UNSOLICITED SUMMER EMPLOYMENT APPLICATION

Assume that you are looking for summer employment but few summer jobs have been advertised. Write to H. L. Winman and Associates or Macro Engineering Inc, asking for a summer job. Use your knowledge of the companies, plus your actual background, to write an interesting letter.

Address your letter to the attention of Tanys Young or George Dunn in Phoenix. If there is an H. L. Winman and Associates branch in your area, you may address your letter to branch manager Vern Rogers.

11

The Technique of Technical Writing

This chapter concentrates on a few writing techniques that will enable you to convey information both quickly and efficiently. I assume you are already proficient in grammar or "English," and can recognize and correct basic writing problems. If you need practice in basic writing, I suggest you refer to a textbook such as the *Prentice-Hall Handbook for Writers*. You can also refer to the glossary of terms in Chapter 12 for information on how to resolve many of the problems that crop up in technical writing (how to form abbreviations and compound adjectives, how to spell problem words, when to use numerals or spell out numbers in narrative, and so on).

THE WHOLE DOCUMENT

Throughout this book I have stressed the need to tell readers immediately what they most want to know. This means structuring your writing so that the very first paragraph (if possible, the first sentence) satisfies their curiosity. Most executives and many technical readers are busy people who have time to read only essential information. By presenting the most important items first, you can help them decide whether they need to read the whole document or whether they should hand it to someone more familiar with the topic.

THE PYRAMID TECHNIQUE

The pyramid technique is so named because it starts at the top with a small morsel of essential information and then substantiates it with a broad base of details, facts, and evidence. In most letters and short reports it has only two

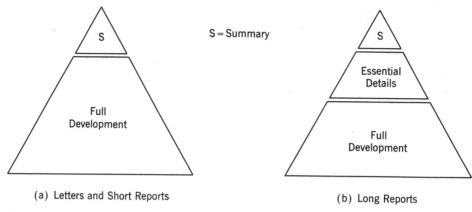

(a) Letters and Short Reports   (b) Long Reports

Figure 11-1. The pyramid technique.

stages: a brief *summary* followed by the *full development,* as shown in Figure 11-1(a). In long reports an additional stage, the *essential details* (see Figure 11-1[b]), is inserted between the summary and the full development.

Readers normally are not consciously aware of the pyramid technique. They simply find that a document in which it is employed is very easy to follow. In the letter report in Figure 11-2, Stanley Roning summarizes in the first paragraph the two things Tina Mactiere wants to know right away: whether the training course was a success and what results were achieved. He uses the remainder of the letter to fill in background details, to state briefly how the course was run, to report on student participation, and to comment on student reaction.

The pyramid technique requires that readers know the main elements of the story by the end of the first paragraph. They can then read the full development more intelligently. If a reader is very busy and the report is long, he or she may choose to stop reading at the end of the summary and continue with the full development later, or may scan the full development quickly and pass the report on to someone else for action.

In the full development you tell the whole story. You amplify what you have already stated in the summary by inserting all the technical details readers need to understand the subject fully. You may develop the subject in whatever order you like, although much of the time you will find that a past-present-future order is the simplest and most effective. This in turn may present a problem, since you will have to maintain continuity between the summary (which frequently ends with a comment on future action) and the first paragraph of the full development (which normally starts by saying how the project began).

To make a good transition from the summary *back* to the background information requires skill. It has been done well in Figure 11-2 because the writer begins paragraph 2 by referring to the training course mentioned in the previous paragraph (he starts the full development by saying: "This was a pilot course . . ."). Further examples of good transitions can be found in the letter reports in Figures 4-3 of Chapter 4 and 5-2 of Chapter 5.

October 20, 19xx

THE RONING GROUP

Communication Consultants

Ms. Tina Mactiere, President
Macro Engineering Inc
600 Deepdale Drive
Phoenix, AZ 85007

Dear Ms. Mactiere:

Results of Pilot Report Writing Course

The report writing course we conducted for members of your engineering staff *essential info*
was completed successfully by 14 of the 16 participants. The average mark
obtained was 63%.

This was a pilot course set up in response to an August 13, 19xx enquiry
from Mr. F. Stokes. At his request, emphasis was placed on giving participants practical experience in writing business letters and technical reports.
Attendance was voluntary, the 16 participants being selected at random from
29 applicants.

Best results were achieved by participants who recognized their writing problems before they started the course, and willingly became actively involved
in the practical work. A few presumably had expected it to be an "information" type of course, and hence were less willing to take part in the heavy
writing program. Our comments on the work done by individual participants
are attached.

Course critiques completed by participants indicate that the course met their
needs from a letter and report writing viewpoint, but that they felt more
emphasis could have been placed on technical proposals and oral reporting.
Perhaps such topics should be covered in a short follow-up course.

We enjoyed developing and teaching this pilot course for your staff, and
particularly appreciated their enthusiastic participation.

Sincerely

Stanley G. Roning.

Stanley G. Roning
President

SGR:rb
enc

Figure 11-2. A letter report written using the pyramid technique.

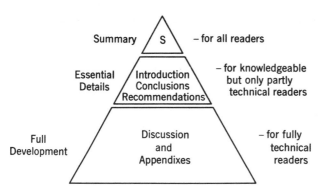

Figure 11-3. Pyramid technique applied to the formal report.

There is no problem in making this transition in the long formal report because the summary and the introduction (which contains the background information) are on separate pages and are often separated by the table of contents. Thus a technical writer can develop a brief but complete story in the summary without having to consider how to build a transition to the next paragraph. Indeed, the organization of many modern formal reports assists writers to adapt their writing to the pyramid technique.

In Figure 11-3 the three main sections of the alternative format for the formal report discussed in Chapter 6 are paralleled with the three stages of development demanded by the pyramid technique. The summary of the formal report is the initial stage of the pyramid; the introduction, conclusions, and recommendations are the essential details; and the discussion is the full development. In effect, the formal report written in this way becomes three reports: the brief but informative **summary** written in general terms that can be understood by any reader; the **essential details** of the introduction, conclusions, and recommendations, written for the knowledgeable but not necessarily technical reader; and the **full development** of the discussion, written for the reader who has the interest and technical capability to understand all the evidence the writer has to present.

What does this technique mean to the writer of scientific and engineering documents? Principally, it provides the opportunity to cater to more than one level of reader within the same document. This is particularly true of long formal reports, less so of short business letters. But simply adopting the two- or three-stage approach of the pyramid technique is insufficient in itself to ensure good writing. It is a mechanical means to effective writing that must be supported by two less tangible but equally important factors: tone and writing style.

TONE

Being able to write for several levels of reader in the same document does not imply that you can omit identifying a specific reader (see Chapter 1). Without this important step you will not be able to set the right tone. The reader you identify is always the person who will be reading the full development. Thus, you should write primarily for your most technically knowledgeable reader, while directing the essential details and the summary to progressively less technically knowledgeable persons.

Whether your writing should be formal or informal will depend on the situation and your familiarity with the reader. Formal reports should adopt a formal tone. (Note, however, that a formal tone is not stiff or pompous; there is no room for writing that makes readers feel uncomfortable because they are not as knowledgeable as you are.) A formal report conveys an image of your company's reputation even before the first page is opened; the words between the covers must maintain that image. Business letters are generally less formal, depending on their importance. For example, a management-level letter proposing a joint venture on a major defense project would be formal, whereas letters between engineers discussing mutual technical problems would be informal. A memorandum report can be informal, since normally it is an in-plant document most often written between persons who know each other.

Varying levels of tone are evident in the following extracts from three separate H. L. Winman and Associates' documents, all written on the same subject.

1. **Extract from a Memorandum.** John Wood's Material Testing Laboratory has compression-tested samples of concrete for Karen Woodford of the Civil Engineering Department. In his memorandum reporting the test results, John writes:

   Informal Tone I have tested the samples of concrete you took from the sixth floor of Tarryton House and none of them meets the 4800 psi you specified. The first failed at 4070 psi; the second at 3890 psi; and the third at 4050 psi. Do you want me to send these figures over to the architect, or will you be doing it?

2. **Extract from a Letter Report.** Karen Woodford conveys this information to the architect in a brief letter report:

   Semiformal Tone Our tests of three samples taken from the sixth floor of Tarryton House show that the concrete at 52 days still was 800 psi below your specification of 4800 psi. We doubt whether further curing will increase the strength of this concrete more than another 150 psi. We suggest, however, that you examine the design specifications before embarking on an expensive and time-consuming remedy.

3. **Extract from a Formal Report.** The architect rechecked the design specifications and decided that 4200 to 4300 psi still would not satisfy the design requirements. He then requested H. L. Winman and Associates to prepare a formal report that he could present to the general contractor and the concrete supplier. Karen Woodford's report said, in part:

   Formal Tone At the request of the architect we cut three 30 × 15 cm diameter cores from the sixth floor of Tarryton House 52 days after the floor had been poured. These cores were subjected to a standard compression test with the following results (detailed calculations are attached at Appendix A):

Core No.	Location	Failed at:
1	0.46 m W of col 18S	4070 psi
2	0.84 m N of col 22E	3890 psi
3	1.42 m N of col 46E	4050 psi

The average of 4003 psi for the three cores is 797 psi below the design specification of 4800 psi. Since further curing will increase the strength of the concrete by no more than 150 psi, we recommend that this concrete pour be rejected.

Although the information conveyed by these three examples is similar, the tone the writer adopts varies sufficiently to help set the scene for each situation.

STYLE

The pyramid technique has no effect on writing style, other than consideration of the reader at each stage. Complexity of subject and technical level of reader have some effect on style, in that you should use careful, simple language when describing a very complex topic to a moderately knowledgeable reader. As the reader's technical level increases, or the complexity of the subject decreases, you can use longer sentences and words, and more complex sentences. Here are some suggestions:

1. When presenting low-complexity background information, and descriptions of nontechnical or easy-to-understand processes, write in an easygoing style that tells readers they are encountering information that does not require total concentration. Use slightly longer paragraphs and sentences, and insert a few adjectives and adverbs to color the description and make it more interesting.
2. For important or complex data, use short paragraphs and sentences. Present one item of information at a time. Develop it carefully to make sure it will be fully understood before proceeding to the next item. Use simple words. This punchy style will warn readers that the information demands their full attention.
3. When writing instructions or step-by-step descriptions, start with a narrative-type opening paragraph that introduces the topic and presents any information that the readers should know or would find interesting; then follow it with a series of subparagraphs each describing a separate step. These subparagraphs should conform to the following general rules:
3.1 Each should develop only one item of the process.
3.2 They should be short; even single sentence subparagraphs are acceptable, particularly for technical instructions.
3.3 They should be parallel in construction, that is, they should generally conform to the same "shape." The importance of parallelism is discussed later in this chapter.
3.4 Each step of an instruction should start with a strong verb that *tells* the reader to do something (see page 227 for suggestions).

WRITING SEQUENCE

If you are to set the right tone throughout, you must write in reverse order, starting with the full development. Writing a report in the order in which it will be read is difficult, if not impossible. An engineering technician who writes the summary before the full development will use too many adjectives and adverbs, big words when shorter words would be more effective, and dull opening statements such as "This report has been written to describe the investigation into defective MN-1 compasses carried out by H. L. Winman and Associates." He or she will be writing without having established exactly what to say in the full development.

The best writing sequence therefore should be the reverse of that shown in the pyramid in Figure 11-3:

Step 1: Write the full development. Direct it to the type of technical reader who will use or analyze your report in depth.

Step 2: Write the essential details (omitted in short informal reports). Make them brief and direct them to a semitechnical reader or person in a supervisory or managerial position. They should comprise:

    a. Introductory information to help the reader understand why the project (and hence the report) was undertaken and what the terms of reference were.

    b. The results of the project, or the main conclusions that can be drawn from the full development; this should follow naturally from the introductory information and satisfy the terms of reference.

    c. A recommendation (if one is to be made) showing what needs to be done next.

Step 3: Write the summary. Direct it to a nontechnical reader who has absolutely no knowledge of the project or the contents of the report. Make it very brief and make sure it tells:

    a. Why the report was written.

    b. What was found out.

    c. What the main conclusions and recommendations are.

PARAGRAPH NUMBERING

The types of industrial documents that most often carry paragraph numbers are military reports, specifications, and technical instructions. Some companies stipulate that all their reports bear paragraph numbers, but they are in the minority. Their reasons for doing so, however, are valid: paragraph numbers are an excellent aid to organization, and they provide a useful means for cross-referencing (both within the report itself and from one report to another).

A paragraph numbering system must be obvious to the reader, simple for you (the writer) to manipulate, and easy for the typist to arrange on the page. The simplest paragraph numbering system starts at 1 and numbers paragraphs

TYPICAL PARAGRAPH NUMBERING SYSTEMS

Figure 11-4. Suggested paragraph and subparagraph numbering systems.

```
1.                A.              1.
2.                  1.            2.
3.                  2.            3.
3.1                   (a)          3.1
3.1.1                 (b)          3.2
3.1.2                   (1)          a)
3.1.2.1                 (2)          b)
                  B.                   (1)
etc., up to       1.                   (2)
6 digits          2.                      etc.
                  etc.

METHOD A      METHOD B      METHOD C
```

consecutively to the end of the document. More complex systems combine numbers, letters, and decimals to allow for subparagraphing. Three possible arrangements are shown in Figure 11-4.

Method A is a numerals-only decimal system, simple and unambiguous, that is often used by the military services and by specification writers. However, the subparagraphing does become difficult to follow in long documents, particularly when indentation cannot be used. (Indention of multiple-digit subparagraphs can quickly result in a very narrow band of information down the right side of the page.)

Method B permits indentation and uses a combination of letters and numerals. Extremely simple, it is popular with many report writers even though it can be ambiguous (there can be several paragraph 1s, 2s, etc, one set under A, another under B, and so on).

The third method, C, combines the decimal and letter-number arrangements for a system with no ambiguity between main paragraphs and subparagraphs. This is the method chosen by Anna King for H. L. Winman and Associates' reports that carry paragraph numbers. She suggests that engineers use normal-size paragraphs for the first two levels, and short paragraphs for the lower-level subparagraphs:

```
1.          }
2.          }   main paragraphs
  2.1       }
  2.2       }   full subparagraphs
    a)      }
    b)      }   short subparagraphs
      (1)   }   very short subparagraphs
      (2)   }   (as one sentence only)
```

This paragraph numbering system combines well with the heading arrangements illustrated in Figure 11-5. For a partial example of this system used in an informal report, turn to Figure 5-6 in Chapter 5.

MAIN CENTER HEADING

SUBSIDIARY CENTER HEADING

The main center heading is always capitalized, and may be underlined. Subsidiary center headings (if used) may be capitalized but not underlined, as shown above, or may be in lower case letters and underlined, as shown immediately below.

Subparagraphing Without Paragraph Numbering

Side Headings

Side headings, as the one immediately preceding this paragraph shows, are usually typed in lower case letters (except for the first letter of each major word), and are under-lined.

Each paragraph is typed level with the same margin as the side heading. In technical writing, the first line of each paragraph is seldom indented.

Subparagraph Headings and Subparagraphing

Subparagraph headings are indented about four or five typewriter spaces.

Each subparagraph is typed as a solid indented block. It should not run back to the left-hand margin because that would make the visible evidence of the subpara-graphing much less obvious.

Secondary Subparagraphing

If further subparagraphing is necessary, the headings and subparagraphs are indented a further four or five typewriter spaces.

<u>Heading Built into the Paragraph.</u> In this lesser-used arrangement, the text continues immediately after the heading. Paragraph headings like these can be used for main paragraphs, subparagraphs, and secondary subparagraphs.

Subparagraphing Combined with Paragraph Numbering

1. <u>Side Headings</u>

   1.1 Normally, when paragraph numbers are used, side headings are assigned simple paragraph numbers, almost like section numbers, as has been done here.

   1.2 Where only a single paragraph follows a side heading or subparagraph heading, it is not assigned a separate paragraph number. But it is typed with its left-hand margin indented slightly, level with the start of the heading above it (see the paragraph immediately following heading 1.3).

   1.3 <u>Subparagraph Headings and Subparagraphing</u>

      If a single paragraph follows the heading, no number is assigned to the para-graph. If more than one paragraph follow the heading, each is assigned an identification number or letter:

      a) This would be the first subparagraph.

      b) This would be the second subparagraph.

      c) Each subparagraph can be further subdivided into a series of very short second-level subparagraphs:

         (1) Here is such a subparagraph.

         (2) Preferably, each such subparagraph should contain no more than one sentence.

      d) Lower level subparagraphs can also be assigned subparagraph headings.

Figure 11-5. Combined system of headings, paragraphs, and paragraph numbers.

ARRANGEMENT OF HEADINGS

There are three main types of headings you are likely to encounter: center headings, side headings, and paragraph headings. Their use is demonstrated in Figure 11-5.

PARAGRAPHS

The role of the paragraph is complex. It should be able to stand alone but normally is not expected to. It must contribute to the document of which it is a part, yet it must not be obtrusive except when called on to emphasize a specific point. It should convey only one idea, although made up of several sentences each containing a separate thought.

Experienced writers construct effective paragraphs almost subconsciously. They adjust length, tone, and emphasis to suit their topic and the atmosphere they want to create. Literary writers, in particular, knowingly stretch and bend the rules of good paragraph construction to obtain exactly the right impact. But even they once had to study and master the techniques of good paragraph writing, although now they let the rhythm of the words guide them far more than the rules.

We are not so fortunate. We have to learn the rules and apply them consciously. But in doing so we must take care not to let our approach become too pedantic, or become so bound by the rules that we write in a stilted manner that is uninteresting and unrhythmic. So when considering the factors designed to help you construct good paragraphs, remember that they are blocks on which to build your writing, not bars to imprison your creativity.

The three elements essential to good paragraph writing are:

Unity
Coherence
Adequate Development

These elements cannot stand alone. All three must be present if a paragraph is to be useful to its reader.

UNITY

For a paragraph to have unity, it must be built entirely around a central idea. This idea is expressed in a topic sentence and developed in supporting sentences. In effect, this permits us to construct paragraphs using the pyramid technique, with the topic sentence taking the place of the summary and the supporting sentences representing the full development (see Figure 11-6).

The paragraph below is strongly unified because its topic is clearly ex-

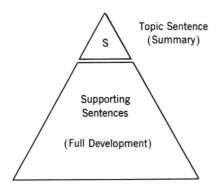

Figure 11-6. Pyramid technique applied to the paragraph.

pressed and the supporting sentences develop it fully (the numbers that precede each sentence are for reference):

> (1) *Joey, when we began our work with him, was a mechanical boy.* (2) He functioned as if by remote control, run by machines of his own powerfully creative fantasy. (3) Not only did he himself believe that he was a machine but, more remarkably, he created this impression in others. (4) Even while he performed actions that are intrinsically human, they never appeared to be other than machine-started and executed. (5) On the other hand, when the machine was not working we had to concentrate on recollecting his presence, for he seemed not to exist. (6) A human body that functions as if it were a machine and a machine that duplicates human functions are equally fascinating and frightening. (7) Perhaps they are so uncanny because they remind us that the human body can operate without a human spirit, that body can exist without soul. (8) And Joey was a child who had been robbed of his humanity.[1]

This opening paragraph of a technical article demonstrates that technical writing does not have to be dull. Let's analyze it, identify why it has unity, and discover why it holds our interest and makes us want to read on.

The paragraph is summarized in sentence (1), the topic sentence, which sets the scene in nontechnical terms that create interest. Sentence (2) amplifies the topic sentence by using slightly more technical terms and introducing the fact that Joey's mechanical aspects are self-created. Then sentence (3) underscores the strength of Joey's imagination by telling us that he caused adult onlookers to feel that he really was a machine. This is carried further in sentence (4), which relates his machinelike actions to normal activities, and sentence (5), which shows how very real his machinelike qualities were, even in moments of inactivity. Then sentences (6) and (7) introduce the fascination (and an element of apprehension) that Joey's condition caused in persons observing him. Sentence (8) restates the idea expressed in sentence (1), but this time in human as opposed to mechanical terms.

---

[1]From Bruno Bettelheim, "Joey, A 'Mechanical Boy.'" Copyright © March 1959 by Scientific American, Inc. All rights reserved.

The topic sentence does not always have to be the first sentence in a paragraph; there are even occasions when it does not appear at all and its presence is only implied. But until you are a proficient writer you would be wise to place your topic sentences right up front, where both you and your readers can see them. Then, when you have experience to support your actions, you can experiment and try placing topic sentences in alternative positions.

The two examples that follow are paragraphs that do not start with a topic sentence. In this descriptive passage on computers the topic sentence is at the end of the paragraph (compare it with the similarly structured technical paragraph describing how exploration crews affected permafrost, p. 225).

> The advantages of a computer have been described by Thelma Forbes, who points out that a small business with a manual inventory system can lose up to 10% of its total business volume each year simply because stock levels are improperly predicted and maintained, and records are tardily updated. In comparison, she contends, a computer system anticipates rather than reacts to a company's needs, and so can increase business volume and replace that lost 10% each year. However, a computer system can just as easily become a liability if it is too small, too large, overly expensive, or improperly programmed. *Plainly, for all the benefits a computer can bring us, not just any system will do.*[2]

In this continuing description of Joey, the "Mechanical Boy," the topic sentence is only implied. If it had been stated it would have read something like this: "Joey's machinelike actions were realistic."

> For long periods of time, when his "machinery" was idle, he would sit so quietly that he would disappear from the focus of the most conscientious observation. Yet in the next moment he might be "working" and the center of our captivated attention. Many times a day he would turn himself on and shift noisily through a sequence of higher and higher gears until he "exploded," screaming "Crash, crash!" and hurling items from his ever present apparatus—radio tubes, light bulbs, even motors or, lacking these, any handy breakable object. (Joey had an astonishing knack for snatching bulbs and tubes unobserved.) As soon as the object thrown had shattered, he would cease his screaming and wild jumping and retire to mute, motionless nonexistence.[3]

COHERENCE

Coherence is the ability of a paragraph to hold together as a solid, logical, well-organized block of information. A coherent paragraph is abundantly clear to its readers; they can easily follow the writer's line of reasoning and have no problem in progressing from one sentence to the next.

Most technical people are logical thinkers and should be able to write

[2]Brian E. Lundeen, *Evaluation of Computer Systems for the Western Farm Implement Company.* Report No. E-28, Antioch Business Consultants, Foothills, Colorado.

[3]Bettleheim, "Joey, 'A Mechanical Boy.'"

logical, well-organized paragraphs. But the organization must not be kept a secret; it must be apparent to every reader who encounters their work. Simply summarizing a paragraph in the topic sentence and then following it with a series of supporting sentences does not make a coherent paragraph. The sentences must be arranged in an identifiable order, following a pattern that helps the reader understand what is being said.

This pattern will depend on the topic and the type of document. Paragraphs describing an event or a process most likely will adopt a sequential pattern; those describing a piece of equipment will probably be patterned on the shape of the equipment or the arrangement of its features. Patterns that can be used for typical writing situations are illustrated in Table 11-1.

Narrative Patterns.   You can write narrative-type paragraphs whenever you want to describe a sequence of steps or events. The past-present-future pattern of a progress report or occurrence report, such as Bob Walton's accident report in Figure 4-3 is a typical example. The pattern should be clearly evident, as in two of the following three paragraphs:

A coherent paragraph (in chronological order)	The accident occurred when D. Friesen was checking in at the Remick Airlines counter. He placed the company Polaroid camera on the counter while he completed flight boarding procedure. When the passenger ahead of him lifted a carry-on bag from the counter, its shoulder strap tangled with the carrying strap of the camera and pulled the camera to the floor. D. Friesen examined the camera and discovered a 1½-inch crack across its back. Remick Airlines' representative K. Trane took details of the incident and will be calling you to discuss compensation.
A much less coherent paragraph (containing the same information but not presented in an identifiable pattern)	The accident occurred when D. Friesen was checking in at the Remick Airlines counter. K. Trane, a Remick Airlines representative, took details of the incident and will be calling you to discuss compensation. The damaged camera received a 1½-inch crack across the back. When the passenger ahead of D. Friesen removed a carry-on bag from the counter, its shoulder strap tangled with the carrying strap of the company Polaroid camera and pulled it to the floor. D. Friesen had placed the camera on the counter while he completed flight boarding procedure.
A coherent paragraph (tracing events from evidence to conclusion)	A mild shimmy at speeds above 52 mph was noticed about 10 days after the new tires had been installed. A visual check of all four wheels revealed no obvious defects. Rotation of the four wheels to different positions on the vehicle did not eliminate the shimmy but seemed to change its point of origin. To pin down the cause I replaced each wheel in turn with the spare wheel, and found that the shimmy disappeared when the spare was in the left front position. The wheel removed from that position was tested and found to have been incorrectly balanced.

Descriptive Patterns.   You can write descriptive paragraphs to describe scenes, buildings, equipment, and any subject having physical features. The pattern can be defined by the shape of a subject, the order in which parts are

Table 11-1 Paragraph Patterns Used in Technical Writing

TOPICS OR SUBJECTS			WRITING PATTERNS	DEFINITIONS AND EXAMPLES
Event Occurrence Situation Process Procedure Method	Generally Narrative		Chronological order	The sequence in which events occurred; e.g. the steps taken to control flooding, how a new product was developed, how an accident happened
			Logical order	Used when chronological order would be too confusing (when describing a multiple-activity process, or several events that occurred concurrently); requires careful analysis to achieve clear narrative
			Cause to effect	The factors (causes) leading up to the result (effect) they produce; e.g., dirty air vents and failure to lubricate bearings cause overheating and eventual failure of equipment
			Evidence to conclusion	The factors, data, or information (evidence) from which a conclusion can be drawn; e.g. how results obtained from a series of tests help identify the source of a problem
			Comparison	A comparison of factors (price, size, weight, convenience) to show the differences between products, processes, or methods
Equipment Scene Appearance Building Location Features	Generally Descriptive	By Shape	Vertical	From top to bottom* of a tall, narrow subject
			Horizontal	From left to right* of a shallow, wide subject
			Diagonal	From corner to corner (for items arranged across a subject)
			Circular	From center to circumference* (for items arranged concentrically) Clockwise* (for items arranged around a subject)
		By Features	In order of: Size	From smallest item to largest item*
			Importance	From most important to least important item*
			Operation	The sequence in which the items are used when the equipment is operated

*Or vice versa.

operated, the arrangement of parts from smallest to largest, or the importance of the various parts. For example:

Excerpt from paragraph (features in order of importance) The most important control on the bombardier's panel is the firing button, which when not in use is held in the black retaining clip at the bottom left corner. Next in importance is the fusing switch at the top right of the panel; when in the "OFF" position it prevents the bombs from being dropped live. Two safety switches, one immediately above the firing button retaining clip and the other to the right of the bank of selector switches, prevent the firing button from being withdrawn from its clip unless both are in the "LIVE" (up) position.

Continuity.   A fully coherent paragraph must also have smooth transitions between the sentences. Smooth transitions give a sense of continuity that makes readers feel comfortable. As they finish one sentence, there is a logical bridge to the next. This can be accomplished by using linking words and by referring back to what has already been said. In the example just quoted there is a natural flow from "The most important . . ." in the first sentence to "Next in importance . . ." in the second; the third sentence then refers back to the firing button to relate the newly introduced safety switches to what has gone before. The transitions are equally good in the first paragraph describing damage to a Polaroid camera, each sentence containing a component that is a development from one of the previous sentences. This is not true of the second paragraph, in which each new sentence introduces a new subject with no reference to what has already been said.

ADEQUATE DEVELOPMENT

The element of adequate development demands good judgment. You must identify your readers clearly enough so that you can look at each paragraph from their point of view. Only then can you establish whether your supporting sentences amplify the topic sentence in enough detail to satisfy their interest.

Simple insertion of additional supporting sentences does not necessarily meet the requirements for adequate development. The supporting sentences must contain just the right amount of pertinent information, all directed to a particular reader. There must never be too little or too much information. Too little results in fragmented paragraphs that offer snippets of information which arouse readers' interest but do not satisfy their needs. On the other hand, too much information can lead to long, repetitive paragraphs that annoy readers. Compare the following paragraphs, all describing the result of exploration crews' first venture with machinery across the Peel Plateau east of Alaska, intended for a reader who is interested in the problems of working in the north but who has never seen what the terrain is like.

Inadequate development

Trails left by tractors look like long narrow scars cut in the plateau. Many of them have been there for years. All have been caused by permafrost melting. They will stay like this until the vegetation grows in again.

Adequate development

To the visitor viewing this far northern terrain from the air, the trails left by tractors clearing undergrowth for roads across the plateau look like long, narrow scars. Even those that have been there for as long as 28 years are still clearly defined. All have been caused by melting of the permafrost, which started when the surface moss and vegetation were removed and will continue until the vegetation grows in again—perhaps in another 30 years.

Over-development

To conservationists, whose main interest is the protection of the environment, viewing this far northern terrain from the air is a heart-rending sight. To them, the trails left by tractors clearing undergrowth for roads across the plateau look like long, narrow scars. The tractors were making way for the first roads to be built by man over an area that until now had been trodden only by Indians indigenous to the area, and the occasional trapper. Some of these trappers had journeyed from Quebec to seek new sources of revenue for their trade. But now, in the very short time span of 28 years, man has defiled the terrain. With his machines he has cut and gouged his way, thoughtlessly creating havoc that will be visible to those that follow for many decades. Those that preceded him for centuries had trodden carefully on the permafrost, leaving no trace of their presence. The new trails, even those that have been there for 28 years, are visible almost as though they had been cut yesterday. And all were caused by melting of the permafrost. . . .

In these examples the descriptive pendulum has swung from one extreme to the other. The first paragraph leaves the reader with questions: What were the tractors doing? How many years? How soon will the vegetation grow in again? The second paragraph develops the topic sentence in just enough detail; it explains why the tractors left semipermanent scars and predicts how long they will remain. The third paragraph, though interesting, is filled with irrelevant information and repetitive statements that detract from the main theme. It might be suitable for a novel but not for a technical report or description.

The first paragraph of the article on Joey, the "Mechanical Boy" also was adequately developed because it provides just sufficient information to support the author's original intent: to introduce the boy. It confines itself to a brief word picture of Joey's problem (but not of Joey himself) and its effect on both the boy and the adults who worked with him. Because its readers would be technical men and women who specialize in other technical fields, the paragraph introduces its subject in general terms that will be understood by and provoke interest in a wide range of readers. The author has not described who Joey is, where he is, and what his surroundings are like. Instead, he has started with a thought-provoking topic sentence and used the whole paragraph to develop the image of the "mechanical" child. To have attempted to introduce names, locations, and similar information would have destroyed the impact of this very effective introductory paragraph.

CORRECT LENGTH

How long should your paragraphs be? The answer is simple: no longer than you need to cover a particular topic for a particular reader. But also keep these points in mind:

1. Variety in paragraph length has a lively visual effect. A series of equal-length paragraphs creates the impression of dullness. If the paragraphs are consistently short, readers will feel cheated because they are not presented with enough information. (Too many short paragraphs create a jackrabbit effect—all starts and stops.) Conversely, a succession of very long paragraphs will make readers feel they are swimming in a sea of information in which they cannot readily identify the main topic; worse, they may subconsciously assume that what you have written is going to be "heavy going," and so start reading your words with a negative bias. Generally, try to limit paragraph length to no more than 10 type-written lines.
2. Readers attach importance to a paragraph that is clearly longer or shorter than those surrounding it. A very short paragraph among a series of generally longer paragraphs attracts attention, just as a longer paragraph among a series of short paragraphs implies that the information it contains is particularly important.
3. Paragraph length needs to be adjusted to suit the complexity of your topic and the technical level of your readers. Generally, complex topics demand short paragraphs containing small portions of information. But their length should be conditioned by your knowledge of specific readers. Even a complex topic can be covered in long paragraphs for persons who are technically knowledgeable.

SENTENCES

Although sentences normally form an integral part of a larger unit—the paragraph—they still must be able to stand alone. While helping to develop the whole paragraph, each has to develop a separate thought. In doing so they play an important part in placing emphasis—in stressing points that are important and playing down those that are not.

Just as experienced writers first had to learn the elements of good paragraph writing, so they also had to learn the elements of good sentence construction. These are:

Unity
Coherence
Emphasis

Unity and coherence perform a similar function in the sentence to their role in a well-written paragraph.

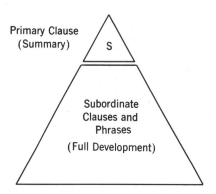

Figure 11-7. Pyramid technique applied to the sentence.

UNITY

Although the comparison is not quite so clearly defined, we can still apply the pyramid technique to the sentence in the same way it is applied to the paragraph and the whole document. In this case the summary is replaced by a primary clause that presents *only one thought,* and the full development by subsidiary clauses and phrases that develop or condition that thought (see Figure 11-7). Thus a unified sentence, as its name implies, presents and develops only a single thought.

Compare these two sentences:

> *A unified sentence that expresses one main thought*
>
> The Amron Building will make an ideal manufacturing plant because of its convenient location, single-level floor, good access roads, and low rent.

> *A complicated sentence that tries to express two thoughts*
>
> The copier should never have been placed in the general office, where those using it interrupt the work being done by the stenographic pool, which has been consistently understaffed since the beginning of the year.

The first sentence has unity because everything it says relates to only one topic: that the Amron Building will make a good manufacturing plant. The second sentence fails to have unity because readers cannot tell whether they are supposed to be agreeing that the copier should have been placed elsewhere, or sympathizing that there has been a shortage of stenographers. No matter how complex a sentence may be, or how many subordinate clauses and phrases it may have, every clause must either be a development of or actively support only one thought, which is expressed in the primary clause.

COHERENCE

Coherence in the sentence is very similar to coherence in the paragraph. Coherent sentences are continuously clear, even though they may have numerous subordinate clauses, so that the message they convey is apparent throughout.

Like the paragraph, they need good continuity, which can be obtained by arranging the clauses in logical sequence, by linking them through direct or indirect reference to the primary clause, and by writing them in the same grammatical form. Parallelism, discussed at the end of this chapter, plays an important role here.

The unified sentence discussing the Amron Building has good coherence because its purpose is continuously clear and each of its subordinate clauses links comfortably back to the lead-in statement (through the phrase "because of its . . ."). The clauses are written in parallel form (they have the same grammatical pattern), which is the best way to carry the reader smoothly from point to point. The first sentence below lacks coherence because there is no logic to the arrangement or form of the subordinate clauses. Compare it with the second sentence, which despite its length is still coherent because it continuously develops the thought of "late" and "damaged" expressed in the primary clause.

An incoherent sentence	The Amron Building will make an ideal manufacturing plant because of its convenient location, which also should have good access roads, the advantage of its one-level floor, and it commands a low rent. (34 incoherent words)
A coherent sentence	Many of the 61 samples shipped in December either arrived late or were damaged in transit, even though they were shipped one week earlier than usual to avoid the Christmas mail tie-up, and were packed in polyurethane as an extra precaution against rough handling. (45 coherent words)

EMPHASIS

Whereas unity and coherence ensure that a sentence is clear and uncluttered, properly placed emphasis helps readers identify the sentence's important parts. By arranging a sentence effectively you can help readers attach importance to the whole sentence, to a clause or phrase, or even to a single word.

Emphasis on the Whole Sentence.   You can give a sentence more emphasis than sentences that precede and follow it by manipulating its length or by stressing or repeating certain words. You can also imply to the reader that all the clauses within a sentence are equally important (without affecting its relationship with surrounding sentences) by balancing its parts.

Although you should aim for variety in sentence length, there may be occasions when you will want to adjust the length of a particular sentence to give it greater emphasis. Readers will attach importance to a short sentence placed among several long sentences, or to a long sentence among predominantly short ones. They will also detect a sense of urgency in a series of short sentences that carry them quickly from point to point. This technique is used effectively by story tellers:

The prisoner huddled against the wall, alone in the dark. He listened intently. He could hear the guards, stomping and muttering. Cursing the cold, probably.

In technical writing we have little occasion to write very short sentences, except perhaps to impart urgency to a warning of a potentially dangerous situation:

> Dangerously high voltages are present on exposed terminals. Before opening the doors:
> 1. Set the master control switch to "OFF."
> 2. Hang the red "NO!" flag on the operator's panel.
> Never cheat the interlocks.

Neither must we err in the opposite direction and write overly long sentences that are confusing. The rule that applies to paragraph writing—adjust the length to suit complexity of topic and technical level of reader—applies equally to sentence writing. If, on average, your sentences tend to exceed 25 words, then probably they are too long.

Similarity of shape can signify that all parts of a sentence are of equal importance. Clauses separated by a coordinate conjunction (mainly *and, or, but,* and sometimes a comma) tell a reader that they have equal emphasis. The following sentences are "balanced" in this way:

> Eight test instruments were used for the rehabilitation project, *and* were supplied free of charge by the Dere Instrument Company.
> The gas pipeline will be 300 miles shorter than the oil pipeline, *but* will have to cross much more difficult terrain.
> The upper knob adjusts the instrument in the vertical plane, *and* the lower knob adjusts it in the horizontal plane.

Emphasis on Part of a Sentence. While coordination means giving equal weight to all parts of a sentence, subordination means emphasizing a specific part and deemphasizing all other parts. It is effected by placing the most important information in the primary clause and placing less important information in subordinate clauses. In each of the following sentences the main thought is italicized to identify the primary clause:

> When the technician momentarily released his grip, *the control slipped out of reach.*
> *The bridge over the underpass was built on a compacted gravel base,* partly to save time, partly to save money, and partly because materials were available on site.
> *He lost control of the vehicle* when the wasp stung him.
> When the wasp stung him, *he lost control of the vehicle.*

Emphasis on Specific Words. Where we place words in a sentence has a direct bearing on their emphasis. The first and last words in a sentence are automatically emphasized. But if we place unimportant words in either of these impact bearing positions, they can rob a sentence of its emphasis:

Emphasis misplaced   Such matters as equipment calibration will be handled by the standards laboratory however.

Emphasis restored   Equipment calibration, however, will be handled by the standard laboratory.

Emphasis misplaced   Without exception change the oil every six days at least.

Emphasis restored   Change the oil at least every six days.

The verbs we use have a powerful influence on emphasis. Strong verbs attract the reader's attention, whereas weak verbs tend to divert it. Active verbs are strong because they tell *who did what*. Passive verbs are weak because they merely pass along information; they describe *what was done by whom*. The sentences below are written using both active and passive verbs. Note that the versions written in the active voice are consistently shorter:

Passive Voice	Active Voice
Elapsed time is indicated by a pointer.	A pointer indicates elapsed time.
The project was completed by the installation crew on May 2.	The installation crew completed the project on May 2.
It is suggested that meter readings be recorded hourly.	I suggest you record meter readings hourly.
The samples were passed first to quality control for inspection, and then to the shipping department where they were packed in polyurethane.	Quality control inspected the samples and then the shipping department packed them in polyurethane.

There are occasions when you will have to use the passive voice because you are reporting an event without knowing who took the action, or prefer not to name a person. For instance you may prefer to write:

The strain gauge was read at ten-minute intervals.

rather than:

Kevin McCaughan read the strain gauge at ten-minute intervals.

Or, if you want to deemphasize Kevin's role:

A technician read the strain gauge at ten-minute intervals.

Unfortunately, many scientists and engineers in industry still believe they should write in the passive voice. They shun the active voice because they feel it is too strong and does not fit their professional character. With the passage of time, this outdated belief is slowly being eroded.

COMPLETENESS

Every day we see examples of incomplete sentences, on television, in magazines, and especially in advertising. Media writers use them to create a crisp, intentionally choppy effect:

SHEER COMFORT!

8200 metres high. Wide seats, just like your living room. Tempting meals. Complimentary refreshments. Only on Remick Airlines. Our Business Class. Try us!

But if we do the same in our business letters and reports, our sentences are likely to be read with raised eyebrows.

A common and very easy error to make, particularly when a piece of writing is going well and you do not want to interrupt your enthusiasm, is to inadvertently form a **sentence fragment.** Normally you correct it later, when you are checking what you have written, but sometimes your familiarity with media writing can cause you not to notice it. These are examples of sentence fragments:

The meeting achieved its objective. Even though three members were absent.
The staff were allowed to leave at 3 p.m. Being the hottest day of the year.

The first sentence in each of these examples is complete (it has a subject-verb-object construction), but the two second sentences are incomplete because, to be understood, each depends on information in the first sentence.

A useful way to check whether or not a sentence is complete is to read it aloud entirely on its own. If it contains a complete thought it will be understood just as it stands. For example, *The meeting achieved its objective* is a complete thought because you do not need additional information to understand it; and so is *The staff were allowed to leave at 3 p.m.* But you cannot say the same when you read these aloud:

Even though three members were absent.
Being the hottest day of the year.

In most cases a sentence fragment can be corrected by removing the period that separates it from the sentence it depends on, inserting a comma in place of the period, and adding a conjunction or connecting word such as *and, but, which, who,* or *because:*

The meeting achieved its objective, even though three members were absent.
The staff were allowed to leave at 3 p.m. *because it was* the hottest day of the year.
Being has been changed to because it was.

Here are two others:

Staff will have to bring bag lunches or go out for lunch from October 6 to 10. While the lunchroom is being renovated.

Change the period to a comma.

All 20-year employees are to be presented with long-service awards. Including three who retired earlier in the year. At the company's annual banquet.

Change both periods to commas.

In particular check sentences which start with a word that ends in "—ing" (referr*ing,* answer*ing,* be*ing*) or an expression that ends in "to" (with reference *to*):

With reference to your letter of June 6. We have considered your request and will be sending you a check.

Change the period to a comma.

Referring to the problem of vandalism to employees' automobiles in the parking lot. We will be hiring a security guard to patrol the area from 8 a.m. to 6 p.m., Monday through Firday.

Although the period could be changed to a comma, a better sentence could be formed by reconstructing the fragment:

To resolve the problem of vandalism to employees' automobiles in the parking lot, we will be hiring. . . .

A similar sentence error occurs if you link two separate thoughts in a single sentence, joining them with only a comma or even no punctuation. The effect can jar a reader uncomfortably. For example:

Ms. Solvason has been selected for the word processing seminar on March 11, she is not eager to attend.

This awkward construction is known as a **run-on sentence.** It can be corrected by:

1. replacing the comma with a period, to form two complete sentences:

. . . on March 11. She is not eager to attend.

2. retaining the comma and following it with *which* or *but:*

. . . on March 11, which she is . . .
. . . on March 11, but she is. . . .

A run-on sentence with no punctuation is even more noticeable:

The customer said he never received an invoice I made up a new one.

Here, either a period or a comma and a linking word can be inserted between *invoice* and *I:*

. . . an invoice. I made up . . .
. . . invoice, so I made up . . .

You will be surprised how easily such simple sentence or punctuation faults can slip by unnoticed (until of course they reach the reader, who immediately spots them!).

WORDS

The right word in the right place at the right moment can greatly influence your readers. A heavy, ponderous word will slow them down; an overused expression will make them doubt your sincerity; a complex word that they do not recognize will annoy them; and a weak or vague word will make them think of you as indefinite. But the right word—short, clear, specific, and necessary—will help them understand your message quickly and easily.

WORDS THAT TELL A STORY

Words should convey images. We have many strong, descriptive words in our individual vocabularies, but most of the time we are too lazy to employ them because the same old routine words spring easily to mind. We write "put" when we would do better to write "position," "insert," "drop," "slide," or any one of numerous descriptive verbs that describe the action better. Compare these examples of vague and descriptive words:

Vague Words	*Descriptive Words*
The technician *got it out* using the MX extraction tool.	The technician *pulled* it out The technician *extracted* it The technician *unlocked* it The technician *unscrewed* it
While the crew was in town they *picked up* spare parts.	they *bought* they *purchased* they *ordered* they *borrowed* they *stole*
The project will *take a long time.*	*require 300 work hours* *employ two persons for two weeks* *last four months*

Story writers use descriptive words to convey active images to their readers. Because we are concerned with *technical* writing does not mean we should avoid seeking colorful words. One descriptive word that defines size, shape, color, smell, or taste is much more valuable than a dozen words that only generalize. (But a word of warning: reserve most of these colorful, descriptive words for your verbs and nouns rather than for adverbs and adjectives.)

Analogies can offer a useful means for describing an unfamiliar item in terms a nontechnical reader will recognize:

> A resistor is a piece of ceramic-covered carbon about the size of a cribbage peg, with a two-inch length of wire protruding from each end.

Specific words tell the reader that you are a definite, purposeful individual. Vague generalities imply that you are unsure of yourself. If you write:

> It is considered that a fair percentage of the samples received from one of our suppliers during the preceding month contained a contaminant.

you are giving the reader four opportunities to wonder whether you really know much about the topic:

1. "It is considered"   Who has voiced this opinion?
2. "a fair percentage"   How many?
3. "one of our suppliers"   Who? One in how many?
4. "contained a contaminant"   What contaminant? In how strong a
                               concentration?

All these generalities can be avoided in a shorter, more specific sentence:

> We estimate that 60% of the samples received from RamSort Chemicals last June were contaminated with 0.5% to 0.8% mercuric chloride.

This statement tells the reader that you know exactly what you are talking about. As a technical person, you should never create any other impression.

LONG VERSUS SHORT WORDS

Big words create a barrier between writer and reader. Some writers use big words to hide their lack of knowledge; others, because they start writing without first defining clearly what they want to say. There are many long scientific words that we have to use in technical writing—we should surround them with short words whenever possible so our writing will not become ponderous and overly complex.

Sometimes we use a certain word or expression because we see another writer use it. Vance Richards, H. L. Winman and Associates' top architect, writes as flamboyantly as he dresses and his draftspersons tend to copy him. Jim Martin models his writing on Vance's, and Jim has replaced the simple, direct style he brought with him as a new employee with wordy writing like this:

> *Vague* Pursuant to the client's original suggestion, Mr. Richards was of the
> *and* opinion that the structure planned for the client would be most suitable
> *Wordy* for immediate erection on the site until recently occupied by the old
> established costume manufacturer known as Garrick Garments. In ac-

cordance with the client's anticipated approval of this site, Mr. Richards has taken great pains to design a multi-level building that can be considered to use the property to an optimum extent.

"Waffle" is the word technical editor Anna King uses to describe such cumbersome writing. By eliminating unnecessary expressions (*pursuant to; of the opinion that; for immediate erection on; in accordance with; can be considered to use; an optimum extent*), Anna was able to cut Jim's 75 words to a much more effective 42 words:

> *Clear*   Mr. Richards believes the type of building the client wants should be
> *and*   erected on the site previously occupied by Garrick Garments. He has
> *Direct*   assumed that the client will approve this site, and has designed a multi-level building which fully develops the property.

LOW INFORMATION CONTENT WORDS

Words of low information content (LIC) contribute little or nothing to the facts conveyed by a sentence. They are untidy lodgers. Remove them and the sentence appears neater and says just as much. The problem is that practically everyone inserts LIC words into sentences, and we become so accustomed to them that we do not notice how they fill up space without adding any information.

Table 11-2 contains some of the words and phrases we must try to delete from our writing. They are difficult to identify because they often sound like good but rather vague prose. In the following sentences the LIC words have been italicized; they should be either deleted or replaced, as indicated by the notes in parentheses.

> The control is actuated by *means of* No. 3 valve. (delete)
>
> Adjust the control *as necessary* to obtain maximum deflection. (delete)
>
> Tests were run for *a period of* three weeks. (delete)
>
> If the project drops behind schedule *it will be necessary to* bring in extra help. (*I, you, he, we,* or *they will* bring in extra help)
>
> By Wednesday we had a backlog of 632 units, *and for this reason* we adopted a two-shift operation. (replace with *so*)
>
> A new store will be opened *in an area* where market research has *given an indication that there actually is a* need for more retail outlets. (delete both expressions; replace the second one with *indicated the*)

Clichés and hackneyed expressions are similar to LIC words and phrases, except that their presence is more obvious and their effect can be more damaging. Whereas LIC words are intangible in that they impart a sense of vagueness, hackneyed expressions can make a writer sound pompous, garrulous, insincere, or artificial. Referring to oneself as "the writer," starting and ending letters with such overworked phrases as "We are in acknowledgement of"

Table 11-2
Examples of Low Information Content (LIC)
Words and Phrases

The LIC words and phrases in this partial list are followed by an expression in parentheses (to illustrate a better way to write the phrase) or by an (X), which means that it should be dropped entirely.

actually (X)	in close proximity to (close to; near)
a majority of (most)	in connection with (about)
a number of (many; several)	in fact, in point of fact (X)
as a means of (for; to)	in order to (to)
as a result (so)	in such a manner as to (to)
as necessary (X)	in terms of (in; for)
at present (X)	in the course of (during)
at the rate of (at)	in the direction of (toward)
at the same time as (while)	in the event that (if)
bring to a conclusion (conclude)	in the form of (as)
by means of (by)	in the neighborhood of; in the vicinity of
by use of (by)	(about; approximately; near)
communicate with (talk to; telephone; write to)	involves the use of (employs; uses)
connected together (connected)	involves the necessity of (demands;
contact (talk to; telephone; write to)	requires)
due to the fact that (because)	is designed to be (is)
during the course of, during the time that (while)	it can be seen that (thus; so)
end result (result)	it is considered desirable (I or we want to)
exhibit a tendency (tend)	it will be necessary to (I, you, or we must)
for a period of (for)	of considerable magnitude (large)
for the purpose of (for; to)	on account of (because)
for this reason (because)	previous to, prior to (before)
in all probability (probably)	subsequent to (after)
in an area where (where)	with the aid of (with)
in an effort to (to)	with the result that (so, therefore)

Note: Many of these phrases start and end with words such as *as, at, by for, in, is, it, of, to,* and *with.* This knowledge can help you to identify LIC words and phrases in your writing.

Table 11-3
Typical Clichés and Hackneyed Expressions

and/or	in the foreseeable future
as a matter of fact	in the matter of
as per	last but not least
attached hereto	needless to say
at this point in time	please feel free to
enclosed herewith	pursuant to your request
for your information	regarding the matter of
(as an introductory phrase)	this will acknowledge
if and when	we are pleased to advise
in our opinion	we wish to state
in reference to	with reference to
in short supply	you are hereby advised

and "please feel free to call me," and using semilegal jargon ("the aforementioned correspondent," "in the matter of," "having reference to") are all typical hackneyed expressions. Further examples are listed in Table 11-3.

Presenting the right word in the correct form is an important aspect of technical writing. Comments on how to abbreviate words, how to form compound adjectives, whether or not you should split an infinitive (generally, you may), and when you should spell out numbers or present them as numerals are contained in a glossary of usage in Chapter 12.

PARALLELISM

Parallelism in writing means "similarity of shape." It is applied loosely to whole documents, more firmly to paragraphs, sentences, and lists, and tightly to grammatical form. Readers generally do not notice that parallelism is present in a good piece of writing—they only know that the sentences read smoothly. But they are certainly aware that something is wrong when parallelism is lacking. Good parallelism makes readers feel comfortable, so that even in long or complex sentences they never lose their way. Its ability to help readers through difficult passages makes parallelism particularly applicable to technical writing.

THE GRAMMATICAL ASPECTS

If you keep your verb forms similar throughout a sentence in which all parts are of equal importance (that is, in which the sentence has coordination, or is balanced), you will have taken a major step toward preserving parallelism. For example, if I write "unable to predict" in the early part of a balanced sentence, I should write "able to convince" rather than "successful in convincing" in a latter part:

Parallelism violated	Mr. Johnson was *unable to predict* the job completion date, but was *successful in convincing* management that the job was under control.
Parallelism maintained (A)	Mr. Johnson was *unable to predict* the job completion date, but was *able to convince* management that the job was under control.
More direct alternative (B) (parallelism maintained)	Mr. Johnson *predicted* no completion date, but *convinced* management that the job was under control.

The rhythm of the words is much more evident in the latter two sentences. In sentence (B) particularly, the parallelism has been stressed by converting the still awkward ". . . unable . . . able . . ." of sentence (A) to the positively parallel "predicted . . . convinced . . .". Two other examples follow:

Parallelism violated	Pete Hansk likes surveying airports and to study new construction techniques.
Parallelism maintained	Pete Hansk likes to survey airports and to study new construction techniques.

or

Pete Hansk likes surveying airports and studying new construction techniques.

Parallelism violated	His hobbies are designing solid-state circuits and stereo component construction.

Parallelism is particularly important when you are joining sentence parts with the coordinating conjunctions *and, or,* and *but* (as in the examples above), with a comma, or with correlatives, such as

either . . . or
neither . . . nor

Each part of a correlative must be followed by an expression with the same grammatical form. That is, if *either* is followed by a verb, then *or* must also be followed by the same form of verb:

Parallelism violated	You may either repair the test set or it may be replaced under the warranty agreement.
Parallelism maintained	You may either repair the test set or replace it under the warranty agreement.

APPLICATION TO TECHNICAL WRITING

Although parallelism is a useful means for maintaining continuity in general writing, in technical writing parallelism has special application. Parallelism can clarify difficult passages and give rhythm to what otherwise might be dull material. When building sentences that have a series of clauses, you can help the reader see the connection between elements by molding them in the same shape throughout the sentence. There is complete loss of continuity in this sentence because it lacks parallelism:

Parallelism violated	In our first list we inadvertently omitted the 7 lathes in room B101, 5 milling machines in room B117, and from the next room, B118, we also forgot to include 16 shapers.

When the sentence is written so that all the item descriptions have a similar shape, the clarity is restored:

Parallelism restored	In our first list we inadvertently omitted 7 lathes in room B101, 5 milling machines in room B117, and 16 shapers in room B118.

Emphasis can also be achieved by presenting clauses in logical order. If they are arranged climactically (in ascending order of importance), or in order of increasing length, the reader's interest will gain momentum right through the sentence:

> No one ever offered to help the navigator struggle out to the aircraft, a parachute hunched over his shoulder, a roll of plotting charts tucked under his arm, a sextant and an astrocompass dangling from one hand, and his precious navigation bag heavy with instruments, log tables, radar charts, and the route plan for the ten-hour flight clenched firmly in the other.

Within the paragraph parallelism has to be applied more subtly. If it is too obvious, the similarity in construction can be dull and repetitive. The verbs often are the key: keep them generally in the same mood and they will help bind the paragraph into one cohesive unit (this is closely tied in with coherence, discussed under "Paragraphs"). This paragraph has good parallelism:

> The effects of sound are difficult to measure. What to some people is simply background noise, to others may be ear-shattering, peace-destroying drumming. The roar of a jet engine, the squeal of tires, the clatter of machinery, the hiss of air-conditioning, the chatter of children, and even the repetitive squeak of an unoiled door hinge, can seriously affect them and create a distinct feeling of uneasiness.

Similarity of shape is most obvious in the third sentence, with its rhythmic

> *roar* of a jet engine
> *squeal* of tires
> *clatter* of machinery
> *hiss* of air-conditioning
> *chatter* of children
> *squeak* of an unoiled door hinge

Here we have words that paint strong images. They not only have the same grammatical form but also are bound together because they relate to the same sense: hearing.

Of course, parallelism can be of particular effect in descriptive writing like this, where the author wants to convey exciting images. In technical writing the subjects are more mundane, giving us less opportunity for imaginative imagery. Nevertheless, you should try to use parallelism to carry your reader smoothly through your paragraphs, as in this description of part of a surveyor's transit:

> Two sets of clamps and tangent screws are used to adjust the leveling head. The upper clamp fastens the upper and lower plates together, while the upper tangent screw permits a small differential movement between them. The lower clamp fastens the lower plate to the socket, while the lower tangent screw turns

the plate through a small angle. When the upper and lower plates are clamped together they can be moved freely as a unit; but when both the upper and lower clamps are tightened the plates cannot be moved in any plane.

APPLICATION TO SUBPARAGRAPHING

Subparagraphing seldom occurs in literature, but is used frequently in technical writing to separate events or steps, describe an operation or procedure, or list parts or components. Subparagraphing always demands good parallelism.

In the following paragraph, a series of tests has been divided into subparagraphs (for clarity, only the initial words of each test are shown here):

> Three tests were conducted to isolate the fault:
>> In the first test a matrix was imposed upon the face of the cathode ray tube and. . . .
>> The second test consisted of voltage measurements taken at. . . .
>> A continuity tester was connected to the unit for test 3 and. . . .

The last subparagraph is not parallel with the first two. To be parallel it must adopt the same approach as the others (it should first mention the test number and then say what was done):

> For the third test, a continuity tester was connected to the unit and. . . .

To be truly parallel the subparagraphs should start in *exactly* the same way:

> In the first test a matrix was. . . .
> In the second test a series of. . . .
> In the third test a continuity tester was. . . .

But this would be too repetitive. The slight variations in the original version make the parallelism more palatable.

If more than three tests have to be described, a different approach is necessary. To continue with "A fourth test showed . . .," "For the fifth test . . .," and so on, would be dull and unimaginative. A better method is to insert a number in front of each paragraph:

> Seven tests were conducted to isolate the fault.
>> 1. A matrix was imposed upon the face of the cathode ray tube and. . . .
>> 2. Voltage measurements were taken at. . . .
>> 3. A continuity tester was connected to the unit and. . . .

Parallelism has been retained, and we now have a more emphatic tone that lends itself to technical reporting.

A third version employs active verbs together with parallelism to build a strong, emphatic description that can be written in either the first or third person:

> In laboratory tests conducted to isolate the fault we:
> 1. *Imposed* a matrix upon the face of the cathode ray tube and. . . .
> 2. *Measured* voltage at. . . .
> 3. *Connected* a continuity tester to the unit and. . . .

To change from the first person to the third person, only the lead-in sentence has to be rewritten:

> In tests conducted to isolate the fault, the laboratory:
> 1. *Imposed.* . . .
> 2. *Measured.* . . .
> 3. *Connected.* . . .

Assignments

Exercise 1. The following sentences and short passages lack compactness, simplicity, or clearness, and particularly contain low-information-content (LIC) expressions. Improve them by deleting unnecessary words, or by partial or complete rewriting.

1. The line printer operates at the rate of 120 characters per second.
2. For your information, the monitor unit will be shipped on or about, but no later than, February 15.
3. Please feel free to contact me at any time if you experience any further problems regarding the disk drive.
4. The international airport that serves the City of Montrose is located in close proximity to the city itself.
5. Check the resistance with the use of an ohmmeter placed in connection with terminals J6 and M9.
6. At this point in time we would like to state that your surveyor's transit has been repaired and is ready for you to pick up at your earliest convenience.
7. It is our pleasure to make an announcement that power of the MX2110 has been increased by as much as 80 kilowatts.
8. The introduction of chemical "X" into the mixture resulted in an increase in the solution's acidity.
9. For the Watson Lake installation, the project manager tentatively planned to operate with a round-the-clock schedule.
10. If you plan on repairing the model 280 multimeter, you would be well advised to get a cost estimate beforehand.

11. In a determined effort to reduce dust contamination, a model 260 ecologizer was installed in the vicinity of test station 22.

12. Repairs to your Microstat are likely to cost in the region of $247.35.

13. We enclose our invoice No. 2820 in the amount of $1834.65, in respect of maintenance services provided during the period October 10 to November 25, 19xx.

14. At the May 10 meeting it was decided by the planning committee that its members were in agreement with and approved Ms. Shalopki's proposal to purchase a high-speed copier for use in the engineering department.

15. An examination of your grain swather shows that it will be necessary to cause the machine's drive mechanism to be brought into a dealer's shop, where repairs can be effected.

16. Our technician did in fact monitor the gauge and found that when the pressure dropped below 120 psi it was indeed necessary for him to reduce the load.

17. Flywheel speed is recorded in terms of revolutions per minute, by means of a Hyperion counter which makes use of digital counters for its display.

18. It is our considered opinion that the new engine should not be operated at a speed in excess of 1200 rpm.

19. From my analysis of the acidity tests, it was apparent that the only conclusion I could draw was that the plants in the growth chamber had been adversely affected by water with a high pH.

20. Without a shadow of doubt, wear in the region of the cotter pin will have an end result which I can foresee will cause complete breakdown of the drive shaft in the not-too-distant future.

Exercise 2. The following sentences offer choices between words that sound similar or are frequently misused. Select the correct word in each case.

1. The designer had to decide (weather/whether) to use a vertical or a horizontal rack.

2. Only 10 of the 40 samples could be tested this week (as/for/because) the model 51 test set developed a readout fault. The (balance/remainder) will be tested next week.

3. To Martin Dawes, Anna King's suggestion that the company purchase three additional Vancourt model 201 microprocessors (appeared/seemed) to be feasible.

4. (Accept/Except) for the modem, all components of the system have been received.

5. The frame has to be (oriented/orientated) so that the grid marker points to 075.

6. The engine's life was (affected/effected) by contaminants in the oil.

7. The (morale/morals) of the employees improved when they heard they were to be included in the company's profit sharing scheme.

8. In 19xx, the (amount/number) of projects completed by systems engineering increased by 13% compared (to/with) the previous year.

9. In his letter to Vern Rogers, Martin Dawes (implied/inferred) that all branch offices are to (confirm/conform) to company regulation DM201.

10. Zero-defects manufacturing means that all items we ship are free (from/of) flaws.

11. This morning's attempt to (elicit/illicit) a response from the satellite station was inhibited by a (defective/deficient) receiver switch, which was not repaired until 1:30 p.m.

12. The signals traveled 123 miles with only 8% attentuation, which was 74 miles (farther/further) than during the 19xx test.

13. The (preventive/preventative) maintenance program instituted last February has reduced operating failures by 23%.

14. The largest items on our (stationary/stationery) order this month are a carton containing 50,000 sheets of $11 \times 8\frac{1}{2}$ in. white bond paper, and a box containing 30,000 kraft 12×9 in. envelopes.

15. The budget reduction means that three (fewer/less) services will be offered next year. This is considerably (fewer/less) than the peak achieved in 19xx.

16. When the data (has/have) been interpreted, the project group will write (its/it's/their) report.

17. Sheila Watson has asked Myrna Carling to read her project report before she issues it, because Myrna is (a disinterested/an uninterested) technician who is not familiar with Sheila's work or the project.

18. The technician was (indiscreet/indiscrete) when he described to the visitors exactly how the company proposed to separate the signals into two (discreet/discrete) channels.

19. The heat exchanger operates (continually/continuously) throughout the winter, switching on automatically whenever the ambient temperature drops below 10°C.

20. The (principal/principle) reason I have included this exercise is that it will introduce you to the glossary of technical usage.

Exercise 3. Rewrite the following sentences to make them more emphatic (in many cases, change them from the passive to the active voice).

1. When the I-beam was examined by the technician, a 2½ inch crack was found at its junction with the east wall.

2. The power outage was caused by excessive ice accumulation on the transmission line cables.

3. The agenda for the Arbutus project meeting was distributed by Andy Rittman on April 20.

4. Harvey Winman was sent a Model 9000 computer handbook from Vancourt Business Systems Inc, and it was forwarded to Tina Mactiere at Macro Engineering Inc, who was asked to evaluate it.

5. When visibility was reduced to less than 80 yards by low cloud, the pilots of all flights coming into Montrose International Airport were instructed by the duty controller to make a landing at Buffalo.

6. After the concrete samples had been tested by the laboratory, it was recommended by the structural engineer that pours No. 7 and 13 be rejected.

7. It was reported by technician Myrna Wolski that patient records in the nuclear medicine laboratory had been damaged by high-pressure steam escaping from a cracked pipe.

8. It was recommended by Wayne Robertson, in a memorandum to Tina Mactiere dated October 28, that production of model A202 megastats should be discontinued on November 30.

9. At MEI, the Vancourt 300 message center has been replaced by an electronic display board developed recently by employee Russ Cowan.

10. It was Fred Stokes's suggestion, in a memo circulated to all staff on April 18, that two engineering technicians be selected by each of the five department heads for attendance at the microprocessor conference in Houston, Texas, on May 7 and 8.

Exercise 4. Improve the parallelism in the following sentences.

1. The Morton 101 modem costs $187.50 if you buy from 1 to 5 units, but if you purchase 6 or more units the price drops to $165 each.

2. All field crew supervisors are to attend a one-day first aid course at the Red Cross Center on May 10. You will be given training in administering basic first aid and how to recognize and deal with heart-attack cases in the field.

3. My inspection shows that the defective video terminal at station L10 has
   - two broken keys,
   - a cracked keyboard case,
   - a frayed cable between the monitor and keyboard, and
   - someone has spilled sweetened coffee or juice on some of the keys, so that eight keys are stuck in the depressed position.

4. Three of the applicants were given promotions and transfers were arranged for the other four applicants.

5. Two additional test sets were purchased, partly to increase manometer production and partly as a means for improving quality control.

6. I have decided to buy the A200 photocopier because it is small, fast, and has both an inexpensive purchase price and low operating costs.

7. The analysis was needed not only to determine permafrost regression but also for planning the proposed pipeline route.

8. H. L. Winman and Associates' medical and dental plan requires that new employees have both a medical examination and a dental check. Head office employees go to room 212 in the Watrous building for the dental check, and for the medical examination they go to the Evergreen Medical Clinic. The plan applies both to head office employees and, at branch offices, employees are sent to local clinics.

Exercise 5. Check that the numbers in the following sentences have been expressed in the proper form (refer to Guideline 4 of Chapter 12 for rules for writing numbers in narrative). Rewrite those that have been written incorrectly.

1. Page eight of the specification, paragraph 28.3, specifies a current drop of no more than .5 mA.

2. You will be required to conduct 2 tests, each lasting ¾ of an hour.

3. The renovations will require four workers for three and a half days.

4. The next meeting of the laboratory standards committee will be held on the afternoon of May 6th, at three o'clock.

5. During the transmission line overhaul we checked 11360 connectors, of which

7041 were satisfactory, 3644 showed signs of oxidation, and 685 had to be replaced.

6. Our price for three model seventy black-and-white twelve-inch video monitors will be one hundred and thirty dollars ($130.00) each.

7. 385 of the connectors were stamped KL and had a shaft diameter of .65 cm, while three hundred were stamped MB and had a shaft diameter of exactly half a centimeter.

8. By the year two thousand, we estimate there will be 1 computer in every 1.2 homes.

Exercise 6.   Correct the following sentences, where necessary, to ensure that each is complete.

1. The vat overheated again on February 16. The third time this year.

2. The fumes in the lab affected the technicians, Mr. Watson sent them home at noon.

3. Because the components were badly corroded, the system had to be shut down while they were replaced.

4. When the lines were reconnected, the power had already been off for 4½ hours. Too late to prevent the samples from freezing.

5. Ms. Fennel has agreed to extend her contract, but only until May 31, she has to leave then to take up a permanent position with Marshland Consultants on June 1.

6. In reference to your memorandum dated December 3. The survey crew will start clearing brush north of highway 136 on January 4, 19xx.

7. Some customers have complained they received two, and in some cases three, invoices last month, I have issued credit notes where applicable. And explained to customers we have had a computer problem.

8. I want to transfer K. Johannsen, M. Triggs, and G. Kelln to the Morgan Avenue plant, two of them have accepted the transfer but G. Kelln says he would prefer not to go. For health reasons. He has produced a doctor's certificate indicating the transfer would be unwise.

Exercise 7.   Abbreviate the terms shown in italics in the following sentences. In some cases you will also have to express numerals in the proper form. (Rules for forming abbreviations appear in Guideline 2 of Chapter 12.)

1. Lake Wabagoon is 2.6 *miles northeast* of Montrose, Ohio.

2. The base of the antenna is 1806 *feet above mean sea level*.

3. By driving a constant 55 *miles per hour* we achieved a fuel consumption of 16.7 *miles per gallon*.

4. Production from *number three* well averages *four hundred barrels per day*.

5. A noise level of 52.6 *decibels* was recorded at 825 *hertz*.

6. We are shipping our voltmeter *serial number* 4257 to you for recalibration.

7. The heater produces *fifty-six thousand British thermal units* of heat per hour.

8. Rotate the control *clockwise* until a *continuous wave* signal is heard in the headset.

9. Please supply *fifteen pounds* of bitumastic compound and *three rolls* of *forty-two inch wide* roofing material.

10. The transformer is rated for operation up to *one hundred and ninety-two degrees Fahrenheit.*

11. A meeting has been called for 10:30 *in the morning on Monday December the fifth, 1987.*

12. Television station DMON of Montrose, Ohio has been allocated a transmitting frequency in the *ultrahigh frequency* band (*three hundred megahertz to three gigahertz*).

13. The municipality's 24-*inch inside diameter* storm sewers are fed into a 48-*inch diameter culvert.*

14. The metric liter is equivalent to 0.264 *United States gallons* or 0.220 *Imperial gallons.* Hence, 10 *United States gallons* equal *38 liters.*

15. Strip *three quarters of an inch* of insulation from both ends of *28 inches* of *number twenty American Wire Gauge* hookup wire.

16. The sun's elevation at *zero eight hours, seventeen minutes and twentysix seconds Greenwich mean time* was calculated to be *twentythree degrees, four minutes, and fourteen seconds* of arc.

Exercise 8.   Select the correctly spelled words among the choices offered below.

1. In his speech to the conference delegates, Harvey Winman (aluded/eluded/alluded/elluded) to his company's experience in civil engineering.

2. Wayne Robertson is an (entepreneur/entrepeneur/entrepreneur).

3. Ian Bailey (recommended/reccomended/reccommended) we rent first class (accomodation/acommodation/accommodation) while staying at the Gatsano site.

4. For the Varlon installation (its/it's) better to use heavy (gauge/guage) cable.

5. In the (forward/forword/foreword) to his report, Mr. Winman (complemented/complimented) his engineers on their comprehensive analysis of the problem.

6. The technicians had difficulty (maneuvering/maneuvring/manoeuvering) the equipment after they had set the (flotation/floatation) gear free in the water.

7. A (seperate/separate) bag of (desiccant/dessicant) was packed with each module in the shipment.

8. Tina Mactiere was (embarased/embarrased/embarassed/embarrassed) by the (exorbitant/exhorbitant) cost of the proposal she had to present to her client.

9. The (auxiliary/auxilliary) units were (paraleled/paralleled/paralelled/parallelled) for maximum circuit protection.

10. The project team was commissioned to (develop/develope) a lubricant which would (supercede/supersede) the product we currently use.

11. Each member of the field team (received/recieved) three different (innoculations/inocculations/inoculations) before leaving for the project in East Africa.

12. The effects of bad spelling are (incalculable/incalculatable). How many words did you (mispel/mispell/misspel/misspell)?

Exercise 9.   Rewrite the following letters and memorandums to improve their effectiveness.

1. To: N. Fielding, Nirvana Dam
   From: G. Wasalusky, Material Control
   Date: September 16, 19xx

Pursuant to your request of December 13, today I took the necessary action. The instruments you requisitioned (see Req #2815) have been shipped to the Nirvana Dam project site. In the event that they have not been delivered to the site by the end of the month, please complete form 1816, attached hereto, and forward it to this office for my attention.

Gary

2. Dear Mr Carstairs:

Your memorandum of Jun 10 received and at hand. You will be glad to know your problem is being studied. We have discovered a defective circuit board in the output stage. Replacement boards are hard to come by: I phoned around. No one has one locally. So I've ordered one from the manufacturer. I've asked them (Miranee Electronics) to ship it air express. I'll telephone you as soon as it comes in. But not for three weeks. They can't ship it before July 6.

Sincerely,
Martita Lopez

3. Dear Mr Sanhoo:

I write with reference to your letter of February 17th. It is this company's honest opinion that the 17 RAM20 video monitors that you reported in your letter had arrived in a damaged condition (of the 24 we shipped to your company on January 22nd) were in effect subjected to overly stressful handling during transportation by Moreton Trucking. If you would kindly call the local branch manager of Transborder Insurance Corporation at your convenience, and arrange for one of their adjusters to make a comprehensive and detailed inspection of the shipping containers, then we will be in a much better position to assess the validity of your claim. In the event that the adjuster's report is to the effect that the video terminals were indeed damaged during transportation from our warehouse to your place of business, then we will be only too happy to acknowledge your request for a credit note. In the interim we have shipped a further 17 video monitors to your company, in response to the request in your letter of February 17th, and trust that these arrangements will meet your needs and prove satisfactory to you, and remain,

Yours very sincerely,
Carl D. Nickersen.

4. To: Annette Lesk
From: Bob Chirlander
Date: March 31, 19xx
Subject: Progress report No. 4

As you will have recognized from my previous progress report (No. 3, dated February 28, 19xx), my subsequent telephone calls regarding replacement personnel, and the results of the project manager's visit to the sight on March 18. The project has now dropped even farther behind schedule and, unfortunately, it is entirely within the realms of possibility that a delayed completion date will occur.

In an effort to speed up the work, Mr. Rossini, the project manager, by telephone on March 23rd, authorized hiring 3 additional installers, two of whom arrived on 25th March and have started work, but not a word has as yet been received about the third replacement, can you advise when she is due to arrive at the cite?

To date, forty % of the modules have been installed and connected to the control unit. By this stage we should have completed installing 120 of the two hundred modules. As you can easily see, this puts us much behind and I cannot see us completing the project until six weeks later than scheduled. Right now I estimate the completion date to be on or about July 20, 19xx. If you can expedite the arrival of the third replacement I would appreciate it very much indeed because then I could shave one week off the date I have predicted. Which will put us only five weeks behind schedule.

Bob

5. April 24, 19xx

Dear Ms. Calhoun:

I refer to your telephone call of April 17, and your letter of April 10 in which I read with considerable concern of the difficulties you have been experiencing with the Model 800 Environment Control Chamber you purchased from us on February 23, 19xx. I noted particularly that you are now requesting warranty service.

It always places us in an awkward position when a piece of industrial equipment fails shortly after the expiration of the warranty period. As I believe you already know, the warranty for your model 800 chamber expired over eight weeks ago, on February 23 to be exact, but the digital timer on the unit did not fail until six weeks later, on April 9 to be exact. Harven Industries—the manufacturer of the model 800—are very adamant that they will accept no warranty work beyond 55 weeks after the purchase date, which would be March 16 in your case, and of course as you already know the failure occurred in the 59th week. Which clearly defines the failure as occurring outside the warranty period.

Nevertheless we believe you have a point when you say in your letter that previous warranty work carried out on January 26 (in the 48th week) could have had an effect on the digital timer circuit. Unfortunately my hands are tied and I am powerless to provide warranty service now without the manufacturer's specific authorization, so I suggest you write directly to Harven Industries at 3120 Circular Road, Harmonsville, Kentucky. I wish you luck in your endeavor and regret profoundly that we can be of such little help in this particular instance.

Regards

Dale J. Freeling

Exercise 10.   Use the following information to write a *single* paragraph. Underline the topic sentence (summary statement).

a. A battery is composed of one or more cells connected in series.

b. Within a battery each cell is a source of chemical energy. This chemical energy is converted into electrical energy—but only when needed.

c. There are two types of cells.

d. One type of cell is the primary cell, which can supply electricity until its chemical energy is exhausted; then you have to throw it away.

e. Some cells can be recharged after use. In this type, an electrical charge can be used to return the chemicals within the cell to their original state (i.e. prior to discharge). This type of cell is known as the secondary cell.

Exercise 11.   Rewrite this one-paragraph notice to make it clearer and more personal.

PROCEDURE RE EXPENSE CLAIMS

Expense claims must be handed to the Accounts Section before ·10:30 a.m. on Wednesday for payment on Friday. Personnel failing to hand in their forms at the proper time, may do so at any time until 4:30 p.m. on Thursday but must wait until Monday for payment. Under no account will a late claim be paid in the same week that it was filed. Claims handed in after Thursday will be processed with the following week's claims and will be paid on the next Friday.

Exercise 12.   Write a single paragraph describing the following situation. Underline your topic sentence (summary statement).

a. Your company has a lot of telephone switching equipment to maintain. Most is old, but some is new.
b. New telephone switching equipment is under warranty for one year.
c. The manufacturer's warranty specifies new equipment must be lubricated with Vaprol.
d. You currently use Vaprol for all telephone switching equipment.
e. Tests by the Engineering Department show Gra-Lub to be a better lubricant.
f. Gra-Lub has a graphite base; Vaprol has not.
g. You want to adopt Gra-Lub for all telephone switching equipment.
h. You don't want to stock and use two types of lubricant for the same equipment.
i. You have written to the manufacturer to request authorization to use Gra-Lub on equipment still under warranty.

12

Glossary of
Technical Usage

A standard glossary of usage contains rules for combining words into compound terms, for forming abbreviations, for capitalizing, and for spelling unusual or difficult words. This glossary also offers suggestions for handling many of the technical terms peculiar to industry. Hence, it is oriented toward the technical rather than the literary writer.

The glossary is preceded by six guidelines for specific aspects of technical writing:

Guideline 1: Combining Words into Compound Terms
Guideline 2: Abbreviating Technical and Nontechnical Terms
Guideline 3: Capitalization and Punctuation
Guideline 4: Writing Numbers in Narrative
Guideline 5: Summary of Numerical Prefixes and Symbols
Guideline 6: Introduction to Metric Units

The glossary is arranged alphabetically and contains a selection of words most likely to be misused or misspelled. These include:

Words that are similar and frequently confused with one another; e.g. *imply* and *infer*, *diplex* and *duplex*, *principal* and *principle*.

Common minor errors of grammar, such as *comprised of* (should be *comprises*), *most unique* (*unique* should not be compared), *liaise* (an unnatural verb formed from *liaison*).

Words that are particularly prone to misspelling; e.g. *desiccant*, *oriented*, *immitance*.

Words for which there may be more than one "correct" spelling; e.g., *programer* and *programmer*.

Where two spellings of a word are in general use (e.g. *symposiums* and *symposia*), both are entered in the glossary and a preference is shown for one of them.

Definitions have been included when they will help you select the correct word for a given purpose, or to differentiate between similar words having different meanings. These definitions are intentionally brief and intended only as a guide; for more comprehensive definitions, consult an authoritative dictionary (I recommend *Webster's New Collegiate Dictionary*).

All entries in the glossary are in lower case letters. Capital letters are used where capitals are recommended for that specific word, phrase, or abbreviation. Similarly, periods have been eliminated except where they form part of a specific entry. For example, the abbreviation for "inch" is *in.*, and the period that follows it has been inserted intentionally to distinguish it from the word "in".

Finally, think of this glossary as a guide rather than a collection of hard and fast rules. Our language is continually changing, so that what was fashionable yesterday may seem pedantic today and a cliché tomorrow. I expect that in some cases your views will differ from mine. Where they do, I hope that the comments and suggestions I offer will help you to choose the right expression, word, or abbreviation, and that you will be able to do so both consistently and logically.

GUIDELINES

GUIDELINE 1: COMBINING WORDS INTO COMPOUND TERMS

One of the biggest problems for technical writers (and for the typists who have to record their work) is knowing whether multiword expressions should be compounded fully, joined by hyphens, or allowed to stand as two or more separate words. For example, should you write:

cross check, cross-check, or crosscheck?
counter clockwise, counter-clockwise, or counterclockwise?
change over, change-over, or changeover?

The tendency today is to compound a multiword expression into a single term. But this bare statement cannot be applied as a general rule because there are too many variations, some of which appear in the glossary.

Most multiword expressions are compound adjectives. When two words combine to form an adjective they are either joined by a hyphen or compounded to form one word. They are usually joined by a hyphen if they are formed from a noun-adjective expression:

Adjective + Noun	As a Compound Adjective
vacuum tube	vacuum-tube voltmeter
cathode ray	cathode-ray tube
high frequency	high-frequency oscillator

But when one of the combining words is a verb, they often combine into a one-word adjective. Under these conditions they normally will compound into a single-word noun:

Two Words	As a Noun	As an Adjective
lock out	lockout	lockout voltage
shake down	shakedown	shakedown test
cross over	crossover	crossover network

Three or more words that combine to form an adjective in most cases are joined by hyphens. For example, *lock test pulse* becomes *lock-test-pulse generator*. Occasionally, however, they are compounded into a single term, as in *counter-electromotive force*. Specific examples are in the glossary.

Obviously, these "rules" cannot be taken at full face value because there are occasions when they do not apply. Useful guides for doubtful combinations are the *Government Printing Office Style Manual*[1] which contains a list of over 19,000 compound terms, and Rudolph Flesch's admirable handbook *Look It Up*.[2]

GUIDELINE 2: ABBREVIATING TECHNICAL AND NONTECHNICAL TERMS

You may abbreviate any term you like, and in any form you like, providing you indicate clearly to the reader how you intend to abbreviate it. This can be done by stating the term in full, then showing the abbreviation in parentheses to indicate that from now on you intend to use the abbreviation. Here is an example:

> Always spell out single digit numbers (sdn). The only time sdn are not spelled out is when they are being inserted in a column of figures.

When forming abbreviations of your own, take care not to form a new abbreviation when a standard one already exists. For example, if you did not

---

[1]*United States Government Printing Office Style Manual,* Superintendent of Documents, Washington, D.C., 20402.

[2]Rudolph Flesch, *Look It Up: A Deskbook of American Spelling and Style* (New York: Harper and Row, 1977).

know that there is a commonly accepted abbreviation for pound (weight), you might be tempted to use *pd*. This would not sit well with your readers, who might resent replacement of their old friend *lb*.

There are three basic rules that you should observe when forming abbreviations:

1. *Use lower case letters,* unless the abbreviation is formed from a person's name:

centimeter	— cm
kilogram	— kg
approximately	— approx
decibel	— dB (the *B* represent *Bell* (Alexander Graham Bell)

2. *Omit all periods,* unless the abbreviation forms another word:

horsepower	— hp
cubic centimeter	— cm^3
cathode ray tube	— crt
foot/feet	— ft
inch	— in.
singular	— sing.

3. *Write plural abbreviations in the same form as the singular abbreviation:*

inches	— in.
pounds	— lb
kilograms	— kg
hours	— h *or* hr

Of course there are exceptions which have grown as part of the language. Through continued use these unnatural abbreviations have been generally accepted as the correct form and now cannot easily be displaced. A few examples follow:

for example *(exempli gratia)*	— e.g.	*(There is a slowly growing trend to write these as* eg *and* ie*)*
that is *(id est)*	— i.e.	
morning *(ante meridiem)*	— a.m.	
afternoon *(post meridiem)*	— p.m.	
inside diameter	— ID	
number(s)	— No.	

Other unusual abbreviations exist in each technical discipline. Although the glossary offers a reasonably comprehensive list of standard abbreviations and some technical abbreviations, for specific technical terms you may need to refer to a list of standard abbreviations compiled by one of the technical societies in your discipline.

GUIDELINE 3: CAPITALIZATION AND PUNCTUATION

Lower case letters should be used as much as possible in technical writing. Too many capital letters cause an untidy appearance and reduce the effectiveness of capital letters inserted for emphasis. The glossary therefore recommends lower case letters except (1) where usage has resulted in general adoption of a capitalized form (as in No., GCA, and the Brinell and Rockwell hardness tests), (2) for proper nouns, and (3) for expressions coined from proper nouns which have not yet been accepted as common terms. (Note, for example, that although we still use capital letters when referring to Doppler's principle, we write of doppler radar in lower case letters.)

The tendency today is to underpunctuate technical writing. This does not mean you may omit punctuation with a "when in doubt, leave it out" attitude. But you need not punctuate every clause and subclause. The intent is to obtain smooth reading; the criterion is to make the message clear. Hence, you should insert punctuation where it is needed rather than rigorously divide the information into tight little formal compartments.

GUIDELINE 4: WRITING NUMBERS IN NARRATIVE

The conventions that dictate whether a number should be written out or expressed in figures differ between ordinary writing and technical writing. In technical writing you are much more likely to express numbers in figures.

The rules listed below are intended mainly as a guide. There will be many times when you have to make a decision between two rules that conflict. Your decision should then be based on three criteria:

Which method will be most readable?
Which method will be simplest to type?
Which method did I use previously under similar circumstances?

Good judgment and a desire to be consistent will help you select the best method each time.

The basic rule for writing numbers in technical narrative is:

Spell out single-digit numbers (one to nine inclusive).
Use figures for multiple-digit numbers (10 and above).

Exceptions to this rule are:
Always Use Figures:
1. When writing specific technical information, such as test results, dimensions, tolerances, temperatures, statistics, and quotations from tabular data.

2. When writing any number that precedes a unit of measurement: 3 in.; 7 lb; 121.5 MHz.

3. When writing a series of both large and small numbers in one passage: During the week ending May 27 we tested 7 transmitters, 49 receivers, and 38 power supplies.

4. When referring to section, chapter, page, figure (illustration), and table numbers: Chapter 7; Figure 4.

5. For numbers that contain fractions or decimals: 7¼; 7.25.

6. For percentages: 3% gain; 11% sales tax.

7. For years, dates, and times: At 3 p.m. on January 9, 1986.

8. For sums of money: $2000; $28.50; 20 dollars; $0.27 (preferred) or 27 cents.

9. For ages of persons.

Always Spell Out:

1. Round numbers that are generalizations: about five hundred; approximately forty thousand.

2. Fractions that stand alone: repairs were made in less than three quarters of an hour.

3. Numbers that start a sentence. (Better still, rewrite the sentence so that the number is not at the beginning.)

Additional rules are:

Spell out one of the numbers when two numbers are written consecutively and are not separated by punctuation: 36 fifty-watt amplifiers or thirty-six 50-watt amplifiers. (Generally, spell out whichever number will result in the simplest or shortest expression.)

Insert a zero before the decimal point of numbers less than one: 0.75; 0.0037.

Use decimals rather than fractions (they are easier to type), except when writing numbers that are customarily written as fractions.

Insert commas in large numbers containing five or more digits: 1,275,000; 27,291; 4056. (Insert a comma in four-digit numbers only when they appear as part of a column of numbers.)

Write numbers that denote position in a sequence as 1st, 2nd, 3rd, 4th, . . . 31st, . . . 42nd, . . . 103rd, . . . 124th, . . .

GUIDELINE 5: SUMMARY OF NUMERICAL PREFIXES AND SYMBOLS

The table below summarizes the numerical prefixes and abbreviations used in the glossary, and conforms to the requirements for writing SI metric units.

MULTIPLE/ SUBMULTIPLE	PREFIX	SYMBOL	MULTIPLE/ SUBMULTIPLE	PREFIX	SYMBOL
10^{18}	exa	E	10^{-1}	deci	d
10^{15}	peta	P	10^{-2}	centi	c

MULTIPLE/ SUBMULTIPLE	PREFIX	SYMBOL	MULTIPLE/ SUBMULTIPLE	PREFIX	SYMBOL
10^{12}	tera	T	10^{-3}	milli	m
10^9	giga	G	10^{-6}	micro	μ
10^6	mega	M	10^{-9}	nano	n
10^3	kilo	k	10^{-12}	pico	p
10^2	hecto	h	10^{-15}	femto	f
10	deca	da	10^{-18}	atto	a

GUIDELINE 6: INTRODUCTION TO METRIC UNITS

For this third edition, the glossary includes terms and symbols within the International System of Units (SI). The trend toward worldwide adoption of metric units of measurement means that for some time both the basic inch/pound system and the metric (SI) system will be in use concurrently. The terms and symbols introduced here are those you are most likely to encounter in your technical reading or may want to use in your technical writing.

The acronym "SI" is used in all languages to represent the name "Système International d'Unités." Both the acronym and the name were adopted for universal usage in 1960 by the eleventh Conférence Générale des Poids et Mesures (CGPM), which is the international authority on metrication. Since then, some of the metric terms the conference established have crept into our language. For example, *Hertz,* the unit of frequency measurement, was first introduced as a replacement for *cycles per second* in the early 1960s; now it is used universally, both in the technical world and by the general public. Other terms already gaining recognition are:

	non-SI	SI
Temperature:	degrees Fahrenheit	degrees Celsius
Length:	miles, yards, feet, inches	kilometers, meters, millimeters
Weight:	tons, pounds ounces	kilograms, grams, milligrams
Liquid volume:	gallons, quarts	kiloliters, liters

The glossary defines the basic SI units and shows how they should be written, either in full or abbreviated as a symbol, and their multiples and submultiples. It does not attempt to explain how the units are derived, nor does it include many of the less common units. The individual glossary entries are supplemented by the guidelines listed below, many of which are equally applicable to non-SI terms and units.

Guidelines.   SI symbols must be written, typed, or printed:

1. in upright type, even if the surrounding type slopes or is in italic letters.
2. in lower case letters, except when the name of the unit is derived from a person's name (e.g. the symbol **F** for *farad* is derived from *Faraday*), in which case the first letter of the symbol is capitalized (e.g. **Wb** for *weber*).
3. with a space between the last numeral and the first letter of the symbol: **355 V, 27 km** (not 355V, 27km).
4. with no "s" added to a plural: **1 g, 236 g.**
5. with no period after the symbol, unless it forms the last word in a sentence.
6. with no space between the multiple or submultiple symbol and the SI symbol: **3.6 kg, 150 mm, 960 kHz.**
7. with a solidus (oblique stroke:/) to represent the word *per:* **cm/s** (centimeters per second). Only one solidus should be used in each expression.
8. with a dot at midletter height (·) to represent that the symbols are multiplied: **1m·s** (lumen seconds).
9. as a symbol, if a number is used with the SI unit: ". . . the tank holds **400 L**"; but spelled out when no number is used with the unit: ". . . capacity is measured in liters" (not: "capacity is measured in L").

Spelling: "-er" or "-re"?   An anomaly exists concerning the spelling of **liter** and **meter**. SI stipulates the **-re** spelling (**litre** and **metre**), but in the United States we have traditionally used the **-er** spelling. I have continued to use the more familiar traditional spelling in *Technically-Write!* but caution readers that the spelling may change in the United States as metric units of measurement become more widely established.

THE GLOSSARY

General abbreviations used throughout the glossary are:

abbr	abbreviate(d); abbreviation
def	definition
lc	lower case
pl	plural
pref	prefer, preferred, preference
SI	International System of Units

A

a; an   use *an* before words that begin with a silent *h* or a vowel; use *a* when the *h* is sounded or if the vowel is sounded as *w* or *y; an hour* but *a hotel, an opinion* but *a European*

aberration

ab initio   def: from the beginning

above-   as a prefix, *above-* combines erratically: *aboveboard, above-cited, aboveground, above-mentioned*

abrasion

abscess

abscissa

absence

absolute   abbr: **abs**

absorb(ent); adsorb(ent)   *absorb* means to swallow up completely (as a sponge absorbs moisture); *adsorb* means to hold on the surface; as if by adhesion

abut; abutted; abutting; abutment

accelerate; accelerator; accelerometer

accept; except   *accept* means to receive (normally willingly); *he accepted the company's offer of employment; except* generally means exclude: *the night crew completed all the repairs except rewiring of the control panel*

access(ible)

accessory; accessories   abbr: **accy**

accidental(ly)

accommodate; accommodation

account   abbr: **acct**

accumulate; accumulator

achieve   means to conclude successfully, usually after considerable effort; avoid using *achieve* when the intended meaning is simply to reach or to get

acknowledg(e)ment   *acknowledgement* pref

acquaint(ance)

acquiesce   def: agree to

acquire; acquisition

acre; acreage

across   not *accross*

actually   omit this word: it is seldom necessary in technical writing

actuator

A.D.   def: Anno Domini

adapt; adept; adopt   *adapt* means to adjust to; *adept* means clever, proficient; *adopt* means to acquire and use

adapter; adaptor   *adapter* pref

adaption; adaptation   *adaption* pref

addendum   pl: *addenda*

adhere to   never use *adhere by*

ad hoc   def: set up for one occasion

adjective (compound)   two or more words that combine to form an adjective are either joined by a hyphen or compounded into a single word; see Guideline 1

adsorb(ent)   see **absorb**

advantageous

advice; advise   use *advice* as a noun and *advise* as a verb: *the engineer's advice was sound; the technician advised the driver to take an alternative route;* spell: **adviser, advisable**

aerate

aerial   see **antenna**

aero-   a prefix meaning of the air; it combines to form one word: *aerodynamics, aeronautical;* in some instances it has been replaced by *air: airplane, aircraft*

aesthetic   alternate spelling of **esthetic**

affect; effect   *affect* is used only as a verb, never as a noun; it means to produce an effect upon or to influence (*the potential difference affects the transit time*); *effect* can be used either as a verb or as a noun; as a verb it means to cause or to accomplish (*to effect a change*); as a noun it means the consequences or result of an occurrence (*the detrimental effect upon the environment*), or refers to property, such as *personal effects*

aforementioned; aforesaid   avoid using these ambiguous expressions

after-   as a prefix, usually combines to form one word: *afteracceleration, afterburner, afterglow, afterheat, afterimage*

agenda   although plural, *agenda* is generally treated as singular: *the agenda is complete*

aggravate   the correct definition of *aggravate* is to increase or intensify (worsen) a situation; try not to use it when the meaning is *annoy*

aggregate

aging; ageing   *aging* pref

agree to; agree with   to be correct, you should *agree to* a suggestion or proposal, but *agree with* another person

air-   as a prefix, normally combines to form one word: *airborne, airfield, airflow;* exceptions: *air-condition(ed) (er) (ing), air-cool(ed) (ing)*

air horsepower   abbr: **ahp**

airline; air line   an *airline* provides aviation services; an *air line* is a line or pipe that carries air

alkali; alkaline   pl: *alkalis* (pref) or *alkalies*

allege(d); alleging

allot; allotted; allotment

all ready; already   *all ready* means that all (everyone or everything) is ready; *already* means by this time; *the samples are all ready to be tested; the samples have already been tested*

all right   def: everything is satisfactory; never use *alright*

all together; altogether   *all together* means all collectively, as a group; *altogether* means completely, entirely; *the samples have been gathered all together, ready for testing; the samples are altogether useless*

allude; elude   *allude* means refer to; *elude* means avoid

almost   never contract *almost* to *most;* it is correct to write *most of the circuits have been tested,* but wrong to write *the circuits are most ready*

alphanumeric   def: in alphabetical, then numerical sequence

alternate; alternative   *alternate(ly)* means by turns: *the inspector alternated among the four construction sites; alternative(ly)* offers a choice between only two things: *the alternative was to return the samples;* although it is incorrect to say *there are three alternatives,* it is becoming common usage

alternating current   abbr: **AC**

alternator

altitude   abbr: **alt**

a.m.   def: before noon (*ante meridiem*)

amateur

ambient   abbr: **amb**

ambiguous; ambiguity

American Wire Gage   abbr: **AWG**

among; between   use *among* when referring to three or more items; use *between* when referring to only two; *amongst* is seldom used

amount; number   use *amount* to refer to general quantity: *the amount of time taken as sick leave has decreased;* use *number* to refer to items that can be counted; *the number of technicians assigned to the project was 30% greater than anticipated*

ampere(s)   abbr: **A** (pref) or **amp**; other abbr: **kA, mA, μA, nA, pA, A/m** (ampere per minute)

ampere-hour(s)   abbr: **Ah** (pref) or **amp-hr** (more common)

amplitude modulation   abbr: **AM**

an   see **a**

anaemic; anemic   *anemic* pref

anaesthetic; anesthetic   *anesthetic* pref

analog; analogous

analyse(r); analyze(r)   *analyze(r)* pref

AND-gate

and/or   avoid using this term; in most cases it can be replaced by either *and* or *or*

angstrom   abbr: **Å**

anion   def: negative ion

anneal(ing)

annihilate; annihilated; annihilation

antarctic   see comment for **arctic**

ante-   a prefix that means before; combines to form one word: *antecedent, anteroom*

ante meridiem   def: before noon; abbr: **a.m.**; can also be written as *antemeridian,* but never as *antimeridian*

antenna   the proper plural in the technical sense is *antennas;* although *antennae* is sometimes seen, its use should be limited to zoology; *antenna* has generally replaced the obsolescent *aerial*

anti-   a prefix meaning opposite or contradictory to; generally combines to form one word: *antiaircraft, antiastigmatism, anticapacitance, anticoincidence, antisymmetric;* if combining word starts with *i* or is a proper noun, insert a hyphen: *anti-icing, anti-American*

antimeridian   def: the opposite meridian (of longitude); e.g. the antimeridian of 96° 30′W is 83° 30′E

anxious   although *anxious* really implies anxiety, current usage permits it to be used when the meaning is simply keen or eager

anybody; any body   *anybody* means any person; *any body* means any object: *anybody can attend; discard the batch if you find any body containing foreign matter*

anyone; any one   *anyone* means any person; *any one* means any single item: *you may take anyone with you; you may take any one of the samples*

anyway; any way   *anyway* means in any case or in any event; *any way* means in any manner; *the results may not be as good as you expect, but we want to see them anyway; the work may be done in any way you wish*

apparatus(es)

apparent(ly)

appear(s); seems(s)   use *appears* to describe a condition that can be seen: *the equipment appears to be new;* use *seems* to describe a condition that cannot be seen: *he seems to be clever*

appendix   def: the part of a report that contains supporting data; pl: *appendixes* (pref) or *appendices*

appreciate   means to value or to cherish; it should not be used as a synonym for *understand,* as in *we appreciate your difficulty in finding spare parts*

approximate(ly)   abbr: **approx**; but *about* is a better word

arbitrary

arc; arced; arcing

architect; architecture

arctic capitalize when referring to a specific area: *beyond the Arctic Circle;* otherwise use lc letters: *in the arctic;* don't omit the first *c*

area the SI unit for area is the *hectare* (abbr: **ha**)

areal def: having area

around def: on all sides, surrounding, encircling; avoid confusing with *round*

arrester; arrestor *arrester* pref

artwork

as avoid using when the intended meaning is *since* or *because;* to write *he could not open his desk as he left his keys at home* is incorrect (replace *as* with *because*)

as per avoid using this hackneyed expression, except in specifications

asphalt

assembly; assemblies abbr: **assy**

assure means to state with confidence that something has been or will be made certain; it is sometimes confused with *ensure* and *insure,* which it does not replace; see **ensure**

as well as avoid using when the meaning is *and*

asymmetrical

asynchronous

athletic not *atheletic*

atmosphere abbr: **atm**

atomic weight abbr: **at. wt**

attenuator

atto def: 10^{-18}; abbr: **a**

audible; audibility

audio frequency abbr: **af** (pref) or **a-f**

audiovisual

aural def: that which is heard; it should not be confused with *oral,* which means that which is spoken

author; writer avoid referring to yourself as *the author* or *the writer;* use *I, me,* or *my*

authoritative

auto- a prefix meaning self; combines to form one word: *autoalarm, autoconduction, autoionization, autotransformer*

automatic frequency control abbr: **afc** (pref) or **AFC**

automatic volume control abbr: **avc** (pref) or **AVC**

auxiliary abbr: **aux**

average see *mean*

avocation def: an interest or hobby; avoid confusing with *vocation*

ax; axe **ax** pref; pl: *axes*

axis the plural also is *axes*

azimuth abbr: **az**

B

back- as a prefix normally combines into one word: *backboard, backdate(d), backlog*

balance: remainder use *balance* to describe a state of equilibrium (*discontinuous permafrost is frozen soil delicately balanced between the frozen and unfrozen state*), or as an accounting term; use *remainder* when the meaning is the rest of: *the remainder of the shipment will be delivered next week*

balk(ed); baulk(ed) *balk(ed)* pref

ball bearing

bandwidth

bare; bear *bare* means barren or exposed; *bear* means to withstand or to carry (or a wild animal)

barometer abbr: **bar.**

barrel; barreled; barreling the abbr of *barrel(s)* is **bbl**

barretter

barring def: preventing, excepting

bases this is the plural of both *base* and *basis*

basically

B.C. def: Before Christ

because; for use *because* when the clause it introduces identifies the cause of a result: *he could not open his desk because he left his keys at home;* use *for* when the clause introduces something less tangible: *he failed to complete the project on schedule, for reasons he preferred not to divulge*

becquerel def: a unit of activity of radionuclides (SI): abbr: **Bq**; other abbr: **PBq, TBq, GBq, kBq**; in SI, the *becquerel* replaces the *curie*

begin(ning)

benefit; benefited; benefiting

beside, besides   *beside* means alongside, at the side of; *besides* means as well as

between   see **among**

bi-   a prefix meaning two or twice; combines to form one word: *biangular, bidirectional, bifilar, bilateral, bimetallic, bizonal*

biannual(ly); biennial(ly)   *biannual(ly)* means twice a year; *biennial(ly)* means every two years; the *bi* of *bimonthly* and *biweekly* means every two

bias; biased; biases; biasing

billion   def: 10^9

billionaire   only one *n*

billion electron volts   although the pref abbr is **GeV,** *beV* and *bev* are more commonly used

Bill of Materials   abbr: **BOM**

bimonthly   def: every two months

binary

binaural

bioelectronics

bionics   def: application of biological techniques to electronic design

birdseye   (view)

biweekly   def: every two weeks

blow-   as a prefix combines to form one word: *blowhole, blowoff, blowout*

blueprint

blur; blurred; blurring; blurry

board feet   abbr: **fbm** (derived from *feet board measure*)

boiling point   abbr: **bp**

bona fide   def: in good faith, authentic, genuine

bonus; bonuses

borderline

brakedrum; brake lining; brakeshoe

brake horsepower; brake horsepower-hour   abbr: **bhp, bhp-hr**

brand-new

break-   when used as a prefix to form a compound noun or adjective, *break* combines into one word: *breakaway, breakdown, breakup;* in the verb form it retains its single-word identity; *it was time to break up the meeting*

bridging

Brinell hardness number   abbr: **Bhn**

British thermal unit   abbr: **Btu**

buoy; buoyant

bureacracy; bureaucrat

bur(r)   *burr* pref

bus(es); bused; busing; bus bar

business; businesslike; businessperson   avoid using *businessman* or *businesswoman* unless referring to a specific male or female person

by-   as a prefix, *by-* normally combines to form one word: *bylaw, byline, bypass, byproduct*

byte   def: a unit of computer memory

C

calendar; calender; colander   a *calendar* is the arrangement of the days in a year; *calender* is the finish on paper or cloth; a *colander* is a sieve

caliber; calibre   *caliber* pref

calking; caulking   *calking* pref

cal(l)iper   *caliper* pref

calorie   abbr: **cal**

calorimeter; colorimeter   a *calorimeter* measures quantity of heat; *a colorimeter* measures color

cancel(l)ed; cancel(l)ing   *canceled, canceling* pref, but always *cancellation*

candela   def: unit of luminous intensity (replaces *candle*); abbr: **cd**; recommended abbr for candela per square foot and square meter are **cd/ft^2** and **cd/m^2**

candlepower; candlehour(s)   abbr: **cp, c-hr;** *candle* has been replaced by *candela*

candoluminescence

cannot   one word pref; avoid using *can't* in technical writing

canvas; canvass   *canvas* is a coarse cloth used for tents; *canvass* means to solicit

capacitor

capacity for   never use *capacity to* or *capacity of*

capillary

capital letters   abbr: **caps.;** use capital letters as little as possible (see Guideline 3)

car-   as a prefix normally combines to form one word: *carload, carlot, carpool, carwash*

carburet(t)or   *carburetor* pref; a third, seldom used spelling is *carburetter;* also: **carburetion**

carcino- as a prefix combines to form one word

case- as a prefix normally combines to form one word: *casebook, caseharden, casework(er);* exceptions: *case history, case study*

cassette

caster; castor pref spelling is *caster* when the meaning is to swivel freely; *castor* is used when the reference is to castor oil, etc.

catalog(ue) *catolog, catologed,* and *cataloging* pref

catalyst; catalytic

category; categories; categorical

cathode-ray tube abbr: **crt** (pref) or **CRT** (commonly used)

cation def: positive ion

-ceed; -cede; -sede only one word ends in *-sede: supersede;* only three words end in *-ceed; exceed, proceed, succeed;* all others end in *-cede:* e.g. *precede, concede*

centerline abbr: **₵** (pref) or **CL**

Celsius abbr: **C;** see **temperature**

census

center-to-center abbr: **c-c**

centi- def: 10^{-2}; as a prefix combines to form one word: *centiampere, centigram;* abbr: **c;** other abbr:

centigram	**cg**
centiliter	**cL**
centimeter	**cm**
centimeter-gram-second	**cgs**
centimeter per second	**cm/s**
square centimeter	**cm²**

centigrade abbr: **C;** in SI, **centigrade** has been replaced by **Celsius;** see **temperature**

centri- a prefix meaning center; combines to form one word: *centrifugal, centripetal*

chairperson avoid using *chairman* or *chairwoman;* if uncomfortable using *chairperson,* try writing **the chair**

chamfer

changeable; changeover

channel; channeled; channeling

chapter abbr: **chap.**

chassis both singular and plural are spelled the same

check as a prefix combines to form one word: *checklist, checkpoint, checkup* (noun or adjective)

checksum def: a term used in computer technology

chisel; chiseled; chiseling

chrominance

cipher

circuit; circuitous; circuit breaker

cite def: to quote; see **site**

climate avoid confusing *climate* with *weather; climate* is the average type of weather, determined over a number of years, experienced at a particular place; *weather* is the state of the atmospheric conditions at a specific place at a specific time

clockwise(turn) abbr: **CW**

co- as a prefix, *co-* generally means jointly or together; it usually combines to form one word: *coexist, coequal, cooperate, coordinate, coplanar* (*co-worker* is an exception); it is also used as the abbr for *complement of* (an arc or angle): *codeclination, colatitude*

coalesce; coalescent

coarse; course *coarse* means rough in texture or of poor quality; *course* implies movement or passage of time: *a coarse granular material; the technical writing course*

coaxial abbr: **coax.**

coefficient abbr: **coef**

collaborate; collaborator avoid writing *collaborate together* (delete *together*)

collapsible

collateral

cologarithm abbr: **colog**

colon when a colon is inserted in the middle of a sentence to introduce an example or short statement, the first word following the colon is not capitalized; when a colon is used at the end of a sentence to introduce subparagraphs that follow, the first word of each subparagraph should be capitalized unless all the subparagraphs are very short (i.e. only one sentence long); a hyphen should not be inserted after the colon.

colorimeter see **calorimeter**

column abbr: **col.**

combustible

comma a comma normally need not be used immediately before *and, but,* and *or,* but may be inserted if to do so will increase understanding or avoid ambiguity; also see Guideline 3.

commence in technical writing, replace *commence* with the more direct *begin* or *start*

commit; commitment; committed; committing

committee

communicate it is vague to write *I communicated the results to the client;* use a clearer verb: *I mailed . . . , I wrote . . . , I telephoned . . . ,* etc

compare; comparable; comparison; comparative use *compared to* when suggesting a general likeness; use *compared with* when making a definite comparison

compatible; compatibility

complement; compliment *complement* means the balance required to make up a full quantity or a complete set; to *compliment* means to praise; *in a right angle, the complement of 60° is 30°; Mr. Perchanski complimented her for writing a good report*

composed of; comprising; consists of all three terms mean "made up of" (specific items); if any one of these terms is followed by a list of items, it implies that the list is complete; if the list is not complete, the term should be replaced by *includes* or *including*

compound terms two or more words that combine to form a compound term are joined by a hyphen or are written as one word, depending on accepted usage and whether they form a verb, noun, or adjective; the trend is toward one-word compounds; see Guideline 1.

comprise; comprised; comprising to write *comprised of* is incorrect, because the verb comprise includes the preposition *of;* also see **composed of**

concur; concurred; concurrent; concurring

condenser

conductor

confer; conferred; conferring; conference; conferee

conform use *conform to* when the meaning is to abide by; use *comform with* when the meaning is to agree with

conscience

conscious

consensus means a general agreement of opinion; hence to write *consensus of opinion* is

incorrect; e.g. *the consensus was that a further series of tests would be necessary*

consistent with never use *consistent of*

consists of; consisting of see **composed of**

conspicuous

contact *contact* should not be used as a verb when *write, visit, speak,* or *telephone* better describes the action to be taken

contend; contention; contentious

continual; continuous *continual(ly)* means happens frequently but not all the time: *the generator is continually being overloaded* (is frequently overloaded); *continuous(ly)* means goes on and on without stopping: *the noise level is continuously above 100 dB* (it never drops as low as 100dB)

continue(d) abbr: **cont**

continuous wave abbr: **cw**

contra- as a prefix normally combines into a single word

contrast when used as a verb, *contrast* is followed by *with;* when used as a noun, it may be followed by either *to* or *with* (*with* pref)

control; controlled; controlling; controller

conversant with never use *conversant of*

converter; convertible

conveyor

cooperate

coordinate; coordinator

copyright not *copywrite*

corollary

correspond *to correspond to* suggests a resemblance; *to correspond with* means to communicate in writing

corroborate

corrode; corrodible; corrosive

cosecant abbr: **csc** (pref) or **cosec**

cosine abbr: **cos**

cotangent abbr: **cot**

coulomb def: a quantity of electricity, electric charge (SI); abbr: **C**; other abbr: **kC, mC, μC, nC, pC, C/m²**

council; counsel a *council* is a group of people; a *counsel* is a lawyer; *to counsel* is to give advice; also: **counseled; counseling; counselor**

counter- a prefix meaning opposite or reciprocal; combines to form one word: *counteract, counterbalance, counterflow, counterweight*

counterclockwise(turn) abbr: **CCW**

counterelectromotive force abbr: **cemf;** also known as *back emf*

counts per minute abbr: **cpm**

course see **coarse**

criterion pl: *criteria*

criticism; criticize; critique

cross- as a prefix, combines erratically: *cross-check, crosshatch, crosstalk, cross-purpose, cross section*

cross-refer(ence) abbr: **x-ref**

cryogenic

crystal abbr: **xtal**

crystalline; crystallize

cubic abbr: **cu** or **3**; other abbr:

cubic centimeter(s)	cm^3 (pref); **cc**
cubic decimeter(s)	dm^3
cubic foot (feet)	ft^3 (pref); **cu ft**
cubic feet per minute	**cfm** (pref); ft^3/min
cubic feet per second	**cfs** (pref); ft^3/sec
cubic inch(es)	$in.^3$ (pref); **cu in.**
cubic meter(s)	m^3
cubic millimeter(s)	mm^3
cubic yard(s)	yd^3 (pref); **cu yd**

curb; kerb *curb* pref

curie abbr: **Ci;** other abbr: **mCi, μCi;** in SI the *curie* is replaced by the *becquerel* (which see)

curriculum pl: *curriculums* (pref) or *curricula*

cursor

cycles per minute abbr: **cpm**

cycles per second abbr: **cps;** although still occasionally used, this term has been replaced by **hertz** (which see)

cylinder; cylindrical abbr: **cyl**

D

daraf def: the unit of elastance

data def: gathered facts; although *data* is plural (derived from the singular *datum,* which is rarely used), it is more often used as a singular noun; *when all the data has been received, the analysis will begin*

date(s) avoid vague statements such as "last month" and "next year" because they soon become indefinite; write as a specific date, using day (in numerals), month (spelled out), and year (in numerals): *January 27, 1986* or *27 January 1986* (the latter form has no punctuation); to abbreviate, reduce month to first three letters and year to last two digits; *Jan 27, 86* or *27 Jan 86;* for SI, use numerals only, in this order: year, month, day: *1986 01 27*

days days of the week are capitalized: *Monday, Tuesday*

de- a prefix that generally combines to form one word: *deaccentuate, deactivate, decentralize, decode, deemphasize, deenergize, deice, derate, destagger;* an exception is *de-ionize*

debug(ging)

decelerate def: to slow down; never use *deaccelerate*

decibel abbr: **dB;** the abbr for decibel referred to 1 mW is **dBm**

decimals for values less than unity (one), place a zero before the decimal point: *0.17, 0.0017*

decimate def: to reduce by one-tenth; can also mean to destroy much of

decimeter abbr: **dm**

declination abbr: **dec**

deductible

defective; deficient *defective* means unserviceable or damaged (generally lacking in quality); *deficient* means lacking in quantity (it is derived from *deficit*), and in the military sense incomplete: *a short circuit resulted in a defective transmitter; the installation was completed on schedule except for a deficient rotary coupler which will not be delivered until June 10*

defense

defer; deferred; deferring; deference

definite; definitive *definite* means exact, precise; *definitive* means conclusive, fully evolved; e.g. *a definite price* is a firm price; *a definitive statement* concerns a topic that has been thoroughly considered and evaluated

degree(s) abbr: **deg** (pref in narrative) or ° (following numerals); see **temperature**

demarcation

demi- a little-used prefix meaning half (generally replaced by *semi-*); combines to form one word: *demivolt*

demonstrate; demonstrator; demonstrable not *demonstratable*

depend; dependence; dependent; dependable

deprecate; depreciate *deprecate* means to disapprove of; *depreciate* means to reduce the value of: *the use of "as per" in technical writing is deprecated; the vehicles depreciated by 50% the first year and 20% the second year*

depth

desiccant; desiccate(d)

desirable

despite def: in spite of; *despite* should never be followed by the word *of*

deter; deterred; deterrence; deterrent; deterring

deteriorate

develop not *develope*

device; devise the noun is *device*, the verb is *devise: a unique device; he devised a new circuit*

dext(e)rous *dexterous* pref

diagram; diagramed; diagraming; diagrammatic

dial; dialed; dialing

dialog(ue) *dialogue* pref

diaphragm

diazo

didn't never use the contraction in technical writing; use **did not**

dielectric

diesel; diesel-electric

dietitian

differ use *differ from* to demonstrate a difference; use *differ with* to describe a difference of opinion

different *different from* is preferred; *different to* is sometimes used; *different than* is better not used

diffraction

diffusion; diffusible

dilemma means to be faced with a choice between two unhappy alternatives; should not be used as a synonym for *difficulty*

diplex; duplex *diplex operation* means the si-

multaneous transmission of two signals using a single feature, e.g. an antenna; *duplex operation* means that both ends can transmit and receive simultaneously

direct current abbr: **dc**

directly def: immediately; do not use when the meaning is as soon as

disassemble never use *dissemble* when the meaning is to take apart

disassociate see *dissociate*

disc; disk both are correct and commonly used; **disk** pref

discernible

discreet; discrete *discreet* means prudent or discerning: *his answer was discreet; discrete* means individually distinctive and separate: *discrete channels; discretion* is formed from *discreet*, not from *discrete*

disinterested; uninterested *disinterested* means unbiased, impartial; *uninterested* means not interested

dispatch; despatch *dispatch* pref

disseminate

dissimilar

dissipate

dissociate; disassociate *dissociate* pref

distil(l) *distil* pref; but always *distilled, distillate, distillation*

distribute; distribution; distributor

don't; doesn't such contractions should not appear in technical writing

donut; doughnut for electronics/nucleonics, use *donut*

doppler capitalize only when referring to the Doppler principle

double- as a prefix combines erratically: *double-barrelled; doublecheck; double-duty, double entry, doublefaced*

down- as a prefix combines into one word: *downrange, downtime, downwind*

dozen abbr: **doz**

drafting; draftsperson avoid writing *draftsman* or *draftswoman*

drawing(s) abbr: **dwg**

drier; dryer the adjective is always *drier*; the pref noun is *dryer: this material is drier; place the others back in the dryer;* also: **dryly**

drop; droppable; dropped; dropping

due to   an overused expression; *because of* pref

duo-   a prefix meaning two; combines to form one word: *duocone, duodiode, duophase*

duplex   see **diplex**

duplicator

dutiable; duty-free

E

each   abbr:**ea**

east   capitalize only if *east* is part of the name of a specific location: *East Africa;* otherwise use lc letters: *the east coast of the U.S.;* abbr: **E**; the abbr for *east-west* (control, movement) is **E-W**; *eastbound* and *eastward* are written as one word

eccentric; eccentricity

echo; echoes

economic; economical   use *economical* to describe economy (of funds, effort, time); use *economic* when writing about economics: *an economical operation* (it did not cost much); *an economic disaster* (refers to economics)

effect   see **affect**

efficacy; efficiency   *efficacy* means effectiveness, ability to do the job intended; *efficiency* is a measurement of capability, the ratio of work done to energy expended; *we hired a consultant to assess the efficacy of our training methods; the powerhouse is to have a high-efficiency boiler*

e.g.   pref abbr for *exempli gratia* (the Latin of *for example*); avoid confusing with **i.e.**; may also be abbr **eg**

eighth

electric(al)   if in doubt, use *electric;* generally, *electric* means produces or carries electricity, whereas *electrical* means related to the generation or carrying of electricity; abbr: **elec**

electro-   a prefix generally meaning pertaining to electricity; it normally combines to form one word: *electroacoustic, electroanalysis, electrodeposition, electromechanical, electroplate;* if the combining word starts with *o*, insert a hyphen: *electro-optics, electro-osmosis*

electromagnetic units   abbr: **emu**

electromotive force   abbr: **emf**

electronic(s)   use *electronic* as an adjective, *elec-*

tronics as a noun: *electronic countermeasures; your career in electronics*

electron volt(s)   abbr: **eV** (pref) or **ev**

electrostatic units   abbr: **esu**

elevation   abbr: **el**

elicit; illicit   *elicit* means to obtain or identify; *illicit* means illegal

eligible; eligibility

ellipse

embarrass; embarrassed; embarrassing; embarrassment

embedded

emigrate; immigrate   *emigrate* means to go away from; *immigrate* means to come into

emit, emitter, emittance; emission; emissivity

enamel; enameled; enameling

encase; incase   *encase* pref

encipher

enclose; inclose   *enclose* pref; *inclose* is used mainly as a legal term

endorse, indorse   *endorse* pref

engineer; engineered; engineering

enquire; inquire   *inquire* pref

enrol; enroll   both are correct, but *enroll* pref; universal usage prefers *ll* for *enrolled, enrolling,* and *enrollment*

en route   def: on the road, on the way; never use *on route*

ensure; insure; assure   use *ensure* (or *insure*) when the meaning is to make certain of: *use the new oscilloscope to ensure accurate calibration;* use *insure* when the meaning is to protect against financial loss: *we insured all our drivers;* use *assure* when the meaning is to state with confidence that something has been or will be made certain: *he assured the meeting that production would increase by 8%*

entrepreneur; entrepreneurial

entrust; intrust   *entrust* pref

envelop; envelope   *envelop* is a verb which means to surround or cover completely; *envelope* is a noun that means a wrapper or a covering

environment(al)

equal   *equality* and *equalize* have only one *l; equally* always has *ll; equaled* and *equaling* preferably have only one *l*, but sometimes are seen with *ll*

equi- a prefix that means equality; combines to form one word: *equiphase, equipotential, equisignal*

equilibrium(s)

equip; equipped; equipping; equipment

equivalent abbr: **equiv**

erase; erasable

errata although *errata* is plural (from the singular *erratum*, which is seldom used), it can be used as a singular or plural noun: both *the errata are ready* and *the errata is complete* are acceptable

erratic

erroneous

escalator

especially; specially *specially* pref when it refers to an adjective (*a specially trained crew*); *especially* should introduce a phrase (*they were all well trained, especially the computer technicians*)

esthetic; aesthetic *esthetic* pref

et al. def: and others

et cetera abbr: **etc**; def: and so forth, and so on; use with care in technical writing; *etc* can create an impression of vagueness or unsureness: *the transmitters, etc, were tested* is much less definite than either *the transmitter, modulator, and power supply were tested,* or (if to restate all the equipment is too repetitious) *the transmitting equipment was tested*

extremely high frequency abbr: **ehf**

everybody; every body *everybody* means every person, or all the persons; *every body* means every single body; *everybody was present; every body was examined for gunpowder scars*

everyone; every one *everyone* means every person, or all the persons; *every one* means every single item: *everyone is insured; every one had to be tested in a saline solution*

exa def: 10^{18}; abbr: **E**

exaggerate

exceed

excel; excelled; excellent; excelling

except def: to exclude; see **accept**

exhaust

exhort

exorbitant

expedite; expediter

explicit; implicit *explicit* means clearly stated, exact; *implicit* means implied (the meaning has to be inferred from the words): *the supervisor gave explicit instructions* (they were clear); *that the manager was angry was implicit in the words he used*

extemporaneous

extracurricular

extraordinary

F

face- as a prefix normally combines to form one word: *facedown, facelift, faceplate;* exceptions: *face-saver, face-saving*

Fahrenheit abbr: **F**; see **temperature**

fail-safe

fallout one word in the noun form

familiarize; familiarization

farad def: a unit of electric capacitance; abbr: **F**; other abbr: **μF, nF, pF**

farfetched; far-out; far-reaching; farseeing; farsighted

farther; further *farther* means greater distance: *he traveled farther than the other salesmen; further* means a continuation of (as an adjective) or to advance (as a verb): *the promotion was a further step in his career plan,* and *to further his education, he took a part-time course in industrial drafting*

fascinate; fascination

fasten(er)

faultfinder; faultfinding

feasible; feasibility

February

feet; foot abbr: **ft**; other abbr:

feet board measure	**fbm**
(board feet)	
feet per minute	**fpm**
feet per second	**fps**
foot-candle(s)	**fc** (pref); **ft-c**
foot-pound(s)	**fp** (pref); **ft-lb**
foot-pound-second	**fps** system
(system)	

femto def: 10^{-15}; abbr: **f**; other abbr:

femtoampere(s)**fA**
femtovolt(s) **fV**

ferri- a prefix meaning contains iron in the

ferric state; combines to form one word: *ferricyanide, ferrimagnetic*

ferro-   a prefix meaning contains iron in the ferrous state; combines to form one word: *ferroelectric, ferromagnetic, ferrometer*

ferrule; ferule   a *ferrule* is a metal cap or lid; a *ferule* is a ruler

fewer; less   use *fewer* to refer to items that can be counted: *fewer technicians than we predicted have been assigned to the project;* use *less* to refer to general quantities: *there was less water available than predicted*

fiber; fibrous; Fiberglas   *Fiberglas* is a trade name

field-   as a prefix normally does not combine into a single-word or hyphenated form: *field glasses, field test, field trip;* but: *fieldwork(er)*

figure numbers   in text, spell out the word *Figure* in full, or abbr it to **Fig:** *the circuit diagram in Figure 26* and *for details, see Fig. 7;* use the abbreviated form beneath an illustration; always use numerals for the figure number

final; finally; finalize

fire-   as a prefix combines erratically: *firearm, fire alarm, firebreak, fire drill, fire escape, fire extinguisher, firefighter, firepower, fireproof, fire sale, fire wall*

first   to write *the first two . . .* (or three, etc.) is better than to write *the two first . . . ;*never use *firstly;* as a prefix, *first-* combines erratically: *first-class, firsthand, first-rate*

fix   in technical usage, *fix* means to firm up or establish as a permanent fact; avoid using it when the meaning is repair

flameout; flameproof

flammable   def: easily ignited; see **inflammable**

flexible

flight   usually combines to form two words: *flight control, flight deck, flight plan*

flip-flop

flotation   this is the correct spelling for describing an item that floats: *flotation gear*

flowchart

fluid   abbr: **fl;** the abbr for fluid ounces is **fl oz**

fluorescence; fluorescent

fluorine; fluoridation

focus; focused; focusing; focuses   pl: *focuses* (pref) or *foci*

foot-   as a prefix normally combines into one word: *footbridge, footcandle, foothold, footnote, footwork;* also see **feet**

for   see **because**

fore-   def: that which goes before; as a prefix normally combines into one word: *foreclose, forefront, foreground, foreknowledge, foremost, foresee, forestall, forethought, forewarn*

forecast   this spelling applies to both present and past tenses

forego; forgo   *forego* and *foregoing* mean to go before; *forgo* means to go without

foreman   avoid using in a general sense, except when referring to a person specifically: *John Hayward, the foreman;* never use *forewoman;* a better choice is *supervisor,* if *foreperson* seems too awkward

foreword; forward   a *foreword* is a preface or preamble to a book; *forward* means onward: *the scope is defined in the foreword to the book; he requested that we bring the meeting date forward*

for example   abbr: **e.g.** (pref) or **eg**

former; first   use *former* to refer to the first of only two things; use *first* if there are more than two

formula   pl: *formulas* (pref) or *formulae*

forty   def: 40; it is not spelled *fourty*

fourth   def: 4th; it is not spelled forth

fractions   when writing fractions that are less than unity, spell them out in descriptive narrative, but use figures for technical details: *by the end of the heat run, nine-tenths of the installation had been completed; a flat case $15\frac{1}{4}$ mm square by $\frac{7}{8}$ mm deep;* use decimals rather than fractions, except when a quantity is normally stated as a fraction (such as $\frac{3}{8}$ *in. plywood*)

free-   as a prefix normally combines into one word: *freehand, freehold, freestanding, freeway, freewheel*

free from   use *free from* rather than *free of: he is free from prejudice*

free on board   abbr: **fob** (pref), **f.o.b.** (commonly used), or **FOB**

frequency   abbr: **freq**

frequency modulation   abbr: **FM**

fulfil(l)   the pref spelling is *fulfill, fulfilled, ful-*

filling, and *fulfillment; fulfil* and *fulfilment* can also be spelled with a single *l.*

funnel; funneled; funneling

further see **farther**

fuse as a verb, means join together or weld; as a noun, means a circuit protection device

fuselage

fuze def: a detonation initiation device

G

gage; gauge both spellings are correct; gage is recommended because it is less likely to be misspelled; *gaging* and *gauging* do not retain the *e*

gallon gallons differ between U.S. (231 in.^{3}; 3.785 dm^3) and Britain (277.42 in.^{3}; 4.546 dm^3); abbr: **gal;** other abbr:

gallons per day	**gpd**
gallons per hour	**gph**
gallons per minute	**gpm**
gallons per second	**gps**

gang; ganged; ganging

gas; gases; gassed; gassing; gaseous; gassy

gauge see **gage**

gearbox; gearshift

geiger (counter)

gelatin(e) *gelatin* pref

geo- a prefix meaning of the earth; combines to form one word: *geocentric, geodesic, geomagnetic, geophysics*

giga def: 10^9; abbr: **G;** other abbr:

gigabecquerel(s)	**GBq**
gigahertz	**GHz**
gigajoule(s)	**GJ**
gigaohm(s)	**GΩ; Gohm**
gigapascal(s)	**GPa**
gigavolt(s)	**GV**

gimbal

glue; gluing; gluey

glycerin(e) *glycerin* pref

gotten never use this expression in technical writing; use *have got* or simply *have*

government capitalize when referring to a specific government either directly or by implication; use lc if the meaning is government generally: *the U.S. Government; the Government specifications; no government would sanction such restrictions*

gram abbr: **g;** abbr for gram-calorie is **g-cal**

grammar; grammatical(ly)

gray def: absorbed dose of ionizing radiation (SI); abbr: **Gy;** other abbr: **mGy, μGy;** in SI, the *gray* replaces the *rad*

Greenwich mean time abbr: **GMT**

grill(e) def: a loudspeaker covering, or a grating; *grille* pref

ground (electrical) abbr: **gnd**

ground crew; ground floor

guage wrongly spelled; the correct spelling is *gauge* or *gage* (pref)

guarantee never *guaranty*

guesstimate

guideline(s)

gyroscope abbr: **gyro**

H

half; halved; halves; halving as a prefix, *half* combines erratically; some common compounds are: *half-hour(ly), half-life, half-month(ly), halftone, half-wave;* for others, consult your dictionary

hand- as a prefix normally combines to form one word: *handbill, handbook, handful, handfuls, handhold, handpicked, handset, handshake*

hangar; hanger a *hangar* is a large building for housing aircraft; a *hanger* is a supporting bracket

hard- as a prefix normally combines into one word: *hardbound, hardhanded, hardhat, hardware;* exceptions: *hard-earned, hard-hitting*

haversine abbr: **hav**

H-beam

head- as a prefix normally combines into one word: *headfirst, headquarters, headset, headstart, headway*

heat- as a prefix, *heat* combines erratically; some typical compounds are: *heat-resistant, heat-run, heatsink, heat-treat;* for others, consult your dictionary

heavy-duty

hectare def: a large unit of area, used in surveying and agriculture; in SI, *hectare* replaces *acre;* abbr: **ha**

height (not *heighth*) abbr: **ht**

heightfinder; heightfinding

helix   pl: *helices* (pref) or *helixes*

hemi-   a prefix meaning half; combines to form one word: *hemisphere, hemitropic*

henry   def: a unit of inductance; abbr: **H;** other abbr: **mH, μH, nH, pH**

here-   whenever possible avoid using *here-* words that sound like legal terms: *hereby, herein, hereinafter, hereof;* they make a writer sound pompous; as a prefix, *here-* combines to form one word.

hertz   def: a unit of frequency measurement (similar to *cycle per second,* which it replaces); abbr: **Hz;** other abbr: **THz, GHz, MHz, kHz**

heterodyne

heterogeneous; homogeneous *heterogeneous* means of the opposite kind; *homogeneous* means of the same kind

hi-fi   def: high fidelity; generally replaced by **stereo**

high-   as a prefix either combines into one word or the two words are joined by a hyphen: *highhanded, highlight, high-power, high-pressure, high-speed*

high frequency   abbr: **hf**

high-pressure   (as an adjective) abbr: **h-p**

high voltage   abbr: **hv** (pref) or **HV**

hinge; hinged; hinging

homogeneous   see **heterogeneous**

horizontal   abbr: **hor**

horsepower   abbr: **hp;** the abbr for horsepower-hour is **hp-hr**

hour(s)   abbr: **hr** or **h** (SI)

hundred   abbr: **C**

hundredweight   def: 112 lb; abbr: **cwt**

hybrid

hydro-   a prefix meaning of water; combines to form one word: *hydroacoustic, hydroelectric, hydromagnetic, hydrometer*

hyper-   a prefix meaning over; combines to form one word: *hyperacidity, hypercritical*

hyperbola   the plural is *hyperbolas* (pref) or *hyperbolae*

hyperbole   def: an exaggerated statement

hyperbolic cosine; sine; tangent   abbr: **cosh, sinh, tanh**

hyphen   in compound terms you may omit hyphens unless they need to be inserted to avoid ambiguity or to conform to accepted usage; *e.g. preemptive* is preferred without a hyphen, but *photo-offset* and *re-cover* (when the meaning is *to cover again*) both require one; refer to individual entries and Guideline 1

hypothesis   pl: *hypotheses*

I

I-beam

ibid.   def: Latin abbr for *ibidem,* meaning in the same place; used in footnoting

ID card

i.e.   pref abbr for *id est* (the Latin of *that is*); avoid confusing with **e.g.;** may also be abbr **ie**

if and when   avoid using this expression; use either *if* or *when*

ignition   abbr: **ign**

ill-   as a prefix combines into a hyphened expression: *ill-advised, ill-defined, ill-timed*

illegible

im-   see **in-**

imbalance   this term should be restricted for use in accounting and medical terminology; use *unbalance* in other technical fields

immalleable

immaterial

immeasurable

immigrate   see **emigrate**

immittance

immovable

impel; impelled; impelling; impeller

imperceptible

impermeable

impinge; impinging

imply; infer   speakers and writers can *imply* something; listeners and readers *infer* it from what they hear or read: *in his closing remarks, Mr. Smith implied that further studies were in order; the technician inferred from the report that his work was better than expected*

impracticable; impractical   *impracticable* means not feasible; *impractical* means not practical; a less-preferred alternative for impractical is *unpractical*

in; into   *in* is a passive word; *into* implies action: *ride in the car; step into the car*

in-; im-; un- all three prefixes mean not; all combine to form one word: *ineligible, impermeable, unintelligible;* if you are not sure whether you should use in-, im-, or un-, use *not*

inaccessible

inaccuracy

inadmissible

inadvertent

inadvisable, unadvisable *inadvisable* pref

inasmuch as a better word is *since*

inaudible

incalculable

incandescence; incandescent

incase *encase* pref

inch(es) abbr: **in.**; other abbr:
inches per second   **ips** (pref); **in./s**
inch-pound(s)   **in.-lb**

incidentally in most cases the word *incidentally* is unnecessary

inclose use **enclose**

includes; including abbr: **incl;** when followed by a list of items, *includes* implies that the list is not complete; if the list is complete, use *comprises* or *consists of* (which see)

incomparable

incompatible

incur; incurred; incurring

index pl: *indexes* pref, except in mathematics (where *indices* is common)

indicated horsepower abbr: **ihp;** the abbr for indicated horsepower-hour is **ihp-hr**

indifferent to never use *indifferent of*

indiscreet; indiscrete *indiscreet* means imprudent; *indiscrete* means not divided into separate parts

indispensable

indorse *endorse* pref

industrywide

ineligible

inequitable

inessential; unessential both are correct; *unessential* pref

inexhaustible

inexplicable

infallible

infer; inferred; inferring; inference also see **imply**

inflammable def: easily ignited (derived from *inflame*); *flammable* is a better word: it prevents readers from mistakenly thinking the *in* of *inflammable* means not

inflexible

infrared

ingenious; ingenuous *ingenious* means clever, innovative; *ingenuous* means innocent, naive; *ingenuity* is a noun derived from ingenious

in-house

inoculate

inoperable not *inoperatable*

inquire; enquire *inquire* pref; also *inquiry*

insanitary; unsanitary both are correct; *insanitary* pref

inseparable

inside diameter abbr: **ID**

in situ def: in the normal position

insofar as

instal(l) *install, installed, installer, installing, installation,* and *installment* pref; a single *l* is acceptable for *instal* and *instalment*

instantaneous

instrument

insure the pref def is to protect against financial loss; can also mean make certain of; see **ensure**

integer; integral, integrate, integrator

intelligence quotient abbr: **IQ**

intelligible

inter- a prefix meaning among or between; normally combines to form one word: *interact, intercarrier, interdigital, interface, intermodulation, interoffice*

intermediate frequency abbr: **if.** (pref) or **i-f**

intermediate-pressure (as an adjective) abbr: **i-p**

intermittent

internal abbr: **int**

interrupt

into see **in**

intra- a prefix meaning within; normally combines to form one word: *intranuclear;* if combining word starts with *a,* insert a hyphen: *intra-atomic*

intractable

intrigue; intrigued; intriguing

intrust *entrust* pref

irrational

irregardless never use this expression; use *regardless*

irrelevant frequently misspelled as *irrevelant*

irreversible

irritate; irritant; irritable; irritation

iso- a prefix meaning the same, of equal size; normally combines to form one word: *isoelectronic, isometric, isotropic;* if combining word starts with *o*, insert a hyphen: *iso-octane*

its; it's *its* means belonging to; *it's* is an abbr for it is: *the transmitter and its modulator; if the fault is not in the remote equipment, then it's most likely in master control;* in technical writing *it's* should seldom be used: replace with *it is*

J

job-holder; jobseeker; job lot

joule def: a unit of energy, work, or quantity of heat (SI); abbr: **J**; other abbr: **TJ, GJ, MJ, kJ, mJ, J/m³, J/K** (joule(s) per kelvin), **J/kg, J/mol** (joule(s) per mole)

journey; journeys

judg(e)ment *judgment* pref

judicial; judicious *judicial* means related to the law; *judicious* means sensible, discerning

juxtaposition

K

kelvin def: the SI unit for thermo-dynamic temperature; abbr: **K**

kerb; curb *curb* pref

key- as a prefix normally combines to form one word: *keyboard, keypunch, keying, keystroke;* but *key word*

kilo def: 10^3; abbr: **k**; other abbr:

kiloampere(s)	**kA**
kilobecquerel(s)	**kBq**
kilocalorie(s)	**kcal**
kilocoulomb(s)	**kC**
kilogram(s), (see **kilogram**)	**kg**

kilohertz	**kHz**
kilohm(s)	**kΩ; kohm**
kilojoule(s)	**kJ**
kiloliter(s)	**kL**
kilometer(s)	**km**
kilometers per hour	**km/h**
kilomole(s)	**kmol**
kilonewton(s)	**kN**
kilopascal(s)	**kPa**
kilosecond(s)	**ks** (pref); **ksec**
kilosiemens	**kS**
kilovolt(s)	**kV**
kilovolt-ampere(s)	**kVA**
kilovolt-ampere(s), reactive	**kVAr**
kilowatt(s)	**kW**
kilowatthour(s)	**kWh** (pref); **kw-hr**

kilogram def: the SI unit for mass; abbr: **kg**; other typical abbr: **Mg, g, mg, μg**; also:

kilogram-calorie(s)	**kg-cl**
kilogram(s) per meter	**kg/m**
kilogram(s) per square meter	**kg/m²**
kilogram(s) per cubic meter	**kg/m³**
kilogram meter(s) per second	**kg·m/s**

knockout as noun or adjective, one word

knot abbr: **kn**

knowledge; knowledgeable

L

label(l)ed; label(l)ing single *l* pref

laboratory abbr: **lab**

laborsaving

lacquer

lambert abbr: **L**; use the abbr **L** with care: it is also the SI abbr for *liter*

lampholder

last, latest, latter *last* means final; *latest* means most recent; *latter* refers to the second of only two things (if more than two, use *last*); it is better to write *the last two* (or three, etc.) than *the two last*

lath; lathe a *lath* is a strip of wood; a *lathe* is a machine

latitude abbr: **lat** or **Φ**

lay- as a prefix generally combines to form one word: *layoff, layout, layover* (all nouns)

learned; learnt *learned* pref

least common multiple abbr: **lcm**

left-hand(ed) abbr: **LH**

lend; loan use *lend* as a verb, *loan* as a noun; to write or say "*loan* me your iron" is wrong, but "*lend* me your iron" is correct

length the SI unit of length is the *metre* (which see), expressed in multiples and submultiples of *kilometers* (**km**), *meters* (**m**), and *millimetres* (**mm**)

lengthy not *lengthly;* also: **lengthening, lengthwise**

less see **fewer**

letter of intent; letter of transmittal pl: *letters of intent* or *transmittal*

level; leveled; leveling

liable to means under obligation to; avoid using as a synonym for *apt to* or *likely to*

liaison liaison is a noun; it is sometimes used uncomfortably as a verb: *liaise*

libel; libeled; libeling; libelous

licence; license *license* pref: *licence* is sometimes used as a noun

light- as a prefix *light-* generally combines to form a single word: *lightface* (type), *lightweight;* but *light-year;* the past tense is *lighted,* but *lit* is common in Canada and Britain

lightening; lightning *lightening* means to make lighter; *lightning* is an atmospheric discharge of electricity

likable

linear abbr: **lin;** the abbr for lineal foot is **lin ft**

lines of communication not *line of communications*

liquefy; liquefaction

liquid abbr: **liq**

liter; litre the SI spelling is **litre,** but in U.S. **liter** is more common; abbr: **L;** other abbr: **kL, mL, μL;** the abbr for *liter(s) per day/hour/minute/second* are **L/d, L/h, L/m, L/s**

loan see **lend**

loath; loathe *loath* means reluctant; *loathe* means to dislike intensely

lock- as a prefix combines into a single word: *locknut, lockout, locksmith, lockstep, lockup, lockwasher*

locus pl: *loci*

logarithm abbr: common—**log;** natural **—ln**

logbook

logistic(s) use *logistic* as an adjective, *logistics* as a noun: *logistic control; the logistics of the move*

long- as a prefix normally combines into a single word or is hyphenated: *long-distance, longhand, longplaying, long-term, long-winded;* but *long shot*

longitude abbr: **long-** or λ

long-play(ing)(record) abbr: **LP**

lose; loose *lose* is a verb that refers to a loss; *loose* is an adjective or a noun that means free or not secured: *three loose nuts caused us to lose a wheel*

louver; louvre *louver* pref

low frequency abbr: **lf**

low-pressure (as an adjective) abbr: **l-p**

lubricate; lubrication abbr: **lub**

lumen def: a unit of luminous flux (SI); abbr: **lm;** other abbr:

lumen-hour(s)	**lm·h** (pref); **lm-hr**
lumens per square foot	**lm/ft²**
lumens per square meter	**lm/m²**
lumens per watt	**lm/W**
lumen-second(s)	**lm·s**
microlumen(s)	**μ lm**
millilumen(s)	**mlm**

luminance; luminescence; luminosity; luminous

lux def: a unit of illuminance (SI); abbr: **lx;** other abbr: **klx**

M

Mach

macro- a prefix meaning very large; combines to form one word: *macroscopic*

magneto pl: *magnetos;* as a prefix, it normally combines to form one word: *magnetoelectronics, magnetohydrodynamics, magnetostriction;* if combining word starts with *o* or *io,* insert a hyphen: *magneto-optics, magneto-ionization*

magneton; magnetron   a *magneton* is a unit of magnetic moment; a *magnetron* is a vacuum tube controlled by an external magnetic field

maintain; maintenance

majority   use *majority* mainly to refer to a number, as in *a majority of 27;* avoid using it as a synonym for many or most; do not write *the majority of technicans* when the intended meaning is *most technicians*

make-   as a prefix normally combines to form one word: *makeshift, makeup*

malfunction

malleable

man   to avoid sexist connotations, replace man as follows:

for:	*write:*
man(ned),(ning)	staff(ed), (ing)
man-hour(s)	work-hour(s)
manpower	labor

manage; managed; managing; manageable

manifesto   pl: *manifestos*

maneuver; maneuvered; maneuvering; maneuverable

manufacturer   abbr: **mfr**

marshal; marshaled; marshaling; marshaler

marvel; marveled; marveling; marvelous

mass   see **kilogram**

material; materiel   *material* is the substance or goods out of which an item is made; when used in the plural, it describes items of a like kind, such as *writing materials; materiel* are all the equipment and supplies necessary to support a project or undertaking (a term commonly used in military operational support)

matrix   pl: *matrices*

maximum   pl: *maximums* (pref) or *maxima;* abbr: **max;** like *mimimize, maximize* can be used as a verb

maybe; may be   *maybe* means "perhaps": *maybe there is a second supplier;* the verb form *may be* means "perhaps it will be" or "possibly there is": e.g. *there may be a second supplier*

meager; meagre   *meager* pref

mean; median   the *mean* is the average of a number of quantities; the *median* is the midpoint of a sequence of numbers; e.g. in the sequence of five numbers 1, 2, 3, 7, 8, the mean is 4.2 and the median is 3

mean effective pressure   abbr: **mep**

mean sea level   *abbr:* **msl** (pref) or **MSL**

mediocre

medium   when *medium* is used to mean substances, liquids, materials, or the means for accomplishing something (such as advertising), the plural is *media,* in all other senses the plural is *mediums*

mega   def: 10^6; abbr: **M;** other abbr:

megacoulomb(s)	**MC**
megaelectronvolt(s)	**MeV**
megahertz	**MHz**
megajoule(s)	**MJ**
meganewton(s)	**MN**
megapascal(s)	**MPa**
megavolt(s)	**MV**
megawatt(s)	**MW**
megohm(s)	**MΩ; Mohm**

memorandum   pl: *memorandums* (pref) or *memoranda;* abbr: **memo** (singular) or **memos** (plural)

merit; merited; meriting

metal   a single *l* is pref for *metaled* and *metaling* (although *ll* also is acceptable); *metallic* and *metallurgy* always have *ll*

meteorology; metrology   *meteorology* pertains to the weather; *metrology* pertains to weights, measures, and calibration

meter; metre   def: metric unit of length; the SI spelling is *metre,* but in U.S. *meter* is more common; abbr: **m;** other typical abbr:

square meter(s)	**m^2**
cubic meter(s)	**m^3**
meters per second	**m/s**
newton meter(s)	**N·m**
newtons per square meter	**N/m^2**
kilogram(s) per cubic meter	**kg/m^3**

micro   def: 10^{-6}; abbr: **μ** (pref) or **u;** other abbr:

microampere(s)	**μA**
microcoulomb(s)	**μC**
microfarad(s)	**μF**
microgram(s)	**μg**
microgray(s)	**μGy**
microhenry(s)	**μH**
microhm(s)	**μΩ; μohm**
microlumen(s)	**μlm**
micromho(s)	**μmho**
micrometer(s)	**μm**

micromole(s)	μ**mol**
micronewton(s)	μ**N**
micropascal(s)	μ**Pa**
microsecond(s)	μ**s** (pref); μ**sec**
microsiemens	μ**S**
microtesla(s)	μ**T**
microvolt(s)	μ**V**
microwatt(s)	μ**W**

as a prefix meaning very small,
micro- normally combines to form one word: *microameter, micrometer, microorganism, microprocessor, microswitch, microwave;* the term *micromicro-* (10^{-12}) has been replaced by **pico** (which see)

microphone abbr: **MIC** (pref) or **mike**

mid- a prefix that means in the middle of; generally combines into one word: *midday, midpoint, midweek;* if used with a proper noun, insert a hyphen: *mid-Atlantic*

mile the word mile is generally understood to mean a statute mile of 5280 ft (1609 m), so the statement *I drove 326 miles* implies statute miles; when referring to the *nautical mile* (6080 ft; 1853 m), always identify it as such: *the flight distance was 4210 nautical miles (or 4210 nmi);* abbr:

statute mile(s)	**mi**
nautical mile(s)	**nmi** (pref); **n.m.**
miles per gallon	**mpg**
miles per hour	**mph**

mileage; milage *mileage* pref

milli def: 10^{-3}; abbr: **m**; other abbr:

milliampere(s)	**mA**
millicoulomb(s)	**mC**
millicurie(s)	**mCi**
millifarad(s)	**mF**
milligram(s)	**mg**
milligray(s)	**mGy**
millihenry(s)	**mH**
millijoule(s)	**mJ**
millikelvin(s)	**mK**
milliliter(s)	**mL**
millilumen(s)	**mlm**
millimeter(s)	**mm**
millimho(s)	**mmho**
millimole(s)	**mmol**
milliohm(s)	**mΩ; mohm**
millinewton(s)	**mN**
millipascal(s)	**mPa**
milliroentgen(s)	**mR**
millisecond(s)	**ms** (pref); **msec**

millisiemens	**mS**
millitesla(s)	**mT**
millivolt(s)	**mV**
milliwatt(s)	**mW**
milliweber(s)	**mWb**

as a prefix, **milli-** combines to form one word: *milliammeter, milligram, millimicron*

millibar def: a unit of pressure (= 100 Pa); abbr: **mbar**

mini- as a prefix combines to form one word: *minicomputer, minireport*

miniature; miniaturization

minimum pl: *minimums* (pref) or *minima;* abbr: **min**

minority use mainly to refer to a number: *a minority by 2;* avoid using it as a synonym for several or a few; to write *a minority of the technicians* is incorrect when the intended meaning is *a few technicians*

minuscule not *miniscule;* def: very small

minute abbr:

time	**min**
angular measure	

mis- a prefix meaning wrong(ly) or bad(ly); combines to form one word: *misalign, misfired, mismatched, misshapen*

miscellaneous

miscible

misspelled not *mispelled*

mitre; mitered; mitering

mnemonic

model; modeled; modeling; modeler

mold; mould *mold* pref

mole def: the SI unit for amount of substance; abbr: **mol**; other abbr: **kmol, nmol, μmol, mol/m**

momentary; momentarily both mean *for a moment,* not *in a moment*

money- as a prefix normally combines to form one word: *moneymaking, moneysaving*

mono- a prefix meaning one or single; combines to form one word: *monopulse, monorail, monoscope*

monotonous

months the months of the year are always capitalized: *January, February,* etc; if abbr, use only the first three letters: *Jan, Feb,* etc; the abbr for *month* is **mo**

moral; morale   often confused; *moral* refers to personal strength of character, the ability to differentiate between right and wrong; *morale* means the general contentedness or happiness of a person or group of people

mortise; mortice   *mortise* pref

mosaic

most   never use as a short form for *almost;* to say *most everyone is here* is incorrect

movable; moveable   *movable* pref

Mr.; Ms.   address men as *Mr.* and women as *Ms.;* use *Miss* or *Mrs.* only if you know the person prefers to be so addressed

multi-   a prefix meaning many; combines to form one word: *multiaddress, multicavity, multielectrode, multistage*

municipal; municipality

N

nano   def: 10^{-9}; abbr: **n**; other abbr:

nanoampere(s)	**nA**
nanocoulomb(s)	**nC**
nanofarad(s)	**nF**
nanohenry(s)	**nH**
nanometer(s)	**nm**
nanosecond(s)	**ns** (pref); **nsec**
nanotesla(s)	**nT**
nanovolt(s)	**nV**
nanowatt(s)	**nW**

naphtha(lene)

nationwide

nautical mile   def: 6080 ft (1853 m); abbr: **nmi** (pref) or **n.m.**; see **mile**

navigate; navigator; navigable

NB   means note well, and is the abbr for *nota bene;* it's more common to use the word *NOTE*

NC   abbr for *normally closed* (contacts)

nebula   pl: *nebulas* (pref) or *nebulae*

negative   abbr: **neg**

negligent; negligence

negligible

nevertheless

newton   def: a unit of force (SI); abbr: **N**; other abbr: **MN, kN, mN, μN, N·m** (newton meter), **N/m** (newtons per meter)

next   it is better to write *the next two* (or next three, etc) than *the two next* (etc.)

nickel

night   never use *nite;* write *nighttime* as one word

nineteen; ninety; ninth   all three are frequently misspelled

NO   abbr for *normally open* (contacts)

No.   abbr for **number** (which see)

noise-cancel(l)ing   single *l* pref

nomenclature

nomogram; nomograph   *nomogram* pref

non-   as a prefix meaning not or negative, normally combines to form one word: *nonconductor, nondirectional, nonnegotiable, nonlinear, nonstop;* if combining word is a proper noun, insert a hyphen: *non-American;* avoid forming a new word with *non-* when a similar word that serves the same purpose already exists (i.e. you should not form *nonaudible* because *inaudible* already exists)

none   when the meaning is "not one," treat as singular; when the meaning is "not any," treat as plural: *none* (not one) *was satisfactory; none* (not any) *of the receivers were repaired*

nonplus(s)ed   *nonplused* pref

no one   two words

norm   def: the average or normal (situation or condition)

normalize

normally closed; normally open   (contacts) abbr: **NC, NO**

normal to   def: at right angles to

north   abbr: **N**; other abbr:

northeast	**NE**
northwest	**NW**
north-south	**N-S** (control, movement)

northbound and *northward* are written as one word; for rule on capitalization, see **east**

no-show

notable

not applicable   abbr: **N/A**

note well   abbr: **NB** (derived from *nota bene*), but *NOTE* is more common

NOT-gate

notice; noticeable; notification

not to exceed   an overworked phrase that

should be used only in specifications; in all other cases use *not more than*

nth   (harmonic, etc.)

nuclear   frequently misspelled

nucleus   the plural is *nuclei*

null

number   although **no.** would appear to be the most logical abbr for number (and is pref), **No.** is much more common (the symbol # is not an abbr for number); the abbr *no.* or *No.* must always be followed by a quantity in numerals; it is incorrect to write *we have received a No. of shipments;* for the difference in usage between *amount* and *number*, see **amount**

numbers (in narrative)   as a general rule, spell out up to and including nine, and use numerals for 10 and above; for specific rules, see Guideline 4

O

oblique; obliquity

oblivious   def: unaware or forgetful; although *oblivious* should be followed by *of*, it is becoming common practice to follow it with *to;* e.g. *she was oblivious of the disturbance* is correct; *she was oblivious to the disturbance*, although acceptable, is less pref

obsolete; obsolescent

obstacle

obtain; secure   use *obtain* when the meaning is simply to get; use *secure* when the meaning is to make safe or to take possession of (possibly after some difficulty): *we obtained four additional samples; we secured space in the prime display area*

occasional(ly)

occur; occurred; occurrence; occurring

o'clock   avoid using; see **time**

of   avoid using in place of *have;* write *we should have measured*, not *we should of measured*

off-   as a prefix either combines into one word, or a hyphen is inserted: *offset, off-center(ed), off-scale, off-the-shelf*

off of   an awkward construction; omit the word *of*

ohm   def: a unit of electric resistance; abbr: Ω or **ohm**; other abbr: **GΩ, Gohm, MΩ, Mohm,**

kΩ, kohm, mΩ, mohm, $\mu\Omega$, uohm; abbr for ohm-centimeter(s) is **ohm-cm;** *ohmmeter* has two *m's*

oilfield; oil-filled

OK; okay   these are slang expressions which should never appear in technical writing

omit; omitted; omission

omni-   a prefix meaning all or in all ways; combines to form one word: *omnibearing, omnidirectional, omnirange*

on; onto   *on* means positioned generally; *onto* implies action or movement: *the report is on Mr Cord's desk; the speaker stepped onto the platform*

once-over

onward(s)   *onward* pref

opaque; opacity

op. cit.   def: Latin abbr for *opere citato*, meaning in the work cited; used in footnoting

operate; operator; operable   not *operatable*

optimum   pl: *optima* (pref), and sometimes *optimums*

oral   def: spoken; avoid confusing with *aural*

orbit; orbital; orbited; orbiting

organize; organizer; organization

OR-gate

orientation   this is the noun; the verb form is *orient, oriented, orienting*

orifice

origin; original; originally

oscillate

oscilloscope   slang abbr: **scope**

ounce(s)   abbr: **oz;** other abbr:
| ounce-foot | **oz-ft** |
| ounce-inch | **oz-in.** |

out-   as a prefix normally combines to form one word: *outbreak, outcome, outdistance;* when *out-* is followed by *of*, insert hyphens if used as a compound adjective (as in *an out-of-date list*), but treat as separate words when used in place of a noun (as in *the printing schedule is out of phase*)

outside diameter abbr: **OD**

outward(s)   *outward* pref

over-   as a prefix meaning above or beyond, normally combines to form one word: *overbunching, overcurrent, overdriven, overexcited, overrun;* avoid using as a synonym for more than, particularly when referring to quantities: *more than 17 were serviceable* is better than *over 17 were serviceable*

overage   means either too many or too old

overall   an overworked word; as an adjective it often gives unnecessary additional emphasis (as in *overall impression*) and should be deleted; avoid using as a synonym for *altogether, average, general,* or *total*

oxidize   *oxidation* is better than *oxidization*

oxyacetylene

P

pacemaker; pacesetter

page; pages   abbr: **p; pp**

paid   not *payed,* when the meaning is to spend

pair(s)   abbr: **pr**

pamphlet

panel; paneled; paneling

paperwork

parabola(s); parabolic; paraboloid

paragraph(s)   abbr: **para**

parallax

parallel; paralleled; paralleling; parallelism; parallelogram   both *parallel to* and *parallel with* are correct

paralysis; paralyses (pl); **paralyze**

parameter; perimeter   *parameter* means a guideline; *perimeter* means a border or edge

paraphernalia

paraplegic

paraprofessional

parenthesis   this is the singular form; pl: *parentheses*

particles   frequently misspelled

partly; partially   use *partly* when the meaning is "a part" or "in part"; use *partially* when the meaning is "to a certain extent," or when preference or bias is implied.

parts per million   abbr: **ppm**

part-time

pascal   def: a unit of pressure (SI); abbr: **Pa;** other abbr: **Gpa, MPa, kPa, mPa, μPa, pPa, Pa·s** (pascal second)

pass-   as a prefix normally combines to form one word: *passbook, passkey, passport, password*

passed; past   as a general rule, use *passed* as a verb and *past* as an adjective or a noun: *the test equipment has been passed by quality control; past experience has demonstrated a tendency to fail at low temperature; in the past . . .*

pay-   as a prefix normally combines to form one word: *paycheck, payload, payroll;* but *pay day*

pencil; penciled; penciling

pendulum   pl: *pendulums*

penultimate   def: the next to last

people; persons   *people* pref: *all the people were present;* use *persons* to refer only to small numbers of people: *three persons were interviewed* (and even here, *people* could be used)

per   in technical writing it is acceptable to use *per* to mean either by or a(n), as in *per diem* (by the day) and *miles per hour;* in literary writing, take care not to use *per* in place of *a* or *an;* avoid using *as per* in all writing

percent   abbr: **%;** use **%** only after numerals: *42%;* use *percent* after a spelled-out number: *about forty percent;* avoid using the expression *a percentage of* as a synonym for *a part of* or *a small part;* also: **percentage** and **percentile**

perceptible

permeable; permeameter; permeance

permissible

permit; permitted; permitting; permittivity

perpendicular   abbr: **perp**

persevere; perseverance

persistent; persistence; persistency

personal; personnel   *personal* means concerning one person; *personnel* means the members of a group, or the staff: *a personal affair; the personnel in the powerhouse;* for *person(s)* see **people**

peta   def: 10^{15}; abbr: **P;** other abbr: **PBq** (petabecquerel)

pharmacy; pharmacist; pharmaceutical

phase   in the nonelectric sense, *phase* means a stage of transmission or development; it should not be used as a synonym for aspect; it is used correctly in *the second phase called for a detailed cost breakdown*

phase-in; phaseout   but use two words for the verb forms: *to phase in, to phase out*

phenolic

phenomenon pl: *phenomena*

photo- as a prefix, normally combines to form one word: *photoelectric, photogrammetry, photoionization, photomultiplier;* if combining word starts with *o*, insert a hyphen: *photo-offset*

pico def: 10^{-12}; abbr: **p**; other abbr:
picoampere(s)	**pA**
picocoulomb(s)	**pC**
picofarad(s)	**pF**
picohenry(s)	**pH**
picosecond(s)	**ps** (pref); **psec**
picowatt(s)	**pW**

piecemeal; piecework

piezoelectric; piezo-oscillator

pilot; piloted; piloting

pint abbr: **pt**

pipeline

plateau pl: *plateaus* (pref) or *plateaux*

plug; plugged; plugging

plumbbob; plumb line

p.m. def: after noon (post meridiem)

pneumatic

polarize; polarization; polarizing

poly- a prefix meaning many; combines to form one word: *polydirectional, polyethylene, polyphase*

polyvinyl chloride abbr: **pvc**

positive abbr: **pos**

post- a prefix meaning after or behind; combines to form one word: *postacceleration, postgraduate, postpaid;* but *post office*

post meridiem def: after noon; abbr: **p.m.**; can also be written as *postmeridian* (less pref)

potentiometer abbr: **pot.**

pound(s) (weight) abbr: **lb**; other abbr:
pound-foot	**lb-ft**
pound-inch(es)	**lb-in.**
pounds per square foot	**psf** (pref); **lb/ft²**
pounds per square inch	**psi** (pref); **lb/in.²**
pounds per square inch, absolute	**psia**

power factor abbr: **pf** or spell out

powerhouse; power line; powerpack

practicable; practical these words have similar meanings but different applications that sometimes are hard to identify; *practicable* means feasible to do: *it was difficult to find a practicable solution* (one that could reasonably be implemented); *practical* means handy, suitable, able to be carried out in practice: *a practical solution would be to combine the two departments*

practice; practise the noun always is *practice;* the verb is *practice* (pref), but can also be *practise*

pre- a prefix meaning before or prior; normally combines to form one word: *preamplifier, predetermined, preemphasis, preignite, preset;* if combining word is a proper noun, insert a hyphen: *pre-Roman*

precede; proceed *precede* means go before; *proceed* generally means carry on or continue: *the dinner was preceded by a brief business meeting; after dinner, we proceeded with the annual presentation of awards;* see **proceed**

precedence; precedent *precedence* means priority (of position, time, etc.): *the pressure test has precedence* (it must be done first); *a precedent* is an example that is or will be followed by others: *we may set a precedent if we grant his request* (others will expect similar treatment)

predominate; predominant; predominantly

prefer; preferred; preference; preferable avoid overstating *preferable* (which states its meaning quite clearly on its own) by using it to make a comparison; e.g. it is wrong to write *more preferable* or *highly preferable*

prescribe; proscribe *prescribe* means to state as a rule or requirement; *proscribe* means to deny permission or forbid

presently use *presently* only to mean soon or shortly; never use it to mean *now* (use *at present* instead)

pressure-sensitive

prestigious

pretense; pretence *pretense* pref

preventive; preventative *preventive* pref

previous def: earlier, that which went before; avoid writing *previous to* (use *before*); see **prior**

principal; principle as a noun, *principal* means (1) the first one in importance, the leader; or (2) a sum of money on which interest is paid: *one of the firm's principals is Mar-*

tin Dawes; the invested principal of $10,000 earned $650 in interest last year; as an adjective, *principal* means most important or chief: *the principal reason for selecting the Arrow microprocessor was its low capital cost; principle* means a strong guiding rule, a code of conduct, a fundamental or primary source (of information, etc); *his principles prevented him from taking advantage of the error*

printout

prior; previous   use only as adjectives meaning earlier: *he had a prior appointment,* or *a previous commitment prevented Mr. Perchanski from attending the meeting;* write *before* rather than *prior to* or *previous to*

privilege

proceed; proceeding; procedure   use *proceed to* when the meaning is to start something new; use *proceed with* when the meaning is to continue something that was started previously

produce; producible; productive

program; program(m)ed; program(m)ing; program(m)er   use of single and double *m* varies widely; double *m* pref

prohibit   use *prohibit from;* never *prohibit to*

prominent; prominence

promissory   (note)

proofread

propel; propelled; propelling; propellant (noun); **propellent** (adjective); **propeller, propellor:** *er* pref

prophecy; prophesy   use *prophecy* only as a noun, *prophesy* only as a verb

proportion   avoid writing *a proportion of* or *a large proportion of* when *some, many,* or a specific quantity would be simpler or more direct

proposition   in its proper sense, *proposition* means a suggestion put forward for argument; it should not be used as a synonym for *plan, project,* or *proposal*

pro rata   def: assign proportionally; sometimes used in the verb form as *prorate: I want you to prorate the cost over two years' operations*

prospectus; prospectuses

protein

proved; proven   use *proven* only as an adjective or in the legal sense; otherwise use *proved; he has been proven guilty; he proved his case*

psycho-   as a prefix normally combines to form one word: *psychoanalysis, psychopathic, psychosis;* if combining word starts with *o,* insert a hyphen: *psycho-organic*

pursuant to   avoid using this wordy expression

purge; purging

Q

quality control   abbr: **QC**

quandary

quantity; quantitative   the abbr of quantity is **qty**

quart   abbr: **qt**

quasi-   a prefix meaning seemingly or almost; insert a hyphen between the prefix and the combining word: *quasi-active, quasi-bistable, quasi-linear*

question mark   insert a question mark after a direct question: *how many booklets will you require?;* omit the question mark when the question posed is really a demand: *may I have your decision by noon on Monday*

questionnaire

quick-   as a prefix normally combines with a hyphen: *quick-acting, quick-freeze, quick-tempered;* exceptions: *quicklime; quicksilver*

quiescent; quiescence

quorum

R

rack-mounted

racon   def: a radar beacon

radian   def: a unit of angular measurement; abbr: **rad**

radiator

radio-   as a prefix, combines to form one word: *radioactive, radiobiology, radioisotope, radioluminescence;* if combining word starts with *o,* omit one of the *o*'s: *radiology, radiopaque;* in other instances *radio* may be either combined or treated as a separate word, depending on accepted usage; typical examples are *radio compass, radio countermeasures, radio direction-finder, radio frequency* (as a noun), *radio-frequency* (as an adjective), *radio range, radiosonde, radiotelephone*

radio frequency   abbr: **rf**

radius   pl: *radii* (pref) or *radiuses*

radix   pl: *radices* (pref) or *radixes*

rally; rallied; rallying

range; ranging; rangefinder; range marker

rare; rarely; rarity; rarefy; rarefaction

ratemeter

ratio; ratios

rational; rationale   *rational* means reasonable, clear-sighted: *John had a rational explanation for the error; rationale* means an underlying reason: *Tricia explained the company's rationale for diversifying the product line*

re   def: a Latin word meaning in the case of; avoid using *re* in technical writing, particularly as an abbr for *regarding, concerning, with reference to*

re-   a prefix meaning to do again, to repeat; normally combines to form one word: *reactivate, rediscover, reemphasize, reentrant, reignition, rerun, reset;* if the compound term forms an existing word that has a different meaning, insert a hyphen to identify it as compound: *re-cover* (to cover again)

reaction   use *reaction* to describe chemical or mechanical processes, not as a synonym for *opinion* or *impression*

reactive kilovolt-ampere; reactive voltampere   see **kilo** or **volt**

readability

readout

realize; realization

recede

receive; receiver; receiving; receivable

recipe; receipt   often confused; *recipe* means cooking instructions; *receipt* means a written record that something has been received

rechargeable

recommend

reconcile; reconcilable

reconnaissance

recover; re-cover   *recover* means to get back, to regain; *re-cover* means to cover again

recur; recurred; recurring; recurrence   these are the correct spellings; never use *reoccur* (etc)

recycle; recyclable

reducible

reenforce; reinforce   *reenforce* means to enforce again; *reinforce* means to strengthen: *Rick Davis reenforced his original instructions by circulating a second memorandum; the Artmo Building required 34,750 tons of reinforced concrete*

refer; referred; referring; referral; referee; reference

referendum   pl: *referendums* (pref); *referenda* (less common)

reiterate   def: to say again

relaid; relayed   *relaid* means laid again, like a carpet; *relayed* means to send on, as a message would be relayed from one person to another

remit; remitted; remitting; remitter; remittance

remodel; remodeled; remodeling

removable

remuneration   def: pay, salary; often misspelled as *renumeration*

rent-a-car

reoccur(rence)   never use; see **recur**

repairable; reparable   both words mean in need of repair and capable of being repaired; *reparable* also implies that the cost to repair the item has been taken into account and it is economically worthwhile to effect repairs

repellant; repellent   use *repellant* as a noun, *repellent* as an adjective; **repeller**

replaceable

reproducible

rescind

reservoir

reset; resetting; resettability

resin; rosin   these words have become almost synonymous, with a preference for *resin;* use *resin* to describe a gluey substance used in adhesives, and *rosin* to describe a solder fluxcore

respective(ly)   this overworked word is not really needed in sentences that differentiate between two or more items; it should be deleted from sentences such as: *pins 4, 5 and 7 are marked R, S, and V respectively*

resumé the correct spelling is *résumé* (with two accents), but the single accent or no accent (*resume*) has become standard

retrieve; retrieval

retro- a prefix meaning to take place before, or backward; normally combines to form one word: *retroactive, retrofit, retrogression;* if combining word starts with *o*, insert a hyphen: *retro-operative*

reverse; reverser; reversal; reversible

revolutions per minute; revolutions per second abbr: **rpm; rps**

rheostat

rhombus pl: *rhombuses* (pref) or *rhombi*

rhythm; rhythmic; rhythmically

ricochet; richocheted; ricocheting

right-handed(ed) abbr: **RH**

rivet; riveted; riveter; riveting

road- as a prefix normally combines to form one word: *roadblock, roadmap, roadside*

roentgen abbr: **R**

role; roll a *role* is a person's function or the part that he plays (in an organization, project, or play); a *roll*, as a technical noun, is a cylinder; as a verb, it means to rotate: *the technicians' role was to make the samples roll toward the magnet*

root mean square abbr: **rms**

rosin see **resin**

rotate; rotator; rotatable; rotary

round def: circular; avoid confusing with **around** (which see)

ruggedize

rustproof; rust-resistant

S

salable; saleable *salable* pref

salvageable

same avoid using *same* as a pronoun; to write *we have repaired your receiver and tested same* is awkward; a better version is *we have repaired and tested your receiver*

sapphire

satellite

saturate; saturation; saturable

save; savable

sawtooth; saw-toothed

scalar; scaler *scalar* is a quantity that has magnitude only; *scaler* is a measuring device

scarce; scarcity

sceptic(al); skeptic(al) *skeptic(al)* pref

schedule

schematic although really an adjective (as in *schematic diagram*), in technical terminology *schematic* can be used as a noun (meaning *a schematic drawing*)

science; scientific(ally); scientist

scissors as a plural word, write *the scissors are;* in the singular form, write *the pair of scissors is*

screwdriver; screw-driven

seamweld

seasonal; seasonable *seasonal* means affected by or dependent on the season; *seasonable* means appropriate or suited to the time of year: *a seasonal activity; seasonable weather*

seasons the seasons are not capitalized: *spring, summer, autumn* or *fall, winter*

secant abbr: **sec**

second as a prefix, *second-* combines erratically: *second-class, second-guess, secondhand, second-rate, second sight;* the abbr for *second* (time) is **sec,** and for *second* (angular measure) it is **"**

secure see **obtain**

-sede *supersede* is the only word to end with *-sede;* others end with *-cede* or *-ceed* (which see)

seem(s) see **appear(s)**

self- insert a hyphen when used as a prefix to form a compound term: *self-absorption, self-bias, self-excited, self-locking, self-resetting;* but there are exceptions: *selfless, selfsame*

semi- a prefix meaning half; normally combines to form one word: *semiactive, semiannually* (every six months), *semi-conductor, semimonthly* (half-monthly), *semiremote, semiweekly* (half weekly); if combining word starts with *i*, insert a hyphen: *semi-idle, semi-immersed*

separate; separable; separator; separation all are frequently misspelled

sequence; sequential

serial number abbr: **ser no.** or **S/N**

series-parallel

serrated

serviceable; serviceman, servicewoman, serviceperson (pref)

servo- as a prefix, combines to form one word: *servoamplifier, servocontrol, servosystem;* as a noun, *servo* is an abbreviation for *servomotor* or *servomechanism*

sewage; sewerage *sewage* is waste matter; *sewerage* is the drainage system that carries away the waste matter

shall *shall* is rarely used in technical writing (*will* is pref), except in specifications when its use implies that the specified action is mandatory

short- as a prefix, most often combines with a hyphen: *short-circuit, short-form* (report), *short-lived, short-term;* in some cases it combines into one word: *shorthand* (writing), *shorthanded, shortcoming, shortsighted*

shrivel(l)ed; shrivel(l)ing single *l* pref

sic a Latin word which means a quotation has been copied exactly, even though there was an error in the original; e.g. *the report stated: "Our participation will be an issental (sic) requirement."*

siemens def: a unit of electric conductance (SI); abbr: **S;** other abbr: **kS, mS, μS**

sight def: the ability to see; see **site**

signal; signaled; signaling; signaler

signal-to-noise (ratio)

silhouette

silverplate; silver-plate use *silverplate* as a noun or adjective, *silver-plate* as a verb

similar not *similiar*

sine abbr: **sin**

singe; singeing the *e* must be retained to avoid confusion with *singing*

singlehanded

siphon not *syphon*

sirup; syrup *syrup* pref

site, sight, cite three words that often are misspelled; a *site* is a location: *the construction site; sight* implies the ability to see: *mud up to the axles became a familiar sight; cite* means quote: *I cite the May 17 progress report as a typical example of good writing*

siz(e)able *sizable* pref

skeptic(al); sceptic(al) *skeptic(al)* pref

skil(l)ful *skillful* pref; note that general usage dictates that *ll* is pref here, contrary to the preference in most *l* and *ll* situations in this glossary

slip- usually combines into a single word: *slippage, slipshod, slipstream;* but *slip ring(s)*

smolder

solder

solely

someone; some one *someone* is correct when the meaning is any one person; *some one* is seldom used

some time; sometimes *some time* means an indefinite time: *some time ago; sometimes* means occasionally: *he sometimes works until after midnight*

sound combines irregularly: *sound-absorbent, sound-absorbing, sound-powered, soundproof, sound track, sound wave*

south abbr: **S;** other abbr:

southeast	**SE**
southwest	**SW**

southbound and *southward* are written as one word; for rule on capitalization, see **east**

space- as a prefix normally combines to form one word: *spacecraft, spaceflight*

spare(s) as a noun, frequently means spare part(s)

specially see **especially**

specific gravity abbr: **sp gr**

specific heat abbr: **sp ht**

spectro- as a prefix, combines to form one word: *spectrometer, spectroscope;* if combining word starts with *o*, omit one *o: spectrology*

spectrum pl: *spectra*

spiral; spiraled; spiraling

split infinitive to split an infinitive is to insert an adverb between the word *to* and a verb: *to really insist* is a split infinitive; although grammarians used to claim that you should never split an infinitive, they now suggest you may do so if rewriting would result in an awkward construction or ambiguity

spotweld

square abbr: **sq** or ^{2}; other abbr:

square foot	**ft^2** (pref); **sq ft**
square inch	**in.^{2}** (pref); **sq in.**
square meter	**m^2**

square centimeter   **cm**^{2}
square millimeter   **mm**^{2}
curies per square meter   **Ci/m**^{2}
milliwatts per square   **mW/m**^{2}
meter

standby; standoff; standstill   all combine into one word when used as a noun or an adjective

standing-wave ratio   abbr: **swr**

state-of-the-art

stationary; stationery   *stationary* means not moving: *the vehicle was stationary when the accident occurred; stationery* refers to writing materials: *the main item in the October stationery requisition was an order for one thousand writing pads*

statute mile   def: 5280 ft (1609 m); see **mile**

statutory

stencil; stenciled; stenciling

stereo-   as a prefix, combines to form one word: *stereometric, stereoscopic; stereo* can be used alone as a noun meaning multi-channel system

stimulus   pl: *stimuli*

stock; stockholder; stocklist; stock market; stockpile

stop-   as a prefix usually combines to form one word: *stopgap, stopnut, stopover* (when used as a noun or adjective); *stoppage, stopwatch*

strato-   a prefix that combines to form one word: *stratocumulus, stratosphere*

stratum   pl: *strata*

structural

stylus   pl: *styluses* (pref) or *styli*

sub-   a prefix generally meaning below, beneath, under; combines to form one word: *subassembly, subcarrier, subcommittee, subnormal, subpoint*

subparagraph   abbr: **subpara;** abbr for *subsubparagraph* is **subsubpara**

subpoena; subpoenaed   *subpena* also used but less pref

subtle; subtlely/subtly   *subtly* pref

succinct

sufficient   in technical writing, *enough* is a better word than *sufficient*

sulfur; sulphur   *sulfur* pref; as a prefix, *sulf-* combines to form one word: *sulfanilamide*

summarize

super   a prefix meaning greater or over; combines to form one word: *superconductivity, superregeneration*

superhigh frequency   abbr: **shf**

superimpose; superpose   *superimpose* means to place or impose one thing on top of another; *superpose* means to lay or place exactly on top of, so as to be coincident with

supersede   see **-sede**

supra-   a prefix meaning above; normally combines to form one word: *supramolecular;* if combining word starts with *a,* insert a hyphen: *supra-auditory*

surfeit   def: to have more than enough

surveillance

susceptible

switch-   *switchboard, switchbox, switchgear*

swivel; swiveled; swiveling

syllabus   pl: *syllabuses* (pref) or *syllabi*

symmetry; symmetrical

sympathize

symposium   pl: *symposia* (pref) or *symposiums*

synchro-   as a prefix combines to form one word: *synchromesh, synchronize, synchroscope; synchro* can also be used alone as a noun meaning synchronous motor

synonymous   use *synonymous with,* not *synonymous to*

synopsis   pl: *synopses*

synthesis   pl: *syntheses*

synthetic

syphon   *siphon* pref

syringe

syrup(y)

systemwide

T

tablespoon   use *tablespoonfuls* rather than *tablespoonsful;* abbr: **tbsp**

tail-   as a prefix normally combines to form one word: *tailboard, tailless, tailwind;* but *tail end, tail fin*

take-   *takeoff; takeover; takeup;* as nouns and adjectives these terms all combine into a single word

tangent abbr: **tan**

tangible

tape deck

taxable; tax-exempt; taxpayer

teamwork

teaspoon use *teaspoonfuls* rather than *teaspoonsful;* abbr: **tsp**

technician

tele- a prefix that means at a distance; combines to form one word: *teleammeter, telemetry, telephony, teletype(writer)*

telecom; telecon *telecom* is the abbr for *telecommunication(s); telecon* is the abbr for *telephone conversation*

television abbr: **TV**

temperature abbr: **temp;** combinations are *temperature-compensating* and *temperature-controlled;* when recording temperatures, the abbr for *degree* (deg or °) may be omitted: *an operating temperature of 85C; the water boils at 100C or 212F;* the pref (SI) unit for temperature is the degree Celsius (°**C**)

tempered

template; templet both spellings are correct; *template* pref; def: a pattern or guide

temporary; temporarily

tensile strength abbr: **ts**

tentative; tentatively

tenuous

tera def: 10^{12}; abbr: **T;** other abbr:

terabecquerel(s)	**TBq**
terahertz	**THz**
terajoule(s)	**TJ**
terawatt(s)	**TW**

terminus pl: *termini* (pref) or *terminuses;* note that this plural is contrary to the pref plurals for most *-us* endings

tesla def: a unit of magnetic flux density, magnetic inductance (SI); abbr: **T;** other abbr: **mT, μT, nT**

that is abbr: **i.e.** (pref) or **ie**

their; there; they're the first two of these words are frequently misspelled, more through carelessness than as an outright error; *their* is a possessive meaning belonging to them: *the staff took their holidays earlier than normal; there* means in that place: *there were 18 desks in the room,* or *put it there; they're* is a

contraction of *they are* and should not appear in technical or business writing

there- as a prefix combines to form one word: *thereafter, thereby, therein, thereupon*

therefor(e) *therefore* pref

thermo- a prefix generally meaning heat; combines to form one word: *thermoammeter, thermocouple, thermoelectric, thermoplastic*

thermodynamic temperature the SI unit is the kelvin (abbr: **K**), expressed in degrees Celsius (°C)

thesis pl: *theses*

thousand abbr: **k**
thousand foot-pound(s) **kip-ft**
thousand pound(s) **kip**

three- when used as a prefix, a hyphen normally is inserted between the combining words: *three-dimensional, three-phase, three-ply, three-wire;* exceptions are *threefold* and *threesome*

threshold

through never use *thru*

tieing; tying *tying* pref

timber; timbre *timber* is wood; *timbre* means tonal quality

time always write time in numerals, if possible using the 24-hour clock: *08:17* or *8:17 a.m., 15:30* or *3:30 p.m.* 24-hour times may be written as *20:45* (pref), *20:45 hr,* or *20:45 hours;* never use the term "o'clock" in technical writing: write *15:00* or *3 p.m.* rather than *3 o'clock*

time- typical combinations are *time base; time-card; time clock; time constant; time-consuming; time lag; timesaving; time-table; time-wasting*

tinplate; tin-plate use *tinplate* as a noun, *tin-plate* as a verb or adjective

to; too; two frequently misspelled, most often through carelessness; *to* is a preposition that can mean in the direction of, against, before, or until; *too* means as well; *two* is the quantity "2"

today; tonight; tomorrow never use *tonite*

tolerance abbr: **tol**

ton; tonne the U.S. ton is 2000 lb and is known as a *short ton;* the British ton is 2240 lb and is known as a *long ton;* the metric ton is 1000 kg (2204.6 lb) and is known as a *tonne* (abbr: **t**); other terms: *tonmile* and *tonnage*

toolbox; toolmaker; toolroom

top-   *top-heavy; top-loaded; top-up*

torque(d); torquing

total; totaled; totaling

toward(s)   *toward* pref

traceable

trade-   as a prefix combines most often into a single word: *trademark, tradeoff;* but *trade-in* and *trade name*

trans-   a prefix meaning over, across, or through; it normally combines to form one word: *transadmittance, transcontinental, transship;* if combining word is a proper noun, insert a hyphen: *trans-Canada* (an exception is *transatlantic,* which through common usage has dropped the capital A and combined into one word); *transonic* has only one *s*

transceiver   def: a transmitter-receiver

transfer; transferred; transferring; transferable; transference

transmit; transmitted; transmitting; transmittal; transmitter; transmission

transverse; traverse   *transverse* means to lie across; *traverse* means to track horizontally

travel; traveled; traveling

tri-   a prefix meaning three or every third; combines to form one word: *triangulation, tricolor, trilateral, tristimulus, triweekly*

trouble-free; troubleshoot(ing)

truncated

tune; tunable; tuneup   (noun or adjective)

tunnel; tunneled; tunneling

turbo-   a prefix meaning turbine-powered; combines to form one word: *turboelectric, turboprop*

turbulence; turbulent

turn-   *turnaround* and *turnover* form one word when used as adjectives or nouns; *turnstile* and *turntable* always form one word; *turns-ratio* is hyphenated

two-   when used as a prefix to form a compound term, a hyphen normally is inserted: *two-address, two-phase, two-ply, two-position, two-wire;* an exception is *twofold*

type-   as a prefix normally combines into one word: *typeface; typeset(ting); typewriter, typewriting*

U

ultimatum   *ultimatums* (pref) or *ultimata* (seldom used)

ultra-   a prefix meaning exceedingly; normally combines to form one word: *ultrasonic, ultraviolet;* if combining word starts with *a,* insert a hyphen: *ultra-audible, ultra-audion*

ultrahigh frequency   abbr: **uhf**

un-   a prefix that generally means not or negative; normally combines to form one word: *uncontrolled, undamped, unethical, unnecessary;* if combining word is a proper noun, insert a hyphen: *un-American;* if term combines to form an existing word that has a different meaning, insert a hyphen: *un-ionized* (meaning not ionized) to avoid confusion with *unionized* (to belong to a union); if uncertain whether to use *un-, in-,* or *im-,* try using *not*

unadvisable; inadvisable   both are correct; *inadvisable* pref

unbalance; imbalance   for technical writing, *unbalance* pref

unbias(s)ed   *unbiased* pref

under-   a prefix meaning below or lower; combines to form one word: *underbunching, undercurrent, underexposed, underrated, undershoot, undersigned*

underage   means a shortage or deficit, or too young

unequal(l)ed   *unequaled* pref; see **equal**

unessential; inessential   *unessential* pref

unforeseen; unforeseeable

uni-   a prefix meaning single or one only; combines to form one word: *uniaxial, unidirectional, unifilar, univalent*

uninterested   def: not interested; avoid confusing with *disinterested* (which see)

unionized; un-ionized   *unionized* refers to a group of people who belong to a union; *un-ionized* means not ionized

unique   def: the one and only, without equal, incomparable; use with great care and never in any sense where a comparison is implied; you cannot write *this is the most unique design;*

rewrite as *this design is unique*, or (if a comparison must be made) *this is the most unusual design*

unmistakable

unnavigable

unparalleled

unpractical *impractical* pref

unsanitary *insanitary* pref

unserviceable abbr: **u/s**

unstable but *instability* is better than *unstability*

untraceable

up- as a prefix combines to form one word: *update, upend, upgrade, uprange, upswing*

upper case def: capital letters; abbr: **uc**

uppermost

up-to-date

use; usable; usage; using; useful

utilize; utilise *utilize* pref; avoid using *utilizes* when *uses* or *employs* would be a better word

V

vacuum

valance; valence *valance* means a cover over a drapery track; *valence* is an electronic or nucleonic term, as in *valence electron*

valve-grind(ing)

vari- as a prefix meaning varied, combines into one word: *varicolored, variform*

variance write *at variance with;* never write *at variance from*

varimeter; varmeter def: a meter for measuring reactive power; *varimeter* pref

vehicle; vehicular

vender; vendor *vendor* pref

versed sine abbr: **vers**

versus def: against; abbr: **vs**

vertex def: top; pl: *vertexes* (pref) or *vertices;* avoid confusing with *vortex*

very high frequency abbr: **vhf**

vice versa def: in reverse order

video- as a prefix normally combines to form one word: *videocast, videocassette, videotape;* the abbr for videocassette recorder is **VCR**

video frequency abbr: **vf**

viewfinder; viewpoint

vise def: a clamp; in U.S. spelled *vise;* elsewhere: *vice*

visor; vizor *visor* pref

viz def: namely; this term is seldom used in technical writing

vocation; avocation *vocation* is a trade or calling; *avocation* means an interest or hobby

voice-over; voiceprint

volt def: electric potential or potential difference; abbr: **V;** other abbr: **MV, kV, mV, μV, nV;** also:

volt-ampere(s)	**VA**
volt-ampere(s), reactive	**VAr**
volts, alternating current	**Vac**
volts, direct current	**Vdc**
volts, direct current, working	**Vdcw**
volts per meter	**V/m**

volt- combines into one word: *voltammeter, voltohmyst*

volume abbr: **vol**

vortex def: spiral; pl: *vortexes* (pref) or *vortices;* avoid confusing with *vertex*

VU-meter

W

waive; waiver/waver *waive* and *waiver* mean to forgo one's claim or give up one's right; *waver* means to hesitate, to be irresolute

walkie-talkie

war- as a prefix, combines to form one word: *warfare, wartime*

warranty

waste, wastage

water- combines irregularly: *watercool(ed); water cooler; waterflow; water level; waterline; waterproof; water-soluble; watertight*

watt def: a unit of power, or radiant flux (SI); abbr: **W;** other abbr: **TW, GW, MW, kW, mW, μW, nW, pW, W/m^2;** the abbr for *watt-hour(s)* is **Wh** (pref) or **W-hr;** as a prefix *watt-* forms *watthourmeter* and *wattmeter*

wave- normally combines to form one word: *waveband, waveform, wavefront, waveguide, wavemeter, waveshape;* exceptions are *wave angle* and *wave-swept*

wavelength abbr: λ

waver   see **waiver**

weather   use only as a noun; never write *weather conditions;* avoid confusing with *climate* and *whether* (which see)

weatherproof

weber   def: a unit of magnetic flux (SI); abbr: **Wb;** other abbr: **mWb**

Wednesday   often misspelled

week(s)   abbr: **wk**

weight   abbr: **wt**

well-   as a prefix normally combines with a hyphen: *well-adjusted, well-defined, well-timed*

west   abbr: **W;** *westbound* and *westward* are written as one word; for rule on capitalization, see **east**

where-   as a prefix, combines to form one word: *whereas, wherefore, wherein;* when combining word starts with *e,* omit one *e: wherever*

whether; weather   *whether* means if; *weather* has to do with rain, snow, sunshine, etc.

while; whilst   *while* pref

whoever

wholly   not *wholely*

wide; width   abbr: **wd**

wideband; widespread

wirecutter(s); wire-cutting; wirewound

withheld; withhold

words per minute   abbr: **wpm**

work-   as a prefix usually combines to form one word: *workbench, workflow, workload, workshop*

working volts, dc   abbr: **Vdcw**

worldwide

wrap; wrapped; wrapping; wraparound

writeoff; writeup   both combine into one word when used as noun or adjective

writer   see **author**

writing   only one *t*

X

x-   *x-axis; X-band; x-particle; x-radiation; x-ray*

Xerox

X-Y recorder

Y

y-   *Y-antenna; y-axis; Y-connected; Y-network; Y-signal*

yard(s)   abbr: **yd**

yardstick

year(s)   abbr: **yr;** typical combinations are *year-end* and *year-round*

your; you're   *your* means belonging to or originating from you: *I have examined your prototype analyzer; you're* is a contraction of *you are* and should not appear in technical or business writing

Z

Z-axis

zero   pl: *zeros* (pref) or *zeroes;* typical combinations are *zero-access, zero-adjust, zero-beat, zero-hour, zero level, zero-set, zero reader*

zip code

zoology; zoological

Index

MARKING CONTROL CHART

This control chart will show you which aspects of your writing need attention and, as time progresses, whether you have successfully corrected your most predominant faults. As each assignment is returned to you, count up the errors indicated as marginal notations by your instructor and enter them on the chart. For instance, if on assignment 1 your instructor enters "F" and "U" once, and "S" three times, in the margin, you are being told you have used the wrong format (F), your work is untidy (U), and you have three spelling errors (S). In column 1 of the chart enter "1" in the squares opposite F and U, and "3" in the square opposite "S".

ASSIGNMENT NUMBER

1	2	3	4	5	6	7	8	9	10	11	12	13	14	15

A — Awkward construction

B — Brevity overdone; too few details

C — Continuity weak; paragraph lacks coherence/unity

D — Development inadequate; support your argument

E — Error! Check your facts, data, information

F — Format incorrect

G — Grammar fault

H — Heavy going; dull; uninteresting

I — Illogical or irrelevant (correct or omit)

J — Jumpy—too many short sentences, reads like primary reader

K — King-size paragraphs, sentences or words (shorten them)

L — Low Information Content words or phrase (delete)

M — Missing words or information

N — No! Never do this; never use slang, contractions, unexplained abbreviations, etc.

O — Organization poor

P — Punctuation error, or punctuation missing

Q — Query: what does this mean? not understood; can't read your writing

R — Repetition

S — Spelling error

T — Tone wrong

U — Untidy, messy, or careless work (improve "presentation")

V — Vague; ambiguous; not clear enough

W — Wishy-washy; weak argument; unconvincing

X — X-out (delete, omit) this unnecessary statement

Y — Yak! Yak! Yak!—too wordy; too many generalities

Z — Lacks continuity; needs better transitions

— Numbers wrongly presented

// — Use parallel construction